Chapter 1 Psychopathology

Perceptions	1	Overvalued idea, obsessions and delusions	4
Illusions	1	Psychopathology in movement	4
Hallucinations	2	Common psychiatric syndromes	5
Psychopathology in thought process	3	OSCE grid	6
Psychopathology in speech	3	Revision MCQs and MEQs	7

Perceptions and related psychopathology

Perception is a complex process involving reception of information from sensory organs, followed by recognition, processing and reorganisation of the information into memory and a subsequent response relating to the information received.

Overview of perceptual phenomena

Criteria	Imagination (Fantasy)	Illusion	Pseudohallucination	Hallucination	Real perception
Voluntary or involuntary	Voluntary	Involuntary	Involuntary (except eidetic imagery)	Involuntary	Involuntary
Vividness	Not vivid	Vivid	Vivid	Vivid	Vivid
Space	Inner subjective space	Inner subjective space	Inner subjective space	Outer space	Outer space
Insight	Intact	Intact	Intact	Impaired	Intact
Psychiatric disorders	Normal experience	Delirium	Depression	Schizophrenia	Normal experience

Illusions

An *Illusion* is an involuntary false perception in which a transformation of real object takes place (e.g. the curtain in the ward is perceived by a patient as a female witch).

Description of common illusions:
- **Complete illusion** – the brain's tendency to fill the missing part of the object and is banished by attention
- **Affective illusion** – occurs in depressed or manic state.
- **Pareidolic illusion** is a type of intense imagery that persists even when the subject looks at a real object in the external environment (e.g. seeing a woman's face when looking at the cloud). This illusion is increased by attention. . The image and percept occur together although and the image is recognised as unreal.
- **Jamais vu** – illusion of unfamiliarity. It occurs in people with temporal lobe epilepsy and exhaustion.
- **Déjà vu** – illusion of familiarity
- **Macropsia** – illusion of exaggeration in size
- **Micropsia** – illusion of reduction in size

Afterimage and palinopsia:
An **afterimage** is a visual illusion that refers to an image continuing to appear in one's vision after the exposure to the original image has ceased. **Palinopsia** is a medical condition referring to the phenomenon of seeing afterimage persistently. And occurs in depression, anxiety, hallucinogen use, Parkinson's disease and migraine

Hallucinations

What is a hallucination? In brief, a hallucination refers to perception of an external object when no such object is present.

Auditory hallucinations	(a) **Simple/Elementary auditory hallucination** (e.g. sounds or musical) • Elementary auditory hallucination can be part of the normal experience. • **Epidemiology**: more common in women; older people. • **Aetiology:** alcohol misuse (alcohol hallucinosis in clear consciousness), deafness, right temporal lobe lesion (b) **Complex auditory hallucination (mood congruent)** Examples: ▪ Stating depressive or manic themes in second person e.g. "You are useless." ▪ Command hallucination: "You are useless and you should hurt yourself." ▪ Occurring in depressive disorder or bipolar disorder. Command hallucinations can be dangerous. (c) **Complex auditory hallucinations (mood incongruent):** Examples: ▪ Voices discussing the person in third person. • Voices giving a running commentary on a person's behaviour. • Thoughts spoken out loud (Thought echo/ Echo de la pense) Complex auditory hallucination occurs in schizophrenia and is usually poorly localised in space. The voice tends to increase if sensory input is restricted or the person is drowsy. Some people with schizophrenia try to distract themselves from the complex auditory hallucinations by listening to music or meaningful speech.
Visual hallucinations	(a) **Simple/Elementary visual hallucination:** (e.g. Flash light) Common in acute organic diseases and lesions are found in the occipital cortex. After head injury, there are simple visual hallucinations such as flashes and stars seen. These simple visual hallucinations are known as positive scotomata. During migraine attack, the person often sees zigzag lines. The underlying pathology is due to transient disturbance in occipital cortex. (b) **Complex visual hallucination** (e.g. involving people and animals) Occurring in organic diseases and lesions are found in the temporal lobe.
Lilliputian hallucination (Visual)	▪ Hallucinated objects, usually people, appear greatly reduced in size (microptic). ▪ Lilliputian hallucinations also occur in psychiatric states associated with febrile or intoxicating conditions (e.g. alcohol). It can be observed in the absence of recognizable organic disorders.
Alcoholic hallucinosis	• It is a form of acute hallucinosis seen in people with alcohol dependence following excess alcohol intake. It can also be part of the withdrawal syndrome. • This phenomenon is rare in women • It is well-localised, derogatory and in second person • Auditory hallucinations occur in a clear intellectual field without confusion or intellectual impairment. • It has slow recovery and responds poorly to antipsychotics...Recurrence is frequent.
HypnoGogic (Going to sleep) and hypnoPompic (waking	• Conscious level fluctuates considerably in different stages of sleep, and both types of hallucination probably occur in a phase of increasing drowsiness when the structure of thought, feelings, perceptions, fantasies and self-awareness become blurred and merged

u<u>P</u>) hallucinations	into oblivion.
	- The perception may be visual, auditory or tactile. It is sudden in occurrence and the subject believes that it woke him up, for example a loud voice in the street outside, a feeling of someone pushing him over the bed, or seeing a man coming across the bedroom. These phenomena may be considered normal even though they are real hallucinations.
Autoscopy (hallucination)	- In autoscopy, the person seeing one's "self" or "double". The double imitates the movement and facial expressions of the original, as if being a reflection in a mirror.
- The double typically appears as semitransparent, and associated auditory, kinaesthetic, and emotional perceptions are frequent.
- The hallucinatory experience rarely lasts longer than a few seconds.
- The most common emotional reactions are sadness and bewilderment.
- It also occurs in organic disorders such as parietooccipital lesions, temporoparietal lobes, epilepsy, schizophrenia and substance misuse. Neurological and psychiatric disorder can occur in 60% of cases
- M: F = 2:1; Mean age = 40 years.
Extracampine hallucination	Extracampine hallucination is experienced outside the limits of sensory field including visual field and range of audibility. It is different from autoscopy as the person often sees others outside the sensory field but not one's self.
Kinesthetic hallucination	Kinesthetic hallucination involves the sense of bodily movement (e.g. a person believes that his elbow joint is rotating involuntarily but there is no such sign on physical examination).
Cenesthetic hallucination	A sensation of an altered state in a body organ without corresponding receptors which can explain the sensation in normal human physiology.

Psychopathology in thought process

Circumstantiality	The speech takes a long time to reach the point because it includes a great deal of unnecessary details. This occurs in schizophrenia, dementia, temporal lobe epilepsy and normal people.
Tangentiality	The stream of thought diverges from the topic and the speech appeared to be unrelated and irrelevant at the end. It is due to loosening and diffuseness of speech. It occurs in people with schizophrenia and mania.
Flight of ideas	A continuous speech where topics jump rapidly from one to another and there is a logical link between topics. It occurs in mania and accompanied by pressure of speech.
Loosening of associations	A speech where topics seem to be disconnected and it is hard for the others to establish a logical link between topics. The speech becomes diffuse and unfocused.
Knight's move thinking	Schizophrenia patient changes from one topic to another between abruptly and there is no logical link between thoughts. The train of thoughts is similar to the knight's movement in chess.
Thought insertion	External thoughts are inserted into the patient's mind which do not belong to the patient.
Thought withdrawal	Patient's thoughts are being taken away by the others.
Thought broadcasting	Patient's own thoughts are known to the others through broadcasting like a radio or TV station.

Psychopathology in speech

Logoclonia	This refers to spastic repetition of syllables. Logoclonia occurs in people with autism, schizophrenia, Alzheimer's disease and Parkinson disease.
Logorrhoea	Logorrhoea refers to excessive verbal production.
Neologism	Neologism refers to the invention of persons' own words which cannot be found in the English language and making condensations of several other words. (e.g. a patient refers 'lambrain' as clouds in the sky because they look like lambs and produce rainfalls) The word has a special meaning for the person. Neologism occurs in people with schizophrenia.
Metonym	Metonym refers to an approximate but related term which is used in an idiomatic way. (e.g. a person refers 'sky sheep' as the clouds in the sky because the clouds look like sheep). The word is found in the English language. Metonym occurs in people with schizophrenia and associated with loosening of associations.
Word salad	Word salad refers to a jumble of words and phrases, tossed around as in a salad bowl, with little

	obvious connection between them. It occurs in people with schizophrenia and associated with loosening of associations.
Mutism	Mutism refers to the state that the person is silent and voiceless. Organic causes of mutism include catatonia, herpes simplex virus encephalitis, locked-in syndrome, myasthenia gravis and polio infection. Psychological causes include conversion disorder and malingering.
Stupor (Akinetic mutism)	The person is mute but he or she does not suffer from aphasia. The person is unresponsive and immobile although he or she is fully conscious. His/her eyes are able to follow an object. The opening and closing of eyes are under voluntary control. The person usually maintains a resting posture and neurological examination often reveals normal reflexes.

Overvalued ideas, obsessions and delusions

Overvalued idea	Possible idea that is pursued beyond normal boundaries and causes distress and functional impairment to the patient. The idea can be challenged (e.g. a patient believes that other people in the train are looking her and she feels very upset. It is possible that the people are staring at her because of abnormal clothing or behaviours).
Obsession	Repetitive and irrational thoughts that the patient recognises those thoughts are indeed originated from his or her own mind and can be challenged. Obsessions often lead to anxiety and compulsions.
Delusion	Firmly maintained false belief which is contradicted by reality. It is idiosyncratic, incorrigible and preoccupying. Partial delusions are found when the person is in the process of recovery. Type of delusion include persecutory (harm by others), nihilistic (the patient ceases to exist), reference (media refers to patient but no supporting evidence), grandiose (exaggerated power and importance), jealousy (believing the partner is unfaithful through pathological reasoning).
Primary delusion	A delusion that arises out of the blue with no explanation. For example, a schizophrenia man suddenly believes that he is from Mars.
Secondary delusion	A delusion which is secondary to other psychopathology such as auditory hallucination in schizophrenia or grandiosity in mania. For example, a manic patient believes that he is the president of a country.
Delusional mood	It consists of a change in mood that precedes the emergence of the delusion. The person feels that something sinister is going to happen.
Delusional perception	A real perception with delusional interpretation: "When the traffic light turns green, God asks me to go to heaven."
Delusional memory	A real memory with delusional understanding: "I had an appendectomy 10 years ago and aliens put an implant in my body to control the world.

Psychopathology in movement

Ambitendency	Repetitive behaviour of cooperation and opposition. It occurs in schizophrenia. It is a form of ambivalence. The person starts to make a movement but before completing it, he/she starts the opposite movement.
Mitgehen	Excessive cooperation and limb movement in response to slight pressure of an applied force even the person is told to resist movement. It occurs in catatonia.
Mitmachen	Limb movement in response to an applied force to any direction without resistance. It occurs in schizophrenia.
Waxy flexibility	Abnormal maintenance of posture in people suffering from catatonia. For example, the person maintains his left arm in the air after the examiner raises his left arm.
Automatic obedience	The person usually suffering from catatonic schizophrenia follows the examiner's instructions blindly without judgement and resistance.
Negativism	The person usually suffers from catatonia and actively performs the opposite action to the instruction from the examiner.
Stereotypy	Non goal – directed repetitive movements. (e.g. rocking forward and backward). It occurs in people with schizophrenia.
Mannerism	Goal – directed repetitive movements. (e.g. a speaker tries to move his hands repetitively to convey his messages).
Echolalia	Repetition of words or phrases spoken by another person. This is seen in autism, schizophrenia

	and dementia
Echopraxia	Repetition of movement demonstrated by another person. This is also seen in catatonia and schizophrenia.
Catalepsy	Abnormal maintenance of postures in catatonia.
Cataplexy	Temporary loss of muscle tone in narcolepsy.
Compulsion	Repetitively and purposeful movements which are resisted by the person as they are senseless. Compulsions aim to neutralise anxiety generated by obsessions.

Common psychiatric syndromes

Charles de Bonnet syndrome	People suffering from this syndrome exhibit complex visual hallucations. This syndrome is associated with central or peripheral reduction in vision. It also occurs in people who are blind or with partial vision. The content of hallucination involves vividly coloured people or animals. There are no delusions and no hallucinations in other modalities. The complex visual hallucinations may last from days to years.
Folie à deux	A delusional disorder where a delusion is transferred from a person with a psychotic disorder to another with whom he or she is in close association, so that they share the same delusion
Capgras syndrome	Delusional misidentification of a familiar person and the patient believes that the familiar person is replaced by an imposter or a double. It occurs in schizophrenia, affective disorder or dementia. This syndrome is more common in women.
Erotomania or DeClerambault's syndrome	This syndrome refers to female patient who loves a man of higher social status but there is no evidence to support their relationship or the patient is known to the man of higher social status.
Fregoli syndrome	Delusional misidentification of an unfamiliar person as a familiar person. It occurs in schizophrenia or dementia.
Ganser's syndrome	In this syndrome, approximate answer interspersed with correct responses, apparent disorientation, clouding of consciousness, vorbeireden (talking past point), pseudohallucination and fluctuation of somatic symptoms. The person may have amnesia for the duration of illness after recovery.
Othello syndrome or morbid jealousy	In this delusional disorder, the patient firmly believes that his or her spouse or partner is unfaithful. While it is possible to have an affair, the way the person reaches the conclusion is pathological. It often associates with confrontation and violence. Treatment involves geographical separation and antipsychotic treatment.
Munchausen syndrome	Aka hospital addiction syndrome and patient wants to be admitted by exaggerating his or her symptoms. Munchausen syndrome by proxy refers to the fact that the mother imposes her child to be admitted and it is a form of child abuse

OSCE grid

OSCE station

A 20-year-old university student was brought in by the counsellor to the emergency department. He was found sitting in the lift of the residential hall the whole day and refused to attend classes. He claims that he hears voices talking to him.

Task: assess his hallucinations.

OSCE grid: approach to hallucinations:

Candidates are advised to ask about hallucinations in other modalities as you will discover more information. This person may have used recreational drugs causing visual or tactile hallucinations. The person can be paranoid and suspicious of doctors. Candidates have to spend time establishing rapport at the beginning.

A) Auditory (second person auditory hallucination in depression and mania)	A1) Introduction When people are under stress, they may have unusual experiences. I understand from your counselor that you have been hearing voices.	A2) Open questions about hallucinations Can you tell me more about the voices? Can you give me an example?	A3) Nature of the voices How do they address you? Do they speak directly to you?	A4) Command hallucinations Do they give you orders? Do you obey them?	A5) Congruence with mood Does your mood influence the content of the voices? For example, when you are sad, you hear the voices say sad things.
B) Auditory (third person auditory hallucination in schizophrenia)	B1) Number of voices B1) Do you hear more than one voice talking about you?	B2) Content of the hallucination What do they say?	B3) Running commentary Do they comment on what you are doing or thinking?	B4) Audible thoughts Are the voices saying out your thoughts aloud?	B5) Echo de la pense Do those voices echo your thoughts after a few seconds?
C) Confirmation of the nature of hallucination (rule out pseudohallucination)	C1) Where do these voices come from? Do you come from external space?	C2) Do the voices come from inside or outside your head?	C3) Do you feel that the voices are real? Are the voices as clear as my voice?	C4) Can you stop them? Can you distract yourself from the voices?	C5) When do these voices occur? Were you falling asleep or waking up? (to rule out hypnagogic and hypnopompic hallucinations)
D) Hallucinations in other modalities and assess insight	D1) Visual hallucinations Have you seen things that other people can't see? What do you see? Can you give me an example?	D2) Olfactory hallucinations Is there anything wrong about the way we smell? Can you tell me more about it? If yes, who sent the gas to you?	D3) Gustatory hallucinations Have you noticed that food or drink seems to have a different taste recently?	D4) Tactile / haptic hallucinations Have you had any strange feelings in your body? How about people touching you? How about insects crawling?	D5) Assess insight Do you have any explanation of above experience? Do you need help (e.g. taking medication to reduce the voices?)

E) Course/ Comorbidity/ risk assessment	E1) Course of hallucinations How long have you experienced those voices?	E2) Assess impact How does it affect your life How is your mood? Has your mood been affected by those voices?	E3) Assess other first rank symptoms Do you worry that those voices are part of a plot? (e.g. harming you or controlling you) Do you feel that your thoughts are being interfered? (e.g. thoughts are being inserted, withdrawn or broadcasted)	E4) Assess substance misuse Do you use recreational drugs or alcohol?	E5) Risk assessment I can imagine that you are stressed by the voices. Some people may want to give up. Do you have thought of ending your life?

Sample MCQs and MEQs

1. A 30-year-old man was brought to the Accident and Emergency Department. He suddenly fell down after hearing a loud sound at a party. There was no loss of consciousness. The psychopathology being described is:
A. Catalepsy
B. Cataplexy
C. Catatonia
D. Posturing
E. Waxy flexibility

Answer: B Cataplexy refers to the temporary paralysis and loss of antigravity muscle tone without loss of consciousness. Cataplexy is often precipitated by emotional excitement and associated with narcolepsy.

2. An elderly Chinese man complained that 'My guts are rotten and blood has stopped flowing to my heart. I am dead.' The psychopathology being described is:
A. Acute intestinal obstruction
B. Delirium
C. Delusion of control
D. Nihilistic delusion
E. Hypochondriasis

Answer: D. Nihilistic delusion or Cotard syndrome is the belief that one is dead or the external world does not exist. It can also take the form of believing that parts of the body do not exist. Nihilistic delusions can be secondary to severe depression, schizophrenia or to an organic disorder.

3. A 50-year-old man is referred from vascular surgeon for pushing sensation in the abdominal aorta. He complains that he feels someone is pushing his abdominal aorta and he is very disturbed by this sensation. The psychopathology being described is:
A. Cenesthetic hallucination
B. Delusional perception
C. Haptic hallucination
D. Kinesthetic hallucination
E. Somatic passivity

Answer: A. Cenesthetic hallucination involves a false perception of an altered stated in body organs without corresponding sensory receptor to explain the sensation. Kinesthetic hallucinations involve the sense of muscles or joints. For example, the patient might feel like their limbs are being twisted. Common causes include schizophrenia and withdrawal state from benzodiazepines or alcohol intoxication.

4. A man sees a blue car driving past him and he realizes that the terrorists are going to kill him. This is most likely to be which of the following:
A. Delusion of hypochondriasis
B. Delusion of passivity
C. Delusional perception
D. Delusion of persecution
E. Visual hallucination

Answer: C. A delusional perception is a normal perception falsely interpreted by the patient and held as being significant to him. It is one of Schneider's first-rank symptoms.

5. Morbid jealousy is not:
A. a misidentification phenomenon
B. associated with erectile dysfunction
C. associated with violence
D. encapsulated
E. more common in men than women

Answer: A. Misidentification syndrome refers to Capgras and Fregoli syndrome. Morbid jealousy is a delusional disorder that the marital or sexual partner is unfaithful, typically accompanied by intense searching for evidence of infidelity and repeated interrogations and direct accusations of the partner that may lead to violent quarrels. Morbid jealousy is more common in men than in women. Morbid jealousy is associated with erectile dysfunction and alcohol misuse.

MEQ
A 20-year-old woman presents with psychotic symptoms such as seeing images and hearing voices which ask to cut her wrist for 6 months. She also complains of mood swings. Her parents divorced and she was abused by her step-father.

Q1. What are the differential diagnoses?
1. Bipolar disorder with psychotic features.
2. Borderline personality disorder
3. Major depression with psychotic features
4. Schizophrenia
5. Substance misuse (e.g. hallucinogens, amphetamine)

Q2. Her mother wants to know how you would differentiate schizophrenia from other differential diagnoses. Your answer is:
1. Schizophrenia patients usually hear third-person auditory hallucinations.
2. Audible thoughts, thought echo, and running commentary usually occur in schizophrenia.
3. Schizophrenia patients have other first rank symptoms such as thought interference, passivity and primary delusions.

EMIS:

Psychopathology

A – Completion illusion
B – Dysmeglopsia
C – Extra-campine hallucinations
D – Haptic hallucination
E – Hypnogogic hallucination
F – Hypnopompic hallucination
G – Kinaesthetic hallucination
H – Liliputian hallucination
I – Pareidolia
J – Synesthesia

Question 1: The experience of one's name been called out when one is about to fall asleep – Ans: Hypnogogic hallucination

Question 2: A young women has been diagnosed with first episode psychosis. She claimed that she is unable to concentrate as she hears her father shouting at her continuously from his flat 2 miles away – Ans: Extra-campine hallucination

Question 3: A child watches the clouds and reports being able to see images of a computer game in the clouds – Ans: Pareidolia

References:
Campbell RJ. *Psychiatric Dictionary*. Oxford: Oxford University Press, 1996.
Puri BK, Hall AD. *Revision Notes in Psychiatry*. London: Arnold, 2002.
Sims A. *Symptoms of the Mind: An Introduction to Descriptive Psychopathology*. London: Saunders, 2003

Notes:

Chapter 2 Clinical interview, formulation and management

Interviewing techniques
Psychiatric history
Mental state examination
Physical examination

Differentials diagnosis and hierarchy of diagnosis
Formulation
Relevant investigations
Management and multi-disciplinary approach

1) **Open-ended questions:** *'Can you tell me why you were admitted?'* Open-ended questions are often used in the initial phase of the interview to produce spontaneous responses from the patient, which are potentially what feel most important to the patient. They convey a sense of genuine interest to the patient.
2) **Closed-ended Questions**: *'Did you attempt to end your life prior to admission?'* Closed-ended questions often follow open-ended questions to efficiently elicit specific details.
3) **Summation** refers to the brief summary of what the person has said so far and is done periodically to ensure the interviewer understands the person correctly. E.g. *'I would like to make sure that I understand you correctly so far. You are saying that you do not think your experience is part of schizophrenia based on your own readings (in a man who does not believe that he suffers from schizophrenia and wants to seek a second opinion)'*
4) **Transition** is a useful technique to gently inform the person that the interview is going onto another topic.
5) **Empathic statement**s convey the message that the psychiatrist finds the patient's concern is important and acknowledge the patient's sufferings. E.g. *'I can imagine that you were terrified when you realized that you could not move half of your body (to a man suffering from post-stroke depression).'*

Psychiatric History

	Purpose	Demonstration
Introduction	Before you start the interview, ensure the person understands the purpose and make that there is no hearing impairment.	"Hello, my name is Dr. Zhang. Has anyone told you about the nature of this interview? Let me explain…"
Identifying Information and demographics	Key demographic data include: 1. Full name. 2. Age. 3. Gender. 4. Marital status: married, divorced, single, widow. 5. Occupation (if the person is unemployed, the interviewer should explore previous job and duration of unemployment; if the person is a housewife, explore her spouse or partner's occupation. 6. Current living arrangement: living alone, homeless, with family. 7. Current status: inpatient or outpatient.	"Before we start the interview, it is important for me to ensure that I got your name right…" "May I know your current age?" Example for presentation: 'Mr. Tan is a 36-year-old taxi driver, married and stays with his family in a 3-room HDB flat. He is currently an inpatient at ward 33, National University Hospital.'

Presenting complaint	The presenting compliant can be part of a first episode of illness or as one of a series of episodes. For patients who are hospitalised, it is important to enquire whether this admission is voluntary or under the Mental Disorder and Treatment Act. Seek the person's view in his or her own words about the admission. List the symptoms in lay term in the order of decreasing severity and state the duration during presentation.	"What have brought you here to this hospital or clinic? Can you tell me what has happened before that?" Example for presentation: 'Mr. Tan presents with an intention to end his life, hopelessness, low mood, poor sleep, poor appetite and poor concentration for 3 months.'
History of present illness	Enquire the precipitating factors, symptoms severity, duration and context of the current episode in a chronological order. Enquire maintaining and protective factors. Assess impact on relationships and functioning. Enquire significant negatives (e.g. psychotic symptoms in severe depression). Seek his or her views towards previous psychiatric treatments, assess efficacy and explore previous side effects. Assess for common psychiatric comorbidity and differential diagnosis associated the history of present illness.	Start with allowing the person to talk freely for 5 minutes and demonstrate eagerness to hear the person's concerns. Questions like "What made you seek treatment this time?" may reveal current stressors; and "What are the problems that your illness has caused you?" assesses functional impairments. "What do you think may have caused you to feel like this?" may reveal patient's perception of symptoms.
Past psychiatric history	Enquire past psychiatric diagnoses, past treatments (medication, psychotherapy or ECT); side effects associated with psychotropic medications, adherence to treatment, previous hospitalizations (including involuntary admissions) and treatment outcomes. History of suicide, self-harm, violence and homicide attempts is essential to predict future risk. It is important to identify the precipitating and maintaining factors of each episode, as this would provide important information in formulation.	Example for presentation: 'This current episode is in the context of 20-year history of depressive illness. 20 years ago, Mr. Tan first consulted a psychiatrist in private practice because of low mood, poor sleep and loss of interest for 6 months. The psychiatrist prescribed a tricyclic antidepressant (amitriptyline). Mr. Tan complained of dry mouth and constipation. Due to financial constraint, he consulted a psychiatrist at the National University Hospital……..'
Past medical history and review of major systems	Enquire past and current medical problems and physical symptoms, in particular pain (e.g. migraine), seizures (e.g. temporal lobe epilepsy), stroke, head injury, endocrine disorders and heart diseases. Indicate medications that the patient has been prescribed for the above problems. Enquire past surgical problems and surgery received. Explore drug allergy (especially to psychotropic medication and clarify the allergic reactions)	Example for presentation, 'Mr. Tan suffers from hypertension and hyperlipidaemia. The polyclinic doctor has prescribed atenolol 50mg OM and simvastatin 40mg nocte. Mr. Tan has no past history of surgery. He has no past history of drug allergy.'

	The physical problems may be due to the medication effect (e.g. prolonged QTc resulted from antipsychotics).	
History of substance misuse	Enquire type (alcohol, benzodiazepine, and recreational drugs), amount, frequency, onset of and past treatment for substance misuse.

Explore biological (e.g. delirium tremens/ head injury) and psychosocial complications (e.g. drunk driving, domestic violence) associated with substance misuse.

Explore the use of tobacco (quantity and frequency).

Look for dual diagnosis e.g. depression with alcohol misuse.

Explore the financial aspect (i.e. the funding of substance misuse). | Maintaining a non-judgmental attitude is essential in enquiring substance misuse.

Normalise the experience of substance misuse, "When people are under stress, they may use recreational drugs. I would like to find out from you whether you have such experience."

Avoid a direct question like, "Do you use drugs?" which may prompt the person to deny any drug use. |
| **Family history** | Enquire the family psychiatric and medical histories.

Look for substance misuse (e.g. alcohol) among family members.

Look for early and unnatural deaths which may indicate suicide.

Briefly assess the quality of interpersonal relationship in the family. | Demonstrate empathy if a close family member suffers from severe psychiatric illness.

Identify the aetiology of psychiatric illness in family members. E.g. "What made you think that your mother may suffer from depression?" |
| **Background history** | Relationship or marital history (current and past); wellbeing of children (e.g. child protection issues); psychosexual history (sexual orientation, STDs if relevant, issues related to infertility, for women, candidate needs to explore the last menstrual period, possibility of pregnancy and method contraception); social history (current living situation, level of expressed emotion, past employment and religion).

Enquire details about current occupation: e.g. working hours, interpersonal relationship in the workplace and stress level. Enquire reasons for changing jobs (e.g. interpersonal problems which reflects patient's personality) Look for potential risk associated with an occupation (e.g. alcohol misuse in a man working in a bar).

Developmental history: birth, childhood development, relationship with parents and siblings, history of physical/sexual abuse, prolonged separation, unhappy childhood. | Seek permission from the patient to explore sensitive issue, 'It is important for me to explore the following aspect as it may be relevant to your case. Some of the issues are sensitive. Would it be alright with you? May I know your sexual orientation?'

Questions to assess premorbid personality:

'How would you describe your character?'

'How do you think your friends or relatives would describe you?'

'Has your character changed since you became unwell?' |

	Education history: details of schooling age that schooling began and stopped. If education is stopped prematurely, enquire the reasons. Education background affects performance of the Mini-Mental State Examination (MMSE). Premorbid personality: Ask the person to describe his or her character, habits, interests, attitude to self or others and coping mechanisms. Forensic history: history of offences, the nature of offences, current status (convicted or pending court case), previous imprisonment and remorse towards the victim.	'Can you tell me your attitude towards other people like colleagues, supervisors and the society?'
Risk assessment	Enquire suicidal and homicidal ideations, access to large amount of medications, sharp objects, firearms and tendency to violence. Risk is not limited to the above. In young people with ADHD, candidates need to assess the risk of accidents due to poor impulse control and hyperactivities. In old people with cognitive impairment, candidates need to assess risk of fall, fire, accidents and exploitation.	People with suicidal or homicidal ideation are often guarded to share with the others on their thoughts. Candidates should be empathetic to the patients and offer reassurance that they are interested in the patients' well being and want to offer help as much as possible. E.g. "I am sorry to hear that you have gone through so much hardship. Have you been thinking about death or ending your life?"
Closing	"We may need to wrap up the interview shortly. I would like to invite you to ask me any questions. I will try my very best to address your concerns."	Always close by thanking the patient.

Mental state examination

Rapport and attitude to the interview	The interpersonal nature of rapport include establishing the mutual experience and awareness of the presence of the other, the mutual sharing of thoughts and feelings, the notion of empathy and the development of mutual trust between the interviewers and patients.

Appearance and behaviours	General appearance includes the observation of dress, grooming, facial expression and ancillary objects (e.g. handcuffs, weapons, etc) often provide information about attitude and other non-verbal communications relevant to rapport.
	
	Behaviours: Observation of non-verbal behaviours includes commenting on the eye contact, gestures and other useful communications.
	    (a) Disinhibition refers to behavioural and social manifestations of undue familiarity, coarseness, and aggressive behaviours.
	    (b) Mannerisms (repetitive behaviours without serving any purpose), stereotypies (repetitive behaviours serving a purpose) and uncontrolled aggression may indicate impairment in reality testing.
	    (c) Childlike behaviour or regression: This may occur in an adult with personality disorder. The person may hold a toy or stuffed animal as transitional object.

Speech	(a) The rate, volume and pressure of speech (e.g. increased rate in mania and decreased rate in depression). (b) Loss of tone is seen in people with Asperger's syndrome. (c) Dysarthria involving slurred speech. (d) Dysphasias (e.g. Wernicke's and Broca's dysphasias). (e) Stuttering and cluttering of speech. (f) Explosive quality with forced vocalization (in agitated and angry patient). (g) Preservative features in speech (e.g. difficulty in changing topics in people with frontal lobe syndrome). (h) Elective mutism. (i) Poverty of speech.
Thought	The flow of thought: 1. Circumstantiality. 2. Tangentiality. 3. Flight of ideas/ racing thoughts (in manic patients). 4. Thought blocking. 5. Loosening of associations. The form of thought 1. Obsessions. 2. Overvalued ideas. 3. Delusions. The content of thought 1. This provides an opportunity to assess the capacity of a person to express his or her experiences and clinical information. 2. Content of phobia: crowded area (agoraphobia), centre of attention (social phobia) or object (e.g. spider) 3. Obsessional themes: contamination, harming the others, intellectual aspects (e.g. purpose in life) Thought interference: thought insertion, withdrawal and broadcasting. Overvalued ideas. Delusions: 1. Primary delusions (e.g. morbid jealousy, persecutory, reference) 2. Secondary delusions (e.g. nihilistic, grandiose) 3. Passivity phenomena or delusion of control 4. Delusional misinterpretation

Affect and mood	General description of affect: happy, euphoric, manic, depressed, tearful, anxious, angry or detached.
	
	Range of affect: it can be within normal range, labile affect, blunted affect, flat affect or apathy.
	    a. Labile affect: rapid fluctuation of affect (e.g. from tearfulness and laughter in a 10-minute interview, seen in patients with bipolar disorder).
	    b. Blunted affect: reduction in variation in mood and emotions.
	    c. Flat affect: almost total absence of variation in mood and emotions.
	    d. Apathy means "freedom from feeling" or a dulled emotional tone. It also conveys the sense of indolence and indifference to what normally excites emotion or interest. Detachment is one of the aspects of apathy. Apathy is a negative symptom in schizophrenia and also seen in patients with dementia.
	2. Other descriptions of affect: stability throughout the interview, congruent with the history of present illness and present complaint.
	3. Mood refers to the subjective emotional state described by a patient. Ask the person to rate his or her mood in a scale of 1 to 10 (1 means very sad and 10 means very happy) and look for congruency with the affect observed by the candidate.

	Other features associated with affect: 4. Alexithymia refers to difficulty in a person to verbalise his or her emotion. 5. Laughter is a universal human behaviour and often a pleasurable relief of tension. Appropriate laughter may be a very useful phenomenon to observe in that it confirms the persons' reactivity of affect, their capacity for exhilaration and mutual enjoyment. 6. Crying may occur in a wide range of normal situations (e.g. grief, relief, pain, fear, shame, guilt or humiliation) as well as in a number of psychiatric disorders (e.g. depressive disorder, adjustment disorder, acute stress disorder).	
Perceptual disturbances	1. Illusions (affect illusions, completion illusions, and pareidolic illusions). 2. Hallucinations (auditory, visual, olfactory, gustatory, tactile and haptic hallucinations). 3. Pseudohallucinations.	
Insight	A psychiatric patient may have impaired insight if: 1. He/she believes that she is not ill. 2. He/she does not believe that his/her signs/symptoms are due to a psychiatric condition. 3. He/she does not believe that psychiatric treatment is helpful.	
Cognitive function	1. Students need to comment on the level of consciousness and orientation. If dementia is suspected, students need to test general knowledge of the person (e.g. the name of prime minister), short-term memory (by recall of an address or 3 objects) and to perform Mini mental state examination (MMSE). 2. Assess orientation to time, place and person. 3. Assess attention and concentration. Attention is the ability of a person to maintain his or her conscious awareness on an external stimulus and to screen out irrelevant stimuli. Concentration refers to sustained and focused attention 4. Level of consciousness refers to alertness and arousal and awareness of self and external environment and stimuli 5. Abstract thinking refers to the manipulation of concepts, constructs, ideas, thoughts or images and link them in a flexible way giving rise to relationships and meaning which were not literal	

Physical examination

Physical examination is important for psychiatric practice and students are advised to look for tell-tale signs for underlying physical conditions.

General inspection	1. General well-being and body size (e.g. cachexia in anorexia nervosa, obesity which may suggest metabolic syndrome). 2. Abnormal eyes (e.g. exophthalmos in Graves' disease; lid lag is a sign in thyrotoxicosis). 3. Atypical facial features may suggest syndromes associated with learning disability (e.g. Down's syndrome). 4. Extremities: scars on the arms may indicate deliberate self-harm in people with borderline personality disorder. Needle tracks may indicate intravenous drug use.
Neurological examination	1. Look for tremor due to thyrotoxicosis, withdrawal from alcohol, extrapyramidal side effects). 2. Look for involuntary movements (e.g. restlessness in lower limbs is also known as akathisia which is induced by antipsychotics). 3. Look for abnormal posturing (e.g. waxy flexibility in catatonia). 4. Look for abnormal gait (e.g. shuffling gait in people with Parkinson's disease; broad-based gait as a cerebellar sign in people who misuse alcohol).
Cardiovascular examination	5. Pulse: heart rate and pattern. 6. Blood pressure and postural change in blood pressure (some antipsychotics such as quetiapine cause postural change in blood pressure). 7. Heart sound and murmur.
Abdominal examination	1. Inspect for scars due to previous operation. 2. Inspect for lanugo hair (associated with anorexia nervosa). 3. Palpate for hepatomegaly.

Diagnosis and hierarchy of diagnosis

Students are advised to provide the diagnosis based on the DSM-IV-TR or ICD-10 criteria. DSM-IV-TR uses the five axis approach (Axis I – main psychiatric illness; Axis II – personality disorder or mental retardation; Axis III – general medical condition; Axis IV – psychosocial and environmental problems; Axis V – Global assessment of functioning (GAF score).

Students are advised to rule out the conditions at the upper part of the pyramid before concluding the conditions at the bottom part of the pyramid. The conditions at the upper part of the pyramid are given the diagnostic priority because these conditions are treatable and reversible by biological treatments.

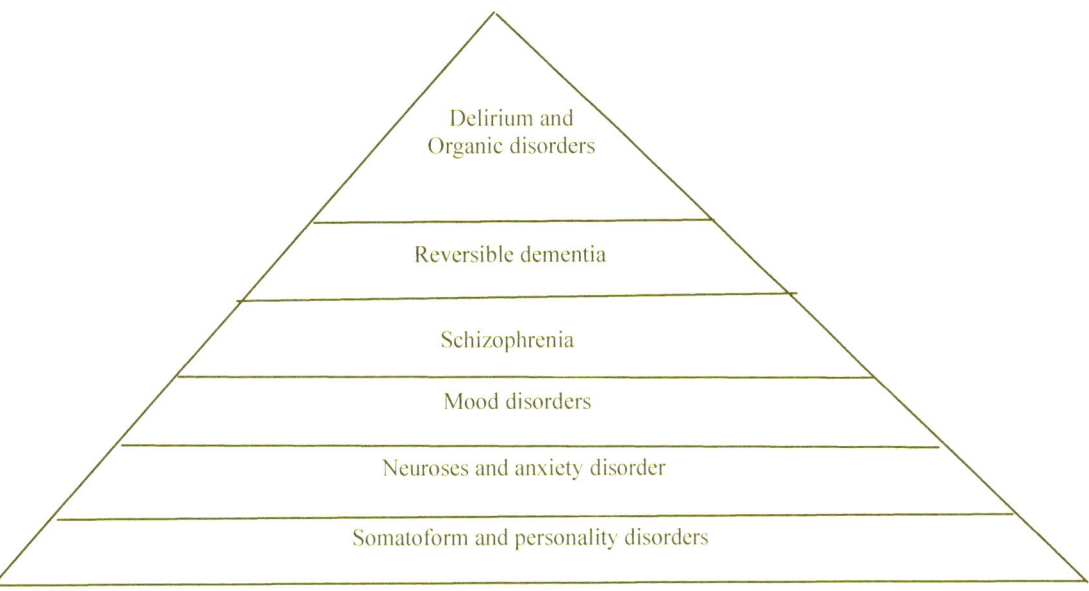

Formulation

Formulation of a case could be based on the under-mentioned model:

	Predisposing factors	**Precipitating factors**	**Maintaining factors**
Biological factors	Genetics (e.g. important for schizophrenia, bipolar disorder, severe depressive episode, obsessive compulsive disorder and Alzheimer's disease)	Examples: - Recent physical illness leading to depression. - Substance misuse leading to psychosis. - Sleep deprivation leading to mania.	Examples: - Chronic medical diseases, - Continued alcohol misuse.
Psychological factors	Personality and temperament.	Examples: - Recent loss and stress. - Non – adherence to medication.	Examples: - The presence of cognitive errors. - Lack of meaningful activities. - Impaired insight.
Social factors	Adversity in socio-economic status; early separation from parents.	Examples: Life events: divorce or unemployment.	Examples: Lack of income or social isolation.

Investigations

1. Routine investigations in psychiatry: FBC, LFTs, U & Es or RFTs, TFTs, Urine drug screen and ECG (to look for prolonged QTc).
2. Further investigations will be required based on patient's history and potential diagnosis, e.g. CT or MRI brain scan, EEG, VDRL, B12 and folate, HIV testing with patient's consent..
3. With the patient's permission, the psychiatrist may obtain collateral history from partner or spouse, friends, family, GPs and other professionals.
4. Investigations do not limit to the clinical setting. The treatment team may consider doing a home visit to understand the interaction between home environment and current psychiatrist illness. Occupational therapist can assess the activity of daily living of an individual while social worker can assess social support.
5. If the patient requires further psychological assessment, the psychiatrist needs to identify a clear goal. For example, a baseline neuropsychological assessment of a person with recent head injury and repeat the assessment after six months of cognitive rehabilitation.

Management and multidisciplinary approach

1. Treatment can be divided as immediate, short term, medium term and long term.

2. **Immediate treatment** includes hospitalisation and close supervision to prevent suicide attempt or prescription of benzodiazepines to prevent alcohol withdrawals. Application of Mental Disorder and Treatment Act to admit the patients who are at risk but refuse admission to the Institute of Mental Health.

3. **Short term treatment** may include biological treatments, risk management and discharge planning.

    a. **Biological treatments:** depending on the diagnosis, comorbidity and drug interactions, psychiatrists may consider antipsychotics, antidepressants, mood stabilisers, benzodiazepine and anti-dementia medication.

    b. **Consulting other specialists:** if necessary, the psychiatrists may obtain input from other experts in medicine, surgery, paediatrics, geriatrics, and obstetrics and gynaecology.

    c. **Risk management** involves modifying and managing risk factors. (E.g. treat the underlying depressive disorder to prevent suicide and to advise caregivers to remove sharp items at home to prevent self-harm). It is important to prevent harm to other people (for example, inform the social agency to ensure the safety of the patient's children).

    d. **Establish therapeutic alliance for future psychological therapy.** Psychoeducation and supportive counselling can be offered while the patient stays in the ward. More sophisticated therapy e.g. cognitive behaviour therapy requires further work after discharge.

    e. **Discharge planning** with active collaboration with patient's relatives, outpatient psychiatrist, case manager (e.g. in Early psychosis programme), GP, and community psychiatric team. The relatives and mental health team can help to identify relapse and plays an active role in the contingency plan

4. **Medium and long term treatment** may include psychological and social treatments.
    a. **Psychological treatments** may involve counselling, supportive psychotherapy, cognitive behaviour therapy or interpersonal therapy. Individuals must explore patient's motivation, preferences and previous response to psychological treatments.

    b. **Social treatments** may involve monitoring by community psychiatric nurses, vocational assessment, supported work schemes, domiciliary self care training, supported accommodation and child protection.

MCQs

1. A 40-year-old woman is referred by her GP because she suffers from depressive disorder. During the interview, she has difficulty to verbalize her emotions. This phenomenon is best described as:

A. Ambivalence
B. Affective flattening
C. Alexithymia
D. Alogia
E. Anhedonia

Answer: C
The difficulty to express one's emotion is known as alexithymia. Primary alexithymia is described as a personality trait, characterized by difficulty in identifying one's emotional state, Secondary alexithymia is a "state" reaction during serious physical illness and severe depressive episode. Defective interhemispheric communication or inhibition of corpus callosum activity may lead to alexithymia.

2. You are assessing a 30-year-old man who is aggressive towards staff in the Accident and Emergency Department. The following are appropriate approaches except:

A. Adopt non-confrontation approach
B. Advise patient to sit near the door
C. Minimize eye contact
D. Leave examination door open
E. Lower your own voice

Answer: B
In this scenario, the psychiatrist should sit near the door to ensure there is an escape route. This will allow the psychiatrist to mobilize resources easily. If the patient sits near to the door, he or she can make the psychiatrist a hostage by blocking the exit.

3. Which of the following statements about mental state examination is correct?

A. During the mental state examination, a psychiatrist should ask a large range of psychopathologies to cover every diagnostic possibility

B. During mental state examination, probe questions about psychopathology are best framed with technical questions

C. Mental state examination is best conducted as formal exercise which follows a schema

D. Serial mental state interrogations are an extremely reliable method of judging the progress of treatment or changes in patient's mental health

E. The best clinical approach to mental state examination is a conversational and informal manner

Answer: E
Unskilled interviewers often conduct the mental state examination is a rigid schema with a lot of technical questions. This will prevent the patient from opening up and volunteering new information. It also explains why option D is incorrect. Option A is incorrect because the psychiatrist should focus on possible psychopathology based on information obtained in history-taking. Option B is incorrect because probe questions about psychopathology are best framed with reference to activities and ideas that are familiar to the patient. Option D is incorrect because state interrogations are unreliable methods of judging the progress of treatment or changes in patient's mental health.

EMIS:
Clinical Interview
A – Circular questioning
B – Clarification
C – Closed questioning
D – Confrontation
E – Empathy
F – Interpretation
G – Open questioning
H – Recapitulation
I – Summation
J – Sympathy

For each of the statements below, please select the most likely clinical interview technique used.

Question 1: I am sorry that you are feeling this way – Ans: Sympathy

Question 2: So, you have continued using sleeping pills despite your former decision to stop using – Ans: Recapitulation

Question 3: I understand that you are very frustrated with your father. I am wondering if you blame her for all that has happened to you – Ans: Interpretation

Reference: Poole R, Higgo R. *Psychiatric Interviewing and Assessment*. Cambridge: Cambridge University Press, 2006.
Notes:

Chapter 3 Schizophrenia, antipsychotics and related disorders

Epidemiology of schizophrenia	Common side effects and alternative antipsychotics
Aetiology	The first generation antipsychotics
Neurobiology of schizophrenia	The second generation antipsychotics
Symptoms of schizophrenia (Based on ICD and DSM)	Extrapyramidal side effects and tardive dyskinesia.
OSCE grid for psychotic disorders	Psychosocial intervention (MOH and NICE guidance)
Physical examination ,investigation and questionnaire	Catatonia
Differential diagnosis	Schizoaffective disorder
Management (MOH guidelines)	Delusional disorders
Antipsychotics: the most and the least	Revision MCQs, MEQs & EMIs

Epidemiology of schizophrenia

	Incidence	Point prevalence	Lifetime risk	Male: Female
Worldwide	15-50 per 100,000 population per year	1%	1%	1:1
Singapore [Chong SA 2004]	10-40 new cases per 100,000 per year	0.6% (23200 cases over 4 million in 2004)	0.7%	1:1

Aetiology

Genetic risk factors	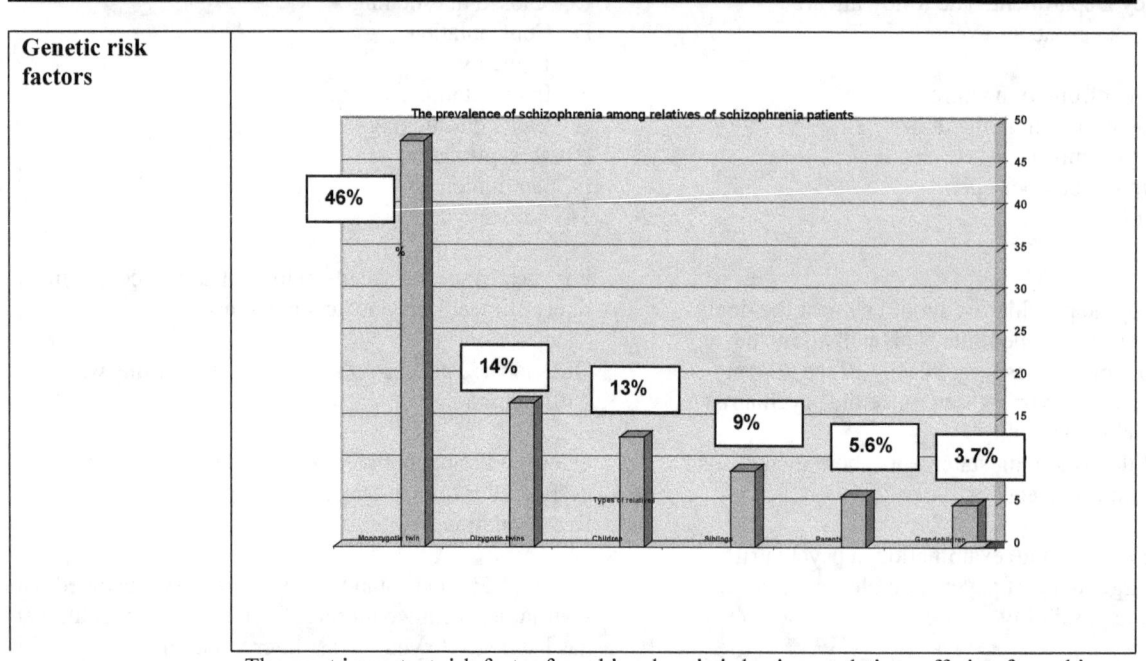
	The most important risk factor for schizophrenia is having a relative suffering from this disorder. The risk to develop schizophrenia for the general population is 0.5 - 1%, the first degree relatives of a schizophrenia patient is 10% and the offspring of two affected parents is 46%. Neuregulin 1 gene on chromosome 8, dysbindin gene on chromosome 8 and chromosome 22q11 (Velo-cardio-facial syndrome) are implicated in the aetiology of schizophrenia.
Antenatal and perinatal risk factors	Second-trimester exposure to influenza infection may increase the risk of the foetus subsequently developing schizophrenia. Associations have also been found with maternal measles and rubella infections. There is a significant association between schizophrenia and premature rupture of

	membranes, preterm labour, low birth weight and use of resuscitation during delivery. Foetal hypoxia during delivery predicts reduced grey matter throughout the cortex in people with schizophrenia but not in controls.
Biological risk factors	**Head injury**: This may lead to paranoid schizophrenia. **Epilepsy and temporal lobe disease**: The most common causative factor that results in both schizophrenia and epilepsy might have developed in the uterus. **Substance misuse**: Cannabis may increase the risk of schizophrenia in people who are homozygous for VAL/VAL alleles in COMT genotypes. Patients suffering from rheumatoid arthritis have lower risk to develop schizophrenia (One-third the risk of the general population).
Demographic risk factors	**Age and gender**: Male schizophrenia patients tend to have more severe disease, with earlier onset, more structural brain diseases and worse premorbid adjustment compared to the female patients. **Advanced paternal age** at the time of birth is a risk factor for the offspring to develop schizophrenia. **Social class**: It is still controversial whether low social class is caused by schizophrenia or is an effect of the course and the nature of the disease. 1. **Breeder hypothesis**: socio-economic adversity precipitates schizophrenia in genetically vulnerable individuals. 2. **Social drift explanation**: people who have an underlying predisposition to schizophrenia are more likely to drift down the social scale. **Rural/urban difference**: The higher prevalence of schizophrenia in urban areas is due to interaction of genetic factors, migration, higher rates of social deprivation and more social problems in the inner city. There is more favourable outcome in non-industrialized countries as compared to industrialized countries. **Ethnicity**: Afro-Caribbean immigrants to the UK have higher risk of schizophrenia even in the second generation.
Psychological risk factors	**Stressful life events** are a precipitant of the first episode psychosis. **High expressed emotion** comprising over-involvement, critical comments and hostility from family members for more than 35 hours per week increase the risk of relapse of schizophrenia.

Neurodevelopmental theory of schizophrenia

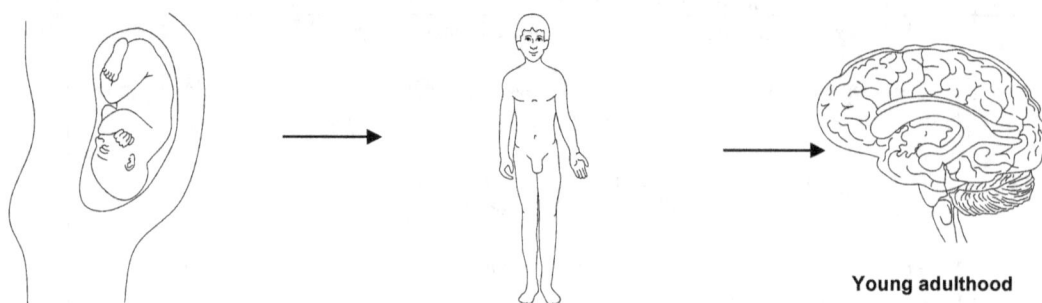

Intrauterine period
Genetic, antenatal and perinatal risk factors cause a disturbance in which the normal pattern of programmed cell death is compromised and leads to a defect in the normal orderly migration of neurons toward the cortical plate. This has serious consequences for the establishment of a normal pattern of cortical connections.

Childhood and adolescence
Constellation of early signs in childhood includes abnormal eye (saccadic) tracking movements, neuropsychological deficits, soft neurological signs (e.g. clumsiness, in-coordination or non-specific EEG changes) and abnormal behaviours.

Young adulthood
Classical schizophrenia symptoms emerge as a result of accumulated intracerebral pathology and spatial disarray of neurons when the human brain reaches certain level of maturity.

Neurobiology of schizophrenia

Gross pathological changes in schizophrenia:
1. Atrophy of the prefrontal cortex and temporal lobe (disturbed neural network in the prefrontal and medial temporal lobes is the core psychopathological feature).
2. Changes in the corpus callosum.
3. Increased ventricular size at commencement of disease (CT changes in the third and lateral ventricles and the temporal horns).
4. Reduction in the overall brain volume.
5. Smaller thalamus.

Histological changes in schizophrenia:
1. Cellular loss in the hippocampus.
2. Reduction of the number of medio-dorsal thalamic neurons.
3. Reduced neuronal density in the prefrontal, cingulate and motor cortex
4. Abnormal patterns of myelination in the hippocampus and temporal lobes as a result of abnormal migration, abnormally sized neurons,

Neurochemical abnormalities:
- **Dopamine**: ↑ dopamine in mesolimbic pathway and the dopamine hypothesis proposes that increased levels of dopamine cause schizophrenia.
- **Serotonin (5 HT)**: There is an interesting relationship between serotonin and dopamine. The two serotonin pathways which are affected in schizophrenia include (1) the projections from dorsal raphe nuclei to the substantia nigra and (2) the projections from the rostral raphe nuclei ascending into the cerebral cortex, limbic regions and basal ganglia. The $5HT_{2A}$ receptor agonism inhibits dopamine release. When there is excess serotonin produced by these two pathways, there is a reduction of the availability of dopamine and that can give rise to the negative symptoms of schizophrenia. Second generation antipsychotics such as risperidone, bind to the D_2, $5-HT_{2A}$ and $α_2$ adrenergic receptors in the brain. Risperidone competes with serotonin and its antagonism at $5HT_{2A}$ receptors cause an increase in dopamine to relieve negative symptoms. As risperidone also blocks D_2 receptors, the positive symptoms are reduced at the same time.

Symptoms of Schizophrenia (Based on ICD-10)

ICD-10 Diagnostic Criteria (Core features – underlined)

General appearance
No specific criteria. Patient may be unkempt and there is deterioration in self care due to decline in **occupation and social function.**

Thoughts:
1. *Thought insertion*
2. *Thought withdrawal*
3. *Thought broadcasting*
4. Formal thought disorders (e.g. neologism)

Affect:
1. Flat or inappropriate affect

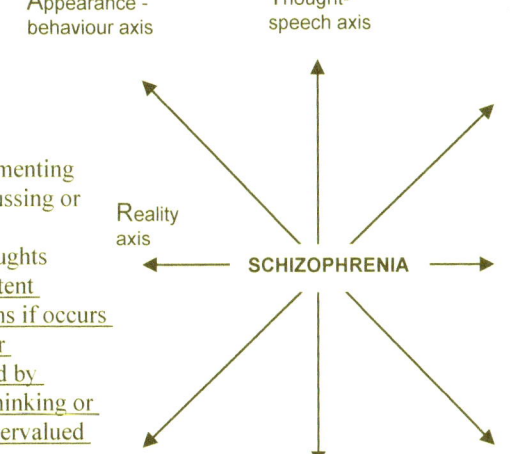

Hallucination:
1. Voices commenting
2. Voices discussing or arguing
3. Audible thoughts
4. Other persistent hallucinations if occurs frequently or accompanied by delusional thinking or sustained overvalued idea (ICD).

Delusion:
1. *Delusional perception*
2. Persistent delusional beliefs are completely impossible. (ICD)

Passivity/ Delusion of control:
1. Made will
2. Made acts
3. Made affect
4. Somatic passivity
5. Thought interference

Interest
1. Loss of interest (avolition), aimlessness, idleness, a self absorbed attitude and social withdrawal over one year (ICD – simple schizophrenia)

Speech
1. Incoherent or irrelevant speech (formal thought disorder)(ICD)
2. Paucity of speech (alogia)(ICD)

Behaviour
1. Catatonic behaviour (ICD)
2. Social impairment (ICD)
3.

Disturbance must last for at least 6 months.
Schneider's first rank symptoms (ABCD):
A – Auditory hallucination (2nd or 3rd person; thought echo; running commentary; voices discussing patient).
B – Broadcasting (thought broadcasting, insertion and withdrawal).
C – Control (Delusion of control and passivity)
D – Delusional perception

DSM-5 Diagnostic Criteria:
Schizophrenia:
Presence of 2 or more of the following symptoms over a 1 month period:
a. Delusions
b. Hallucinations
c. Abnormal speech
d. Disorganized behavior
e. Negative symptoms
(at least 1 of which must be a, b or c)
Such that an individual's premorbid level of functioning is affected in several major domains of life.
There must be continuous impairment over a period of at least 6 months, during which the individual might experience either active or residual symptoms.
These symptoms must not be due to the effects of substance usage or an underlying medical condition

	Further Overview of DSM – 5
Number of symptoms	The DSM-5 specified that for an individual to fulfill this diagnosis, he/she must have at least 2 or more of the following symptoms: a. Hallucinations b. Delusions c. Incoherent and disorganized speech d. Disorganized or catatonic behaviour e. Diminished emotional expression / negative symptoms Of which at least one of (a), (b) or (c) is present Symptoms must be continuous for at least 1 month in duration over a 6 months period of marked impairment in premorbid functioning.
Deterioration in occupational and social function	It is a compulsory criterion.
Duration of symptoms	The minimum duration of disturbance is at least 6 months. The minimum duration of symptoms is at least 1 month.
Inclusion of simple schizophrenia and schizophreniform disorder	Subtypes of schizophrenia have been removed in DSM V due to limited diagnostic stability, validity and reliability. Schizophreniform disorder: The DSM-5 specified that for an individual to fulfill this diagnosis, he/she must have, for a duration of between 1 to 6 months, the following symptoms: a. Hallucinations b. Delusions c. Disorganized or catatonic behaviour d. Disorganized speech e. Diminished emotional expression / negative symptoms The clinician should have considered and ruled out the differentials of schizoaffective disorder, depressive or bipolar disorder with psychotic features.
Other types of schizophrenia	DSM-5 has removed the subtypes of schizophrenia which were previously present in DSM-IV TR. The clinically relevant subtypes of schizophrenia which one may still encounter as based on DSM IV TR includes: 1. Paranoid type (best prognosis)

2. Disorganized type

3. Catatonic type

4. Undifferentiated type
5. Residual type (mainly the presence of negative symptoms)

Catatonia associated with another mental disorder

The DSM-5 specified that the diagnosis of catatonia is made when at least 3 or more of the following symptoms are present:

Disorders of Movement:

a. Stupor

b. Catalepsy

c. Waxy Flexibility

d. Negativism

e. Posturing

f. Mannerism

g. Stereotypy

h. Grimacing
i. Echopraxia

j. Agitation, not influenced by external stimuli

Disorders of Speech:

a. Mutism

b. Echolalia

Other specified schizophrenia spectrum and other psychotic disorder

DSM-5 reserved this category of diagnosis for individual presenting with symptoms characteristic of schizophrenia, but has not met the full diagnostic criteria.

Examples include:

a. Sub-threshold psychotic syndrome

b. Delusions with associated mood disorders

c. Persistent auditory hallucinations in the absence of any other psychotic features

Brief psychosis	The DSM-5 specified that for an individual to fulfill this diagnosis, he/she must have, for a duration of between 1 day to 1 month the following symptoms: a. Delusions b. Hallucinations c. Disorganized speech d. Grossly disorganized or catatonic behavior of which at least 1 of these must be (a), (b) or (c) The individual should be able to return to premorbid level of functioning after the course of the illness. Clinicians need to exclude differentials like major depression or bipolar depression with psychotic features, schizophrenia and exclude the possibility of the symptoms being due to underlying substance use or medical conditions. Brief psychotic disorder could occur either in the presence or absence of stressors.
Post-schizophrenia Depression	Not mentioned
Schizotypal disorder	This disorder is considered to be a personality disorder but DSM-5 also lists this disorder under schizophrenia. It must fulfil 2 core criteria for personality disorder: impairment in self and interpersonal functioning.

OSCE: Interview a patient to establish a diagnosis of schizophrenia			
A 22 – year – old university student is brought by his counsellor to the Accident and Emergency Department. He was seen by the psychiatrist at the University clinic and diagnosed to suffer from schizophrenia. You are the resident working in the Accident and Emergency Department. Task: Take a history to elicit first rank symptoms and other related symptoms to establish the diagnosis of schizophrenia.			

A. Introduction and assess hallucinations.	A1. Introduction 'I would like to ask you some questions. Some of them *may appear a bit strange*. Is that all right with you? I gather that you had been through a lot of stress recently. When under stress sometimes people have *certain unusual experiences*. Have you had such experiences?'	A2. Assess auditory hallucinations By unusual experience, I mean, for example, some people hearing voices when no one around. If yes,.......... 'Do the voices speak directly to you?' (2nd person) 'Do they speak among themselves?' (3rd person) 'What sort of things do the voices say?' **Mnemonics: EAR** **E**choes of own thoughts: 'Do the voices repeat your thoughts?' **A**rguing voices involve at least 2 voices. **R**unning commentary on patient: 'Do these voices describe or comment upon what you are doing or thinking?' Do the voices tell you to do things? (command hallucinations)	A3. Other hallucinations Visual hallucinations: 'Have you ever had experiences during which you saw things or shadows that others could not see?' Tactile hallucinations: 'Do you feel that there are strange sensations within you, as if something is crawling within your body?' Olfactory / gustatory hallucinations: 'Have you ever had experiences during which you smell or experience strange tastes that others do not experience?'
B. Assess thought interferences.	B1.Thought insertion 'Is there any interference with your thoughts?' 'Do others put or force their thoughts into your mind?'	B2. Thought withdrawal 'Could someone take your thoughts out of your head?'	B3. Thought broadcasting 'Do other people know what you think in your mind?' 'Do you feel that your thoughts are broadcasted to other people?'
C. Delusions insight, mood and substance misuse.	C1. Delusion of control or passivity experience. 'Has there been any difficulty with feelings, actions, or bodily sensation? Is there someone or something trying to control you in the following areas:' **Mnemonic – WEA:** W- Will (impulses) E – Emotions (Feelings or affect)	C2 Other delusions. 'When under stress some people find that someone is playing tricks on them. Have you had any such experiences?' Delusion of Persecution: 'Are there some people who try to harm you or make your life miserable?' Delusion of reference: 'Do you think that someone is watching, following or spying on you?'	C3 Assess insight and mood. 'What do you think is the cause for these experiences?' 'Could you suffer from an illness in your mind?' 'Do you think treatment would help to reduce those experiences?' 'How do you describe your mood? Do you feel sad?'

	A – Actions (volitional)	Delusion of grandeur: 'Do you have any special powers or abilities that others don't have?' Delusion of guilt: 'Do you feel like you deserve punishment for mistakes you made in the past? Can you tell me the nature of the mistakes and punishment you deserve?'	'When you feel sad, do you have thought of harming yourself?' 'I encounter some students use recreational drugs when they go for overseas exchange or party. Have you used recreational drugs recently?'
D. Assess negative symptoms and academic disturbances (Note: alogia/lack of speech and flatten affect is observed during interview)	D1. Apathy (lack of motivation) 'Do you encounter any difficulty in looking after yourself? How often do you tend to take a shower or a bath?' 'Has anyone complained about the state the flat is in? Is it difficult to stay tidy or to keep the flat the way you would like it?'	D2. Anhedonia (lack of interests) 'Have you spent any time with friends lately?' 'Do you find it difficult to feel close emotionally to others?' 'Do you have any activity you enjoy to do nowadays?' 'What were your main interests or hobbies in the past?'	D3. Assess social and academic deterioration. 'How do you find your academic performance recently?' 'Can you concentrate in your study?' 'It seems that your academic performance is not as good as in the past. How long has it been?'

Risk assessment and schizophrenia

Violence in people with schizophrenia is uncommon but they do have a higher risk than general population. Prevalence of recent aggressive behaviour among outpatients with schizophrenia is around 5%. The types of violence and aggression are classified as follows: verbal aggression (around 45%), physical violence towards objects (around 30%), violence towards others (around 20%) and self-directed violence (around 10%). Family members are involved in 50% of the assaults with strangers being attacked in 20%. Doctors need to be competent in identify patients at risk and protect both patients and others.

Physical examination, investigation and questionnaire

Physical examination	1) Vital Signs (HR, temperature). 2) BMI to rule out metabolic syndrome and guide choice of medication. 3) Neurological examination. 4) Physical signs suggesting substance misuse (e.g. injection marks).
Investigation	1) FBC, LFT, U&Es, thyroid function tests, fasting lipids and glucose, β-HCG (to rule out pregnancy for women), toxicology screen and ECG (check for prolonged QTc interval).. 2) Optional: Prolactin (if there is galactorrhoea). 3) CT or MRI brain. 4) Urine drug screen to rule out recent use of recreational drugs. 5) EEG if suspects of temporal lobe epilepsy.
Questionnaire	**Positive and negative syndrome scale for schizophrenia (PANSS)** The **PANSS** is a 30-item semi-structured interview to assess positive symptoms (7-items), negative symptoms (7-items) and global psychopathology (16 items). The severity of individual items are scored according to the manuals.

Differential diagnosis

The initial management involves establishing diagnosis and ruling out psychoses that could be secondary to physical morbidity or substance use (MOH guidelines 2011).

1. Misuse of substances such as alcohol, stimulants, hallucinogens, glues or sympathomimetics.
2. <u>Medications include steroids, anticholinergics and antiparkinson drugs.</u>
3. General medication condition including: CVA, CNS infection, CNS tumours, temporal lobe epilepsy, metabolic abnormalities (vitamin B_{12} deficiencies, thiamine deficiencies), head injury, SLE, acute intermittent porphyria, endocrine abnormalities related to thyroid and adrenal glands.
4. Severe depression or mania with psychotic features.
5. Delusional disorders.
6. Dementia and delirium.
7. Personality disorders – Paranoid / schizotypal personality disorders

Management (included recommendations from MOH guidelines)

Acute phase management	Initial treatment (MOH guidelines 2011)	Stabilisation phase	Maintenance phase
1) Prevent harm by hospitalisation. 2) Reduce aggression and threat by rapid tranquilisation (oral lorazepam 1 to 2mg stat, olanzapine zydis 10mg stat or risperidone quicklet 1-2mg stat; if patient refuses oral medication, consider IM lorazepam 2mg stat, and/or IM haloperidol 5-10mg stat). 3) Reduce acute symptoms by regular oral antipsychotics. Start at low dose and titrate upwards over 2 weeks. The choice of antipsychotics is based on risk and benefit ratio and patient's preference after explanation of various options. Close monitoring for 2 months to assess effectiveness.	1) The patient's social supports, functioning and relative risk of self-harm or harm to others must be evaluated for choice of treatment setting. 2) People newly diagnosed with schizophrenia should be offered oral antipsychotic medication. The recommended optimal oral dose of antipsychotics is 300–1,000 mg chlorpromazine equivalents daily for an adequate duration of 4–6 weeks. 3) If there is inadequate response by 4–6 weeks or if patient develops intolerable side effects, the medication should be reviewed and another typical or atypical antipsychotics should be used. 4) Long-acting depot antipsychotics should not be used for acute episodes because it may take 3–6 months for the medications to reach a stable steady state.	1) Offer psychoeducation to enhance knowledge of illness. 2) Minimize the likelihood of relapse by ensuring compliance to medications. Long-acting depot (e.g. IM fluanxol, clopixol), antipsychotics may be indicated in patients in whom treatment adherence is an issue or when a patient expresses a preference for such treatment (MOH guidelines) Reduce expressed emotion by family intervention. 3) Enhance adaptation and coping to social and occupational disturbances by rehabilitation and occupational therapy. 4) Facilitate continued reduction in symptoms and	1) Ensure symptom remission or control by the lowest effective dose of antipsychotics, which should not be lower than half of the effective dose during the acute phase [MOH guidelines, 2011]. 2) Monitor and manage adverse effects related to antipsychotics. 3) Regular follow-up with a psychiatrist on a regular basis. 4) For patients with poor social support, refer to the community psychiatric team for home visit. 5) Oral antipsychotics should be used as first-line treatment for patients with an acute relapse of schizophrenia (MOH guidelines 2011). 6) Patients receiving atypical antipsychotics

	5) Electroconvulsive therapy should be considered for patients who have not responded to an adequate trial of antipsychotics and for patients with life threatening symptoms such as catatonia and prominent depressive symptoms.	promote the process of recovery by psychological interventions e.g. cognitive behaviour therapy and problem solving therapy. 5) Antidepressants should be considered when depressive symptoms emerge during the stable phase of schizophrenia (post-psychotic depression). Antidepressants should be used at the same dose as for treatment of major depressive disorder (MOH guidelines 2011)	should be regularly monitored for metabolic side effects (MOH guidelines 2011). 7) Treatment options for schizophrenia patients who are pregnant should be individualised, with consideration of severity of previous episodes, previous response to treatment and the woman's preference. Abrupt cessation of medications should be avoided. (MOH guidelines 2011)

The 'most' and the 'least' of the first generation and second generation of antipsychotics.

Antipsychotics	The most likely / High risk ↑	The least likely / Low risk ↓
1st generation		
First generation antipsychotics	1) In acute dystonia, young men at most risk. 2) In tardive dyskinesia, elderly women are at most risk.	Nil
Haloperidol	1) High risk for EPSE 2) High risk for galactorrhoea	1) Haloperidol - low risk of weight gain, postural hypotension and sedation.
Trifluoperazine	Nil	1) Trifluoperazine: low risk of weight gain and postural hypotension.
Sulpiride	Nil	1) Sulpiride carries low risk of weight gain. 2) Sulpiride carries low risk of postural hypotension. 4) Sulpiride carries the low risk of sedation.
Clozapine	1) Most common side effect is sedation till the next morning. 2) Second most common side effect is hypersalivation. 3) Clozapine (like olanzapine) carries the highest risk of weight gain.	1) Clozapine is least likely to cause tardive dyskinesia.
2nd Generation		
Risperidone	High risk for EPSE and galactorrhoea compared to other 2nd generation.	1) Risperidone carries low risk of sedation.
Olanzapine	Olanzapine carries the highest risk of weight gain among all antipsychotics.	1) Olanzapine carries low risk of EPSE. 2) Olanzapine carries low risk of hyperprolactinaemia. 3) Olanzapine carries low risk of QTc prolongation.
Quetiapine	1) Quetiapine has high affinity for muscarinic receptors and cause anticholinergic effects.	1) Quetiapine carries the lowest risk for EPSE. 2) Quetiapine carries the lowest risk of sexual dysfunction.

		3) Quetiapine carries low risk of hyperprolactinaemia.
Aripiprazole	Nil	1) Aripiprazole carries the lowest risk of QTc prolongation. 2) Aripiprazole carries low risk of sexual dysfunction. 3) Aripiprazole carries low risk for EPSE. 4) Aripiprazole carries low risk of dyslipidaemia, weight gain and glucose intolerance. 5) Aripiprazole carries low risk of hyperprolactinaemia. 6) Aripiprazole carries low risk of postural hypotension. 7) Aripiprazole carries the low risk of sedation.

Psychogenic polydipsia

5-20% of schizophrenia patients suffer from psychogenic polydipsia.

	Urine volume	Urine osmolality	Sodium
Nephrogenic diabetes insipidus	High	Low	Serum: high Urine: low
Psychogenic polydipsia	High	Low	Serum: low Urine: low
Syndrome of Inappropriate AntiDiuretic Hormone (SIADH)	Low	High	Serum: low Urine: high

Common side effects of antipsychotics and best alternatives.

Antipsychotics associated with each side effect:

Sedation: Chlorpromazine, olanzapine and quetiapine.

Tardive dyskinesia: First generation antipsychotics.

Galactorrhoea: Haloperidol and risperidone.

Postural hypotension ($\alpha 1$ antagonism): Chlorpromazine and quetiapine.

Weight gain ($5HT_{2C}$ and H_1 antagonism): Clozapine and olanzapine.

Sexual dysfunction: Haloperidol, risperidone, sulpiride, chlorpromazine (priapism).

EPSE: First generation antipsychotics e.g. haloperidol and trifluoperazine.

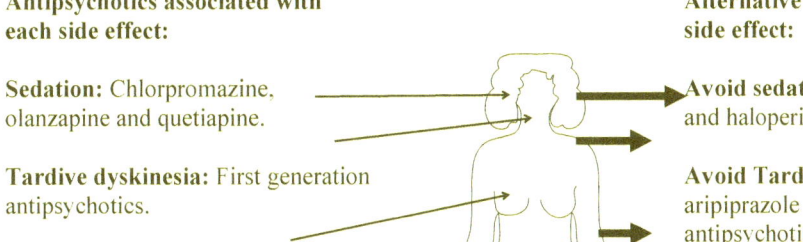

Alternative antipsychotics to avoid each side effect:

Avoid sedation: Aripiprazole, sulpiride and haloperidol.

Avoid Tardive dyskinesia: clozapine, aripiprazole and second generation antipsychotics.

Avoid galactorrhoea: aripiprazole, olanzapine and quetiapine. Galactorrhoea can be treated by bromocriptine.

Avoid postural hypotension: Aripiprazole, sulpiride, haloperidol and trifluoperazine.

Avoid weight gain: aripiprazole, sulpiride, haloperidol, trifluoperazine, diet modification and exercise. There is evidence that metformin can reduce weight gain associated with antipsychotics.

Avoid sexual dysfunction: aripiprazole and quetiapine

Avoid EPSE: quetiapine, olanzapine and aripiprazole.

The first generation antipsychotics

Indications and contraindications	Mechanism of action	Side effects	Drug interactions
Indications: Schizophrenia. Schizoaffective disorder. Substance induced psychosis. Personality disorder with psychotic features. Affective disorders. Tourette's syndrome Huntington's disease Nausea, emesis and hiccups. **Contraindications** Parkinson disease. Lewy body dementia. Elderly who are prone to develop extrapyramidal side effects.	**Symptom control:** mesolimbic dopamine blockade is thought to be the most important for control of positive psychotic symptoms. **Receptor occupancy:** PET studies have shown 65-90% occupancy of brain D2 receptors after normal antipsychotic doses.	**Neurological adverse effects:** **Extrapyramidal symptoms** are due to blockade of D_2 receptors in the basal ganglia. **Tardive dyskinesia** is due to D_2 receptor hypersensitivity. **Drowsiness** is due to antihistamine activity. **Secondary negative symptoms** (indifference to the environment, behavioural inhibition and diminished emotional responsiveness) **are** due to dopamine antagonism in the mesocortical pathway. **Memory impairments** attributable to antimuscarinic effects and dopamine blockade in the cortex. **Impairments in cognitive and psychomotor functions** occur after acute treatment in both healthy volunteers and patients. Chronic treatment does not cause any significant impairment on psychometric tests or psychomotor performance. **Fine motor incoordination** is due to nigrostriatal blockade. **Hormonal side effects:** **Galactorrhoea** is due to dopamine antagonism in the tuberoinfundibular pathway. Plasma neuroleptic levels correlate with prolactin increases. **Other endocrine effects include:** 1) False pregnancy test 2) Weight gain 3) Secondary amenorrhoea 4) Unilateral gynaecomastia. **Allergic side effects:** 1) Contact dermatitis 2) Opacities in the lens 3) Cholestatic jaundice 4) Optic neuritis 5) Aplastic anaemia	**Drug interactions of first generation antipsychotics e.g. phenothiazines:** They potentiate the depressant action of antihistamine, alcohol, GA and benzodiazepine. They increase analgesic effects of opiates. They cause a marked increase in intracellular lithium. They antagonise the dopaminergic effect of anti-parkinsonian drug. Phenothiazines are protein bound, care must be taken when administered with other highly protein-bound medications (e.g. warfarin, digoxin, theophylline) and potent 2D6 inhibitors (e.g. fluoxetine, paroxetine, cimetidine and erythromycin).

		Haematological side effects: 1) Transient leucopenia 2) Agranulocytosis	

Chlorpromazine (Phenothiazines with aliphatic side chain) [Dose: 100mg to 800mg/day; 2012 price: $0.12/mg]

Indications and contraindications	Mechanism of action	Side effects
Chlorpromazine is sedative antipsychotics, indicated for schizophrenia. Chlorpromazine is contraindicated in cholestatic jaundice, Addison's disease, myasthenia gravis, glaucoma and bone marrow depression.	High level of anticholinergic, anti-α-adrenergic, and antihistaminergic actions. At a dose of 100 mg twice daily, chlorpromazine exhibits 80% dopamine D2 receptor occupancy. $T_{1/2}$ = 16-30 hours	Most sedative first generation antipsychotics. Cataract (↑ risk by 3-to-4 fold), miosis, weight gain, increased duration of SWS, galactorrhoea, haemolytic anaemia or agranulocytosis, leucocytosis or leucopenia, cholestatic jaundice (hypersensitivity reaction), quinidine like side effect (prolonged QTc interval, ↓ST and ↓T wave blunting, photosensitive rash and hyperglycaemia. **Priapism:** alpha-receptor antagonism unopposed by cholinergic stimulation may be the underlying mechanism. Summary: Antihistaminergic side effects> anticholinergic side effects = extrapyramidal side effects (EPSE).

Trifluoperazine (Phenothiazines – piperizine side chain) [Dose range: 5mg to 15mg/day; 2012 price: $0.12/mg]

Indications	Mechanism of action	Side effects
Less sedative antipsychotics	Increased D_2 blockade and a lower affinity to muscarinic α-adrenergic and histaminergic receptors. $T_{1/2}$ = 10-20 hours	More likely to cause EPSE. Summary of side effects: EPSE > anticholinergic = antihistaminergic side effects.

Haloperidol (Butyrophenones) [Dose range: 5mg to 10mg/day; 2012 price: $0.10/mg (tablet), Injection 5mg $1.87/mg, Drops $26.66/ ml]

Indications	Mechanism of action	Side effects
Indications: 1) Positive symptoms of schizophrenia. 2) Delirium. 3) Rapid tranquillisation (IM) 4) Available in liquid form (covert medication) 5) Tourette's syndrome Contraindications: 1) Parkinson's disease. 2) Lewy body dementia.	Very potent D_2 blocker Lower level of activity with the nigrostriatal pathway. Little antimuscarinic, antihistaminergic and anti-adrenergic activity. $T_{1/2}$ = 10-30hours	High doses often lead to EPSE, akathisia and akinesia.

Sulpiride (Substituted benzamines) [Dose range: 200mg to 800mg/day; 2012 price: $0.24/mg]

Indications and	Mechanism of action	Side effects

contraindications		
Schizophrenia patients who cannot tolerate EPSE associated with other first generation antipsychotics.

Contraindication: renal failure, acute porphyria. | Selective antagonist at D_2 and D_3 receptors.

$T_{1/2}$ = 7 hours | Hyperprolactinaemia.

Dry mouth, sweating, nausea. |

Flupenthixol (Thioxanthenes) [Dose range: IM 20mg once per month; 2012 price: $0.12/ mg]

Indications	Mechanism of action	Side effects
Depot antipsychotics

For people with schizophrenia who are non – compliant.

Not a recommended as a first-line treatment for depression (some patients found it useful). | Antipsychotic effect.

Low doses may have antidepressant effects.

$T_{1/2}$ = 10-20 hours | Acute dystonia

EPSE

Long term usage may lead to tardive dyskinesia. |

The second generation antipsychotics

Risperidone (Benzisoxazole) [Dose range: 1mg to 6mg/day; 2012 price: $0.40/mg]
Riperdal consta (IM depot) [Dose range: IM 37.5mg to 50mg every 2 weeks; 2012 price: $155.06/ 25 mg]

Indications and contraindications	Mechanism of action	Side effects
Schizophrenia: 1-2mg is the minimum effective dose for the first episode and 4mg for relapse.

Bipolar disorder – mania.

Behavioural problems in dementia and autism.

For covert administration (Risperidone droplets).

Contraindications: patients reported EPSE with risperidone in the past. | High-affinity antagonist of 5-HT$_2$, D$_2$-like.

α-Adrenergic antagonism

It allows more dopaminergic transmission than conventional antipsychotic because the normal inhibitory action of serotonin on dopamine neurons is inhibited due to antagonism of the 5HT$_{2A}$ heteroreceptor.

The affinity of risperidone for D$_2$ receptors is approximately 50-fold greater than that of clozapine and 20%-50% of haloperidol.

$T_{1/2}$ = 3-20 hours | Hyperprolactinaemia and EPSEs at higher doses. (Reduction of EPSE may be confined to a low dose range, of 2-6 mg, with higher doses giving a profile that approaches that of a first-generation agent).

Other side effects include: insomnia dizziness, anxiety, menstrual disturbances and weight gain. |

Olanzapine (Thienobenzodiazepine) [Dose range: 5mg to 15mg /day; 2012 price: $5.69 / 5mg tablet]

Indications and contraindications	Mechanism of action	Side effects
Indications: Schizophrenia, Schizoaffective disorder. Bipolar mania. **Olanzapine zydis (rapidly dissolvable form) for rapid tranquillisation.** **Contraindications** Stroke patients Patients with diabetes. Obese patients. Narrow angle glaucoma due to anticholinergic effect	High affinity for D_2 and $5HT_{2A}$ but low affinity for D_1 receptors. Similar chemical structure to clozapine. Clozapine selectively binds to many different dopamine receptors, whereas olanzapine is only partially selective for D_2 receptors. $T_{1/2}$ = 21-54 hours	Weight gain. High risk of diabetes. Appetite increase. Sedation. Anticholinergic side effects: dry mouth and constipation. Dizziness. EPSE under olanzapine are not absent all together but if they occur tend to be mild, at relatively high levels of D_2 occupancy. This occurs in association with high anticholinergic effect which may contribute to mitigation of EPSE. The annual rate of tardive dyskinesia is 0.5%.

Quetiapine (Dibenzothiazepine) [Dose range: 200mg to 800mg/day; 2012 price: $2.10 / 200 mg tablet]

Indications	Mechanism of action	Side effects
Schizophrenia patients who cannot tolerate EPSE. Parkinson disease patients with psychotic features after taking levodopa.	Has a high affinity for muscarinic receptors. Quetiapine has high affinity for $5-HT_{1A}$ receptors may increase dopamine levels in the hypoactive mesocortical dopaminergic pathway and improve negative symptoms. It has lower affinity for all receptors than clozapine. $T_{1/2}$ = 6 hours	Sedation (17.5%) Dizziness (10%) Constipation (9%) Postural hypotension. No difference from placebo in terms of EPSE and prolactin level. Less weight gain compared to olanzapine and clozapine (clozapine = olanzapine > risperidone > quetiapine > ziprasidone).

Ziprasidone [Dose range: 20 mg BD to 80mg BD; 2012 price: This drug is not available at NUH]

Indications	Mechanism of action	Side effects
Schizophrenia patients who cannot tolerate weight gain and people with schizophrenia taking warfarin.	Potent $5HT_{2A}$ and D_2 antagonist The effects of ziprasidone on negative symptoms and possibly cognitive symptoms may also be related to its potent antagonism for $5-HT_{2A}$ receptors. It also exhibits $5-HT_{1A}$ agonism and inhibits the reuptake of noradrenaline and serotonin. $T_{1/2}$ = 7 hours	The overall effect of ziprasidone on movement disorder is no difference from placebo. Ziprasidone produced only modest weight gain in short term (4- to 6-week) trials, with a median weight gain of 0.5kg. ECGs revealed a modest prolongation with ziprasidone treatment in short-term (4- and 6-week).

Aripiprazole [Dose range: 15mg – 30mg/day; 2012 price: $10.86 / 15mg tablet]

Indications	Mechanism of Action	Side effects
Schizophrenia patients who develop weight gain, metabolic syndrome, galactorrhoea, EPSE, QTc prolongation associated with other antipsychotics.	D_2 and $5-HT_{1A}$ partial agonist. $5HT_{2A}$ antagonist. High affinity for D_3 receptors; moderate affinity for D_4, $5HT_{2C}$, $5-HT_7$, adrenergic, and histaminergic receptors. There are no significant differences in outcomes compared to 1st and 2nd generation antipsychotics. $T_{1/2}$ = 74h to 94h (due to active metabolites)	Excellent safety and tolerability profile. Most common side effects include: Headache Insomnia Agitation Anxiety. Aripiprazole is less likely to cause elevation in prolactin compared to other antipsychotics.

Paliperidone [Dose range: 6mg to 12mg per day; 2012 price: $6.38 / 6 mg tablet]

Indications and contraindications	Mechanism of action	Side effects
Indicated for schizophrenia, schizoaffective disorder especially patients with Tourette's syndrome with liver impairment. Contraindication: renal impairment. This drug is known to be substantially excreted by the kidney.	Potent $5HT_{2A}$ and D_2 antagonist paliperidone is the major active metabolite of Risperidone $T_{1/2}$ = 23 hours	EPSE Akathisia QTc prolongation Hyperprolactinaemia, metabolic syndrome and increase in risk of seizure

Clozapine (Dibenzodiazepine) [Dose range: 200mg to 450mg daily; 2012 price: $0.26 / 25 mg]

Indications and contraindications	Mechanism of Action	Side effects
Clozapine should be offered to patients whose illness has not responded adequately to treatment despite the sequential use of adequate doses and duration of at least two different antipsychotics (MOH guidelines 2011). Patients require full blood count on a weekly basis for the first 18 weeks, then fortnightly until the end of the first year. Then patient requires full blood count on a monthly basis after 1 year. **Contraindications:** Potential lethal combinations if patients take clozapine and carbamazepine or sulphonamide. This will lead to blood dyscrasia. Patients take lithium and clozapine increase the risk of seizure, confusion, dyskinesia and neuroleptic malignant syndrome (NMS)	Higher antagonist affinity for non-dopamine than for dopamine receptor subtypes); D_2 receptors: moderate affinity for D_2 receptors and a high affinity for $5\text{-}HT_{2A}$ receptors. $5\text{-}HT_{1A}$ partial antagonism with D_2-like antagonism this may contribute not only to mitigation of EPSE but also to enhancement of prefrontal dopamine release and putative therapeutic effects. Hypersalivation caused by clozapine is due to antagonism of α_2-adrenergic receptors and agonism at the M_4 receptor. $T_{1/2}$ = 6-26 hours	**Pulmonary embolism:** 1 in 4500 **Myocarditis:** 1 in 1300. **Agranulocytosis:** 1 in 10,000 Neutropenia and agranulocytosis (agranulocytosis is defined as an absolute neutrophil count of < 500/mm^3). Neutropenia is not dose related and occurs in 1-2 % of patients. If the temperature is over 38.5.°C, consider withholding clozapine until the fever subsides. Weight gain, metabolic syndrome. Hypotension, tachycardia and ST segment changes. Hypersalivation, constipation, and urinary incontinence. The incidence of seizures in people with schizophrenia taking clozapine at more than 600 mg per day is roughly 15%.(first seizure requires a reduction in dose and second seizure requires an addition of anticonvulsant such as sodium valproate)

Extrapyramidal side effects (EPSE) and tardive dyskinesia (TD)

Pseudo-parkinsonism. (Lesions in nigrostriatal pathway).	**Epidemiology:** Incidence (20%) **Risk factors:** more common in older women, particularly those with neurological damage and persist for many months. **Onset:** After a few weeks of usage and develop gradually. **Symptoms:** mimicking Parkinson's disease, with akinesia (generalized slowing and loss of movements, particularly the involuntary movements of expression), rigidity and tremor. Rigidity and akinesia develop more frequently than tremor. **Management:** Gradually reducing the dose can reduce the symptoms. Otherwise change to second generation antipsychotic. Anticholinergic agents (e.g. benzhexol or artane 2mg BD) have been shown to be effective in reducing the severity of EPSEs and may be prescribed to patients experiencing these side effects (MOH guidelines 2011).

Acute dystonia. (Lesions in nigrostriatal pathway).	**Epidemiology:** Incidence (10%). **Risk factors:** Young men are at highest risk. High potency antipsychotic in schizophrenia patients who are antipsychotic naïve is also a risk factor. **Onset:** Occur within a few hours of antipsychotic administration. **Symptoms:** The classic example includes oculogyric crisis (fixed upward or lateral gaze), but torsion dystonia and torticollis occur, as well as spasms of the muscles of the lips, tongue, face and throat. Acute dyskinesia (involuntary movements), with grimacing and exaggerated posturing and twisting of the head, neck or jaw, can also occur. Trismus refers to the dystonic reaction to antipsychotic medication affecting the jaw muscles **Management:** Intramuscular anticholinergics (e.g. IM congentin 2mg stat) in oculogyric crisis and torsion dystonia.
Akathisia. (Lesions in nigrostriatal pathway).	**Epidemiology:** Incidence (25 - 50%) **Risk factors:** acute forms are related to rapid increases in antipsychotic dose. **Onset:** most commonly after the 5th day of initiation of dopamine receptor antagonists. **Symptoms:** The patient may become irritable or unsettled, complains of needing to go out or may try to leave for no clear reason. **Management:** Acute akathisia may respond to anticholinergics. Chronic akathisia responds poorly to anticholinergics but may respond to benzodiazepines. The best-established treatment for either form of akathisia is propranolol.
Tardive dyskinesia. (Lesions in nigrostriatal pathway).	**Epidemiology: 5%** after one year of treatment; 20% after chronic treatment Risk factors: 1) Elderly women, those with affective disorder, organic brain disorder and history of EPSE. 2) Long exposure to antipsychotics. 3) Precipitated by anticholinergic. **Pathology:** TD is caused by supersensitivity of D_2 receptors. **Symptoms:** Typical TD includes: lip smacking, chewing, 'fly catching' tongue protrusion, choreiform hand pill rolling movements and pelvic thrusting. **Management:** Change to second generation antipsychotics. Vitamin E.

Psychosocial interventions (NICE (UK)/ MOH (Singapore) guidelines)

Interventions	Principles
Psychoeducation	Early psycho-education and family intervention should be offered to patients with schizophrenia and their families (MOH guidelines 2011). 1. The main objective is to provide the patient with information about the illness, the range of treatments available and the effect of using recreational drugs such as amphetamines. The patients are informed on the choices. 2. Psychoeducation for individuals with first episode of psychosis or schizophrenia should encourage blame-free acceptance of illness. 3. Develop strategies to promote control of illness by recognising and responding to early warning signs and seek professional advice.
Crisis intervention	To support and assist the patient to recover and reorganize at times of relapse or major life events which overwhelm the patient's capacity to cope.
Grief work on losses	To work through losses both from prior to the illness onset and also losses arising from the disruption, disorganisation and disability associated with schizophrenia.
Supportive psychotherapy	Refinements of individual supportive psychological treatments i.e. targeted psychological treatments for specific components of illness or symptoms developed for affected individual. E.g. coping techniques to deal with psychotic symptoms.
Cognitive behavioural therapy	Psychological therapy, in particular Cognitive Behaviour Therapy (CBT), administered in combination with routine care should be considered for patients with schizophrenia, particularly those with persistent negative and positive symptoms (MOH guidelines 2011). **Components of CBT should involve:** 1. Advise the patients to keep a record to monitor their own thoughts, feelings or behaviours with respect to their symptoms or recurrence of symptoms. 2. Promoting alternative ways of coping with the target symptoms. 3. Reducing stress. 4. Improving functioning. 5. Reducing negative symptoms by activity scheduling. **Specific techniques targeting at auditory hallucination:** 1. Distraction method: Wearing headphones to focus attention away so the hallucinations are extinguished with decreased reactivity. 2. Desensitization: Describe, record and recognize the connection between stressors and hallucinations and explore what the voices mean to them. The NICE guidelines (UK) also recommend the therapist to deliver CBT on a one-to-one basis over at least 16 planned sessions. The therapist should follow a treatment manual to help the patient to establish links between their thoughts, feelings or actions and their current or past symptoms. This will help them to re-evaluate their perceptions, beliefs and reasoning behind the target symptoms.
Family intervention	MOH guidelines: Patients and their family members should be educated about the illness, its course and prognosis as well as the efficacy of the various medications, the anticipated side effects and costs. Family interventions should also incorporate support, problem-solving training and crisis intervention. The NICE guidelines (UK) recommend the following: 1. The therapist should include the service user if possible and offer at least 10 planned

	sessions over a period of 3 months to 1 year. 2. Single-family intervention focusing on "here and now" and the family systems boundaries, coalitions, triangulation) is recommended and take into account the relationship between the main carer and the service user. 3. Therapist should establish a working relationship with family and provide structure and stability. 4. Cognitive techniques can be used to challenge unhelpful attributions e.g. guilt. 5. Behavioural approaches include goal setting and problem-solving.
Arts therapy	The NICE guidelines (UK) recommend art therapies conducted by arts therapists registered by the Health Professions Council. The objectives of arts therapy include: 1. Helping people with schizophrenia to experience themselves differently and develop new ways of relating to others. 2. Expressing themselves and organising their experience into a satisfying aesthetic form. 3. Accepting and understanding feeling which may have emerged during the creative process when doing the art work.
Other techniques	Managing comorbidity e.g. substance abuse, organised support groups in supporting affected individual and family.

Prognosis of schizophrenia

Good prognosis	Poor prognosis
Demographics: women, married,	Men, single
Aetiology: precipitated by stressful life events, no past psychiatric history, family history of affective illness	Past psychiatric history, family history of schizophrenia
Nature of illness: late onset of symptoms, paranoid type, positive symptoms.	Early onset, negative symptoms.
Treatment: good response to treatment, short duration of untreated psychosis.	Poor response to treatment, long duration of untreated psychosis.

Catatonia

Causes of catatonia: schizophrenia, severe depressive disorder, bipolar disorder, organic disorders e.g. CNS infections, CNS tumour, cerebrovascular accident, severe intoxication of recreational drugs and lethal catatonia.

Clinical features: Ambitendency, automatic obedience (mitgehen, mitmachen), waxy flexibility / catalepsy, negativism, stereotypy, mannerism, echolalia and echopraxia.

Investigations: FBC, RFT, LFT, TFT, blood glucose, CK, urine drug screen, ECG, CT, MRI, EEG, urine and blood culture, syphilis screeen, HIV, heavy metal screen, auto-antibody screen and lumbar puncture.

Management strategies (non-pharmacological):
Hydration, early mobilization, close monitoring, transferral to ICU if patient deteriorates.

Medications:
1. Benzodiazepines (e.g. IM lorazepam up to 4mg per day).
2. If benzodiazepine does not work and symptoms are severe, ECT is an option.

Prognosis: Two-third of patients improve after treatment.

Schizoaffective disorder

Epidemiology

Lifetime prevalence	M:F
0.05% - 0.08%	Women > men

ICD-10 and DSM-5 criteria

ICD-10	DSM-5
Schizophrenia and affective symptoms are prominent and occur in the same episode or within a few days of each other. **Manic type**: mania (e.g. elation, irritability, aggression, high energy, flight of ideas, grandiosity); psychotic symptoms (e.g. thought interference, passivity, delusions of persecution) **Depressive type**: depression (e.g. low mood, psychomotor retardation, insomnia, guilt, hopelessness and suicidal thought); psychotic symptoms. **Mixed type**: mixed depression and manic symptoms, psychotic features.	The DSM-5 specified that for an individual to fulfill the diagnostic criteria, there must be the presence of solely hallucinations or delusions for at least 2 weeks in the absence of an affective episode, throughout the whole duration of the psychiatric illness. There must also have an uninterrupted period where there are prominent affective symptoms concurrent with symptoms of schizophrenia (Criterion A). Individuals should have symptoms fulfilling the diagnosis of an affective disorder for most of the duration of the illness. DSM-5 has specified 2 subtypes of schizoaffective disorder, which are: a. Bipolar type - Whereby a manic episode is part of the entire course of the illness b. Depressive type - Whereby a major depressive episode is part of the entire course of the illness

Investigations: same as schizophrenia.

Treatment:
- Psychotic symptoms: antipsychotics (e.g. olanzapine has good mood stabilising effects).
- **Manic subtype:** Mood stabiliser e.g. **lithium** or **carbamazepine.**
- **Depressive subtype:** Antidepressant, usually a SSRI.
- **Poor response to pharmacological treatments:** ECT.
- **Psychosocial treatments:** similar to schizophrenia.

Prognosis:
- The outcome for schizoaffective disorder is intermediate between schizophrenia and affective disorders.
- Manic subtype has a better prognosis than depressive subtype.

Brief or acute/transient psychotic disorder

Epidemiology

Age of Onset	Gender
20-30 years.	More common in women.

Aetiology:
1) Acute stressful life event e.g. disaster, bereavement or severe psychological trauma.
2) Underlying personality disorders: borderline, histrionic, paranoid and schizotypal.

3) Family history of mood disorders or schizophrenia.

ICD-10 and DSM-5 criteria:

	ICD 10 criteria	DSM-5 criteria
Onset	Acute onset if psychotic features occur within 2 weeks of the stressful event. Abrupt onset if psychotic features occur within 48 hours of the stressful event.	The DSM-5 specified that for an individual to fulfill this diagnosis, he/she must have, for a duration of between 1 day to 1 month the following symptoms: a. Delusions b. Hallucinations c. Disorganized speech d. Grossly disorganized or catatonic behaviour Clinicians need to exclude differentials like major depression or bipolar depression with psychotic features, schizophrenia and exclude the possibility of the symptoms being due to underlying substance use or medical conditions.
Precipitant	With or without stressful life event.	With or without stressful life event, and with postpartum onset
Symptoms	Sudden change of a person's mental state from normal to psychotic. 1) Acute polymorphic psychotic disorder (variable hallucinations, delusions or emotions) without symptoms of schizophrenia. 2) Acute polymorphic psychotic disorder (variable hallucinations, delusions, emotions) with symptoms of schizophrenia (less than 1 month of symptom). 3) Acute schizophrenia-like psychotic disorder (stable hallucinations and delusions but less than 1 month symptom). 4) Acute and predominantly delusional psychotic disorders (duration < 3 months)	Present with delusions, hallucinations, disorganised speech, disorganised or catatonic behaviour. Return to premorbid function. Sub classified into: 1) Brief reactive psychosis with marked stressor. 2) Without marked stressor. 3) With postpartum onset.
Exclusion	Schizophrenia, mania, depression, delirium, dementia, intoxication of drugs or alcohol.	Mood disorder with psychotic features, schizoaffective disorder, schizophrenia.

Treatment:
1) Short-term use of low dose antipsychotic e.g. risperidone 1 to 2mg daily to control psychotic symptoms.
2) Short-term use of low dose benzodiazepine e.g. lorazepam 0.5mg for sleep.
3) Problem solving or supportive psychotherapy.

Prognosis:
Complete recovery usually occurs within 2-3 months. Relapse is common. The more acute/abrupt the onset, the better the long term outcome.

Delusional disorders

Incidence	Point Prevalence	Lifetime risk	Mean Age of onset	Gender ratio
1-3 per 100,000	0.03%	0.05-0.1%	35 years for men 45 years for women	More common in women. Erotomania is more common in women.

Aetiology:
1) Genetic risk factors such as family history of schizophrenia. Delusional disorder and paranoid personality disorder.
2) The main neurotransmitter involved is excessive dopamine.
3) The key neuroanatomical areas involved are the basal ganglia and the limbic system.
4) The cognitive theory proposes that delusions are caused by cognitive deficits, resulting in misinterpretation of external reality.
5) Organic diseases such as CNS disorders (e.g. Parkinson's disease, Huntington's disease, sub-arachnoid haemorrhage, brain tumour), degenerative disorders (e.g. Alzheimer's disease), infectious diseases (e.g. AIDS, neurosyphilis, encephalitis), metabolic diseases (e.g. hypercalcaemia, hyponatraemia, hypoglycaemia, uraemia, hepatic encephalopathy), endocrine diseases (e.g. syndrome, hypothyroidism, hyperthyroidism, panhypopituitarism) and vitamin deficiencies (e.g. vitamin B_{12}, folate)
6) Other factors: sensory impairment, isolation, migration with cultural barrier.

ICD-10 and DSM 5 criteria

	ICD-10	DSM-5
Duration	Delusions lasting ≥ 3 months.	At least 1 month's duration.
Other features	1) Delusions are persistent and can be life-long.	DSM-5 diagnostic criteria states that individuals need to have fixed, firmed and unshakeable beliefs (delusions) for a minimum duration of at least 1 month. These delusional beliefs must not have a marked impairment on an individual's level of functioning. Individuals might experience hallucinations at times, but the content of the hallucinations are usually in relation to the delusional beliefs. Clinicians need to distinguish between delusional disorder and schizophrenia.
Exclusion	1) Schizophrenia 2) No mood disorder 3) People present with first rank symptoms. 4) No long-term organic disorder. 5) No substance misuse.	No significant impairment of functioning. The symptoms must not be explained by obsessive compulsive disorder or body dysmorphic disorder with absent insight/delusional beliefs.
Type	1) Persecutory (most common type, being followed, harmed, poisoned or malevolently treated) 2) Jealous type (the patient accuses the partner to be unfaithful but the reasoning is absolutely illogical or impossible, may associate with violence)	DSM-5 specified several subtypes of delusional disorder, which are as follows: a. Erotomanic type - characterized as individuals believing that others are in love with them b. Grandiose type - characterized as

	3) Erotomanic type (usually a female patient believes that a man of higher status (e.g. CEO, President) falls in love with her but it is absolutely illogical or may associate with stalking) 4) Hypochondriasis [infestation (or infected by parasites), dysmorphophobia (concerned one body part is unattractive) and foul body odour] 5) Capgras syndrome 6) Fregoli syndrome.	individuals believing that they possess unique abilities c. Jealous type - characterized as individuals believing that their loved ones are not faithful d. Persecutory type - characterized as individuals believing that others are out there to harm, cheat or even poison them e. Somatic type - characterized as individuals believing that there are some abnormalities pertaining to bodily functions. f. Mixed type - when no major delusional theme could be identified g. Unspecified type Delusions can be also be sub-classified into those with or without bizarre content (previously in DSM IV-TR, delusions were deemed only as non-bizarre).

Management
1) Hospitalisation may be appropriate if the patient has high risk of suicide or self-harm (e.g. high risk for self-operation in delusion of dysmorphophobia); high risk of violence or aggression (e.g. a patient with morbid jealousy is using violence to interrogate the spouse) and there is a need to apply Mental Disorder and Treatment Act to treat the patient during compulsory admission.
2) Pharmacological treatment: similar to schizophrenia, antipsychotics and benzodiazepine. Patient may require covert antipsychotics (e.g. administering liquid antipsychotics through patient's food in patient with very poor insight and refuses oral treatment). The decision of covert medication is determined by a consultant psychiatrist with detailed discussion with family members and after analysis of risks and benefits.
3) Psychosocial interventions include cognitive therapy targeting at delusions, family therapy and provide shelter or alternative accommodation to the spouse of a patient with morbid jealousy.

Revision MCQs

1. Which of the following symptoms of schizophrenia is not a first-rank symptom?

A. Audible thoughts
B. Delusional perception
C. Formal thought disorder
D. Thought insertion
E. Voices discussing or arguing

Answer: C
First-rank symptoms are summarised by an acronym, 'ABCD'. 'A" stands for auditory hallucinations (e.g. third-person, audible thoughts, thought echo). 'B' stands for broadcasting (e.g. thought broadcasting, thought insertion and thought withdrawal). 'C' stands for control or passivity (e.g. delusion of control). 'D' stands for delusional perception. Hence, formal thought disorder is not a first-rank symptom.

2. A 40-year-old woman suffering from schizophrenia and she has been taken haloperidol for the past 10 years. Her QTc is 500 ms. She wants to find out the potential medical complications if she continues to take haloperidol. Which of the following complications is least likely?

A. Atrial fibrillation
B. Palpitations
C. Sudden cardiac death
D. Torsade de pointes
E. Ventricular fibrillation

Answer: A
A prolonged QTc interval mainly affects the ventricles but not the atrium. Torsade de pointes is a form of irregular heartbeat that originates from the ventricles and causes ventricular fibrillation. A prolonged QTc interval is also associated with palpitations, sudden cardiac death and ventricular fibrillation.

3. A 50-year-old woman suffers from schizophrenia and her QTc is 500 ms. Which of the following antipsychotics is least likely to lengthen her QTc interval?

A. Aripiprazole
B. Olanzapine
C. Quetiapine
D. Risperidone
E. Sulpiride

Option: A
Aripiprazole is least likely to prolong the QTc interval and sulpiride carries low risk in QTc prolongation.

4. A junior medical student is interested about haloperidol and asks you which class of antipsychotics it belongs to. Your answer is:

A. Butyrophenones
B. Phenothiazine
C. Piperidines
D. Piperazines
E. Thioxaothenes

Answer: A
Haloperidol belongs to the class of butyrophenones. Chlorpromazine belongs to the class of phenothiazine. Thioridazine belongs to the class of piperidine. Fluphenazine belongs to the class of piperazine.

5. A 30-year-old develops schizophrenia and you have prescribed an antipsychotic agent. His wife wants to know the chance for him to show a complete response. Your answer is:

A. 10%
B. 20%
C. 30%
D. 40%
E. 50%

Answer: C.
Only 30% of patients will show complete response. 60% of patients will respond to some degree. 10% of patients will not respond to any antipsychotic agents.

6. A 38-year-old father is diagnosed with schizophrenia. His wife does not have any psychiatric illness. He wants to know what the risk of his son developing the disorder. Your answer is:

A. 5%
B. 10%
C. 15%
D. 20%
E. 25%

Answer: B.
His son has 10% risk of developing schizophrenia.

7. The Ministry of Health wants to develop a new mental health service targeting at immigrants who suffer from schizophrenia. Which of the following statements is true about the risk of developing schizophrenia among immigrants?

A. Biological factors have a greater aetiological role in comparison to sociological factors
B. Migrants from the inner city area have a higher

risk
C. Migrants have the same risk as the population in their native countries
D. Migrants are less likely to be hospitalized
E. The service should focus on the first generation immigrants because they have higher risk of developing schizophrenia

Answer: B
@ Among immigrants, sociological factors are more important than biological factors in the aetiology of schizophrenia. Migrants have higher risk to develop schizophrenia and are more likely to be hospitalized. The first generation and subsequent generation of immigrants have higher risk to develop schizophrenia and the service should not just focus on the first generation of immigrants.

8. A 20-year-old man presents with the first episode of psychosis in his life. Which of the following factors is the most important predicting factor for schizophrenia?

A. Alcohol misuse
B. Duration of quasi-psychotic symptoms
C. History of schizophrenia in the relative
D. Impairments in social functioning
E. Unemployment

Answer C

Family history of schizophrenia or genetic factor is the most important predicting factor among the above options. Recent population genetics studies have shown that schizophrenia has high heritability. Genome-wide association studies have identified a number of risk loci at the levels of significance.

9. Which of the following risk factors is associated with the highest risk of developing agranulocytosis in schizophrenia patients taking clozapine?

A. Afro-Caribbean descent
B. Young age
C. Female gender
D. High dose of clozapine
E. Long duration of clozapine usage

Answer: C
Female gender, Ashkenazi Jewish descent and older age are associated with high risk of developing agranulocytosis. The risk of developing agranulocytosis is not directly proportional to dose and duration of treatment.

Revision MEQs
A 30-year-old man is brought in by his sister because he has been hearing voices, having a spiritual warfare with Satan and complains that his thoughts are being interfered. He believes that Satan can take his thoughts away. He has been unemployed for 10 years as a result of the above symptoms and refuses to seek treatment. His wife has left him few years ago. Mental state examination shows that his affect is blunted and preoccupied with psychotic experiences.

1. What is your provisional diagnosis and differential diagnosis?
Provisional diagnosis: Schizophrenia
Differential diagnosis: drug-induced psychosis, organic causes (e.g. temporal lobe epilepsy), severe depression with psychotic features.

2. What are the characteristics identified in this patient associated with poor prognosis of your provisional diagnosis?
Male gender, long duration of untreated psychosis, lack of support and poor insight.

3. During the interview, he suddenly ran away from your clinic room and hid in the car park. His sister has difficulty to manage him at home. What would you do next?
1. Advise patient to be admitted.
2. If he refuses admission, refer him to the Institute of Mental Health and admit him under the Mental Disorder and Treatment Act.

4. The patient was admitted to the hospital and stabilised with oral antipsychotics. After discharge, he refused to take medication again. What would you offer to this patient?
1. Psychoeducation about schizophrenia and the role of antipsychotics.
2. Motivational interviewing: analyse the pros and cons of not taking medication and hopefully the patient understands there are more disadvantages than advantages to stop antipsychotics.
3. Depot antipsychotics e.g. IM fluanxol 20mg once per month.

5. Six months after discharge, you see the patient in the clinic. He challenges the diagnosis of schizophrenia and he believes that he suffers from depressive illness. His wife asks you how to differentiate the presentation of schizophrenia from depressive illness. Your answer is:

1. Schizophrenia patients exhibit mood incongruent psychotic features while depressed patients exhibit mood congruent psychotic features.
2. Schizophrenia patients show blunted affect while depressed patients are more reactive in their affect.
3. First rank symptoms usually do not occur in people with depression.
4. In general, patients with depression have better global and occupational functioning in comparison to schizophrenia patients.

EMIS:

Please select the most appropriate answer for each question.
Qn1. Disorders of action

A – Somatic Passivity
B – Made volition
C - Made affect
D – Made impulse
E – Somatic hallucination
F – Compulsion
G – Delusional perception
H – Referential delusion
I – Delusional misrepresentation

1. One of the patients said that someone is using telepathy to cause my hands to tremble though no obvious tremors are noted – Ans: Somatic passivity
2. One of the patient said that "I know it's wrong to do it and in fact I do not want to do it. But despite my resistance, they force me to do it." – Ans: Made impulse
3. "When I heard the train coming, I came to know that there is a plot to finish me off" – Ans: Delusional perception
4. One of the patient vocalized that he feels unpleasant due to insects crawling under his skin – Ans: Somatic hallucination

Qn 2. Please select the most appropriate answer for each question.

A – Voices commanding the patient
B – Voices discussing among themselves
C – Voices echoing patient's own thoughts
D – Voices giving commentary on the patient
E – Voices giving repeated feedback as and when the patient does an act

Which of the following is not consider a first rank symptom of schizophrenia – A

References

Chong SA, Lee C, Bird L, Verma S.(2004) A risk reduction approach for schizophrenia: the Early Psychosis Intervention Programme.
Ann Acad Med Singapore. 2004 Sep;33(5):630-5

Verma S, Chan LL, Chee KS, Chen H, Chin SA, Chong SA, Chua W, Fones C, Fung D, Khoo CL, Kwek SK, Ling J, Poh P, Sim K, Tan BL, Tan C, Tan CH, Tan LL, Tay WK; MOH Clinical Practice Guidelines Workgroup on Schizophrenia. (2011) Ministry of Health clinical practice guidelines: schizophrenia. Singapore Med J. Jul;52(7):521-5.

Lieberman JA, Murray RM (2001) Comprehensive care of schizophrenia. A Textbook of Clinical Management. London: Martin Dunitz.

Castle D, Copolov D, Wykes T (2003) Pharmacological and psychosocial treatments in schizophrenia. London: Martin Dunitz.

Castle D.J., McGrath J, Kulkarni J (2000) Women and schizophrenia. Cambridge: Cambridge University Press.

Yung A, Philips L, McGorry PD (2004) Treating schizophrenia in the prodromal phase. London: Taylor and Francis.

Taylor D, Paton C, Kapur S (2009) The Maudsley prescribing guideline. London: Informa healthcare.

Fleming M. Implementing the Scottish schizophrenia guidelines through the use of integrated care pathways in Hall J, Howard D (2006) Integrated care pathways in mental health. Churchill Livingstone: London.

Liddle PF (2007) Schizophrenia: the clinical picture in Stein G & Wilkinson G (2007) Seminars in general adult psychiatry. Gaskell: London.

Mellor C (2007) Schizoaffective, paranoid and other psychoses in Stein G & Wilkinson G (2007) Seminars in general adult psychiatry. Gaskell: London: 167 – 186.

World Health Organisation (1992) ICD-10 : The ICD-10 Classification of Mental and Behavioural Disorders : Clinical Descriptions and Diagnostic Guidelines. World Health Organisation: Geneva.

American Psychiatric Association (1994) DSM-IV-TR: Diagnostic and Statistical Manual of Mental Disorders (Diagnostic & Statistical Manual of Mental Disorders) (4th edition). American Psychiatric Association Publishing Inc: Washington.

American Psychiatric Association (2013) DSM-5: Diagnostic and Statistical Manual of Mental Disorders (Diagnostic & Statistical Manual of Mental Disorders) (5th edition). American Psychiatric Association Publishing Inc: Washington.

Gottesman II & Shields J (1972) Schizophrenia and genetics: A twin study vantage point. London: Academic Press.

Harvey PD, Sharma T (2002) Understanding and treating cognition in schizophrenia. A Clinician's Handbook. Martin Dunitz: London.

Sham P, Woodruff P, Hunter M and Leff J (2007) In Stein G & Wilkinson G (2007) Seminars in general adult psychiatry. Gaskell: London.

Watt RJ (2001) Diagnosing schizophrenia in Lieberman JA, Murray RM (2001) Comprehensive care of schizophrenia. A Textbook of Clinical Management. London: Martin Dunitz.

[15] King DJ (2004) Seminars in Clinical Psychopharmacology. 2nd edition Gaskell: London.

Anderson IM & Reid IC Fundamentals of clinical psychopharmacology. Martin Dunitz 2002.

Sadock BJ, Sadock VA. Kaplan and Sadock's Comprehensive Textbook of Psychiatry. 9th ed. Philadelphia (PA): Lippincott Williams and Wilkins, 2003.

Johnstone EC, Cunningham Owens DG, Lawrie SM,Sharpe M, Freeman CPL. (2004) Compansion to Psychiatric studies. 7th edition. Churchill Livingstone: London.

Schatzberg AF & Nemeroff CB (2004) Textbook of Psychopharmacology. 3rd edition. American Psychiatric Publishing: Washington.

Stahl SM. Psychopharmacology of Antipsychotics. Martin Dunitz: London, 1999.

Puri BK, Hall AD. Revision Notes in Psychiatry. London:

Arnold, 1998.

Beer MD, Pereira SM, Paton C. Psychiatric Intensive Care. Greenwich Medical Media Limited London: 2001.

Goodman R, Scott S. Child Psychiatry. Oxford: Blackwell Science, 1997.

Semple D, Smyth R, Burns J, Darjee R, McIntosh A (2005) Oxford Handbook of Psychiatry. Oxford University Press: Oxford.

Chapter 4 Mood disorders

Depressive disorder	Bipolar disorder
Epidemiology	Epidemiology
Aetiology	Aetiology
OSCE grid – Assess depressive disorder	ICD-10 and DSM-5 criteria
ICD-10 and DSM-5 criteria	OSCE grid – Assess manic symptoms
Differential diagnosis, investigation and questionnaire	Investigations and questionnaire
Management	Management (MOH and NICE guidelines)
Antidepressants	Mood stabilisers
Electroconvulsive therapy	OSCE video – Explain lithium side effects
OSCE video – Explain ECT	Suicide and deliberate self harm
Psychotherapy	Bereavement, abnormal grief, seasonal affective disorder.
Course and prognosis	Revision MCQs, MEQs & EMIs

Depressive disorder

Epidemiology

	Incidence	Lifetime prevalence	Point prevalence	Age of onset	Gender ratio
International	14.0 per 1000 persons	Overall: 10-20%; 1 in 4 women and 1 in 10 men have depressive disorder in their lifetime.	2-5%	25-45 years	F:M = 2:1
Singapore	As above	As above	5.6%	As above	As above

Aetiology

Genetics:
1) Family studies show that a person has 40-70% chance to develop depressive episode if a first degree relative suffer from depressive episode.
2) Twin studies show that the concordance rate for monozygotic twins is 40 – 50% and for dizygotic twins is 20%.
3) Adoption studies show that the risk to develop depressive disorder of adoptees with family history of depressive disorder is twice as high as in adoptees without family history of depressive disorder.

Organic causes:
1) Physical illnesses include Cushing's syndrome, Addison's disease, Parkinson's disease, stroke, epilepsy, coronary arterial disease and hypothyroidism
2) Medications: Corticosteroids, oral contraceptive pills, beta-blockers, clonidine, metoclopramide, theophylline and nifedipine

Psychosocial factors:
(a) Adversity in childhood
- Maternal loss and disruption of bonding.
- Poor parental care and over-protection among parents.
- Childhood physical and sexual abuse.

(b) Adversity in adulthood
- Women: Absence of a confiding relationship, having more than 3 children under the age of 14 and unemployment (Brown and Harris' social origins of depression, 1978).
- Men: Unemployment, divorce (e.g. unable to pay for maintenance fees and loss of custody).

(c) Recent life events
- Loss of a child.
- Death of a spouse.
- Divorce.
- Martial separation.
- Imprisonment.
- Recent death of a close family member.
- Unemployment.

Neurobiology of depressive disorder:

1. Monoamine theory states that depressed patients have decreased levels of noradrenaline, serotonin and dopamine. Evidence include the finding that the 5-HIAA levels are reduced in the CSF of depressed patients who committed suicide. 5-HIAA is a metabolite of serotonin. The tricyclic antidepressants increase noradrenaline levels. The selective serotonin reuptake inhibitors increase the serotonin levels. The antidepressant bupoprion may increase dopamine levels.
2. Other neurotransmitters include raised acetylcholine levels (associated with depressive symptoms such as anergia, lethargy, psychomotor retardation) and decreased levels of gamma-aminobutyic acid (GABA).
3. Neuroendocrinology: Elevated CRF, ACTH and cortisol in blood and CSF in depressed patients. Non-suppression in dexamethasone suppression test (DST) is greatest in people with severe depression and reversed with antidepressant treatment. Non-suppression of DST is a result of increased hypothalamic CRF release. Depression reduces the level of inhibitory hormone, somatostatin and increases the level of growth hormone. Decreased levels of thyroid hormone (T_4) are associated with depressive symptoms.
4. Neuroimaging: Ventricular enlargement, sulcal widening and reduction in size in the frontal lobe, cerebellum, basal ganglia, hippocampus and amygdala.

ICD-10 and DSM-5 criteria

Summary of ICD-10 F32.0 Mild, F32.1 Moderate, F32.2 *Severe (Bold + Italic)* criteria) diagnostic criteria of depressive disorder:

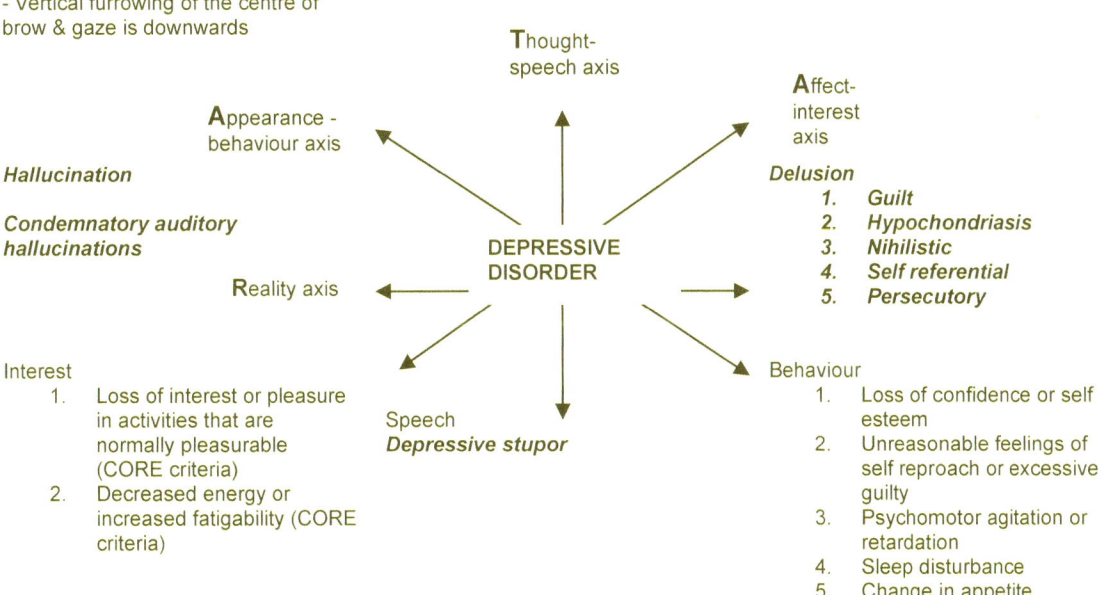

General appearance :
- Not specified by ICD – 10 or DSM-5
- Neglect of dressing & grooming
- Turning downwards of the corners of the mouth
- Vertical furrowing of the centre of brow & gaze is downwards

Thoughts :
1. Recurrent thoughts of death or suicide
2. Diminished ability to think and concentrate

Affect:
1. Depressed mood most of the day, almost every day for 2 weeks (CORE criteria)

Hallucination

Condemnatory auditory hallucinations

Delusion
1. *Guilt*
2. *Hypochondriasis*
3. *Nihilistic*
4. *Self referential*
5. *Persecutory*

Interest
1. Loss of interest or pleasure in activities that are normally pleasurable (CORE criteria)
2. Decreased energy or increased fatigability (CORE criteria)

Speech
Depressive stupor

Behaviour
1. Loss of confidence or self esteem
2. Unreasonable feelings of self reproach or excessive guilty
3. Psychomotor agitation or retardation
4. Sleep disturbance
5. Change in appetite
6. Somatic syndrome

For ICD – 10, mild depressive disorder requires 2 out of 3 core criteria + 2 remaining criteria; moderate depressive disorder requires 2 out of 3 core criteria + 4 remaining criteria; severe depressive disorder requires 3 out of 3 core criteria + 6 remaining criteria ± psychotic features.

Summary of the other types of depressive disorder in DSM-5

	DSM-5
Severity of depressive illness	The DSM-5 states that for individuals to fulfill the diagnostic criteria, they need to have at least 5 of the following symptoms for a minimum duration of 2 weeks: a. Low mood for most of the days (core feature) b. Diminished interest in almost all activities (core feature) c. Weight loss of more than 5% of body weight within a month's duration d. Sleep difficulties characterized as either insomnia (↓ total sleep time, ↓ Random Eye Movement (REM) latency, ↑density of REM sleep and ↑nocturnal wakening) or hypersomnia e. Psychomotor changes characterized as either agitation or retardation f. Generalized feelings of low energy nearly everyday g. Feeling worthless, or with excessive guilt h. Attention and concentration difficulties i. Recurrent passive or active ideations of self harm and suicide These symptoms have caused marked impairments in terms of premorbid functioning. In addition, DSM-5 states that for significant losses (bereavement, financial losses, disability etc), clinicians should carefully consider comorbid major depressive disorder, in addition to usual responses to the losses. There are several subtypes specified by DSM-5, which include: 1. With **anxious distress** - characterized by the presence of at least 2 of the following symptoms: Feeling restless, keyed up, difficulties with concentration, worries that something awful would happen, fear of losing control 2. With **mixed features** – characterized by presence of 3 or more of the following symptoms: elevated mood, grandiosity, increased speech, flight of ideas, increased energy, increased risky behavior, decreased need for sleep 3. With **melancholic features** - characterized by the presence of either (a) Diminished enjoyment in most activities or (b) Unable to react to enjoyable stimulus; and at least 3 or more of the following symptoms: (a) Feelings of excessive guilt, (b) Decreased appetite (c) Psychomotor changes, (d) Early morning awakening (at least 2 hours in advance), (e) Low mood especially in the morning, (f) Distinctively low mood. 4. With **atypical features** - characterized by the following symptoms: (a) Ability of mood to react according to stimulus, (b) Significant increment in appetite or weight, (c) Increased duration of sleep, (d) Heave sensations in arms or legs, (e) Being sensitive to interpersonal rejection. 5. With **mood congruent psychotic features** 6. With **mood-incongruent psychotic features** 7. With **catatonia** 8. With **peripartum onset** – characterized by mood symptoms occurring during pregnancy or in the 4 weeks following delivery 7. With **seasonal pattern**- characterized by regular association between mood symptoms and particular seasons in a year. Individuals should have full remission during the other seasons. At least 2 major depressive episodes must have demonstrated correlation with seasonality in the last 2 years for this diagnosis to be fulfilled.
Recurrent depressive disorder	Major depressive episode – recurrent episode 1. The presence of at least 2 major depressive episodes. 2. To be considered separate episodes, there must be an interval of at least 2 consecutive months in which criteria are not met for a major depressive episode.

	3. Specifier: mild, moderate and severe without psychotic features/with psychotic features, in partial/full remission, unspecified
Cyclothymia	The DSM-5 diagnostic criteria states that individuals would fulfill the diagnosis if: over the time duration of at least 2 years (or 1 year for children and adolescents), there have been numerous episodes (at least half the time) with hypomanic symptoms and depressive symptoms. It must be noted that these episodes do not meet the full diagnostic criteria for hypomania or depression. The individual must not have been without the symptoms for more than 2 months in duration each time.
Dysthymia	**Persistent Depressive Disorder (Dysthymia)** The DSM-5 specifies that an individual must have pervasive depressed mood for most part of the days, for a total duration of at least 2 years to qualify for the above diagnosis (1 year for children or adolescents). Apart from depressed mood, the individual should have at least 2 of the following signs and symptoms: a. Reduction or excessive oral intake b. Difficulties associated with sleep - either insomnia or hypersomnia c. Marked reduction in energy levels d. Reduced self confidence e. Attention and concentration difficulties f. Feelings that life is worthless and hopeless
Other mood disorders	**Disruptive Mood Dysregulation Disorder** The DSM-5 states that this diagnosis should be made for individuals between ages of 6 to 18 years old. Individuals should have the following symptoms for at least 12 months and these symptoms should be present in at least 2 different situational settings. The symptoms include: 1. Significant outbursts of temper manifested verbally or physically, not in keeping with the situational context 2. Temper outbursts are not consistent with developmental level 3. These outbursts occur on average at least 3 times per week 4. In between these temper outbursts, the individual's mood is persistently irritable. **Premenstrual Dysphoric Disorder** The DSM-5 states that the onset of the symptoms should be at least in the week prior to the onset of menstruation, and the symptoms should improve within a few days after the onset of menstruation. The intensity of the symptoms would either become minimal or resolve post-mensuration. DSM-5 specified that an individual needs to have at least 5 of the following signs and symptoms: a. Mood swings b. Increased interpersonal relationship conflicts c. Feelings of low mood associated with hopelessness d. Anxious e. Reduction in interest in usual activities f. Difficulties with concentration g. Marked reduction in energy levels h. Changes in appetite i. Sleep difficulties j. Sense of losing control k. Physical symptoms such as breast tenderness or swelling, muscular pain and a sensation of bloating or weight gain.
Melancholic depression	With melancholic features - characterized by the presence of either (a) Diminished enjoyment in most activities or (b) Unable to react to enjoyable stimulus; and at least 3 or more of the following symptoms: (a) Feelings of

		guilt, (b) Decreased appetite,(c) Psychomotor changes
Atypical features		With atypical features: characterized by the following symptoms: (a) ability of mood to react according to stimulus, (b) Significant increment in appetite or weight increase, (c) Increased duration of sleep, (d) Heave sensations in arms or legs, € Being sensitive to interpersonal rejection.

OSCE grid: Assess depression.

You are the resident working in the AED. A 30 – year - old teacher is referred by polyclinic for management of depression. He cannot cope with the workload and he also has interpersonal problems with the school principal.

Task: To a history to establish the diagnosis of depressive disorder.

Please note that forgetting to have a brief assessment of suicidal risk in a depressed patient may result in a failure.

A. Assess core symptoms of depression.	A1) Assess mood. 'During the past month, how often have you been bothered by feeling down or depressed?' 'Can you rate your current mood from a scale of 1 to 10? 1 means very depressed and 10 means very happy.' 'Which part of the day is the worst?' (Elicit diurnal variation of mood)	A2) Assess energy level. 'Have your energy levels been recently?' 'Do you feel tired most of the time?'	A3) Assess interest. 'Can you tell me more about your interests and hobbies before the current depressive episode?' 'During the past month, how often have you been bothered by having little interest or pleasure in doing things?'
B. Assess biological symptoms of depression.	B1. Assess sleep. 'How has your sleep been lately?' 'Can you fall asleep? If not, how long does it take?' 'How many times do you wake up in the middle of the night? (exclude urination)' 'What time do you wake up in the morning? (look for early morning wakening). If you wake up, can you fall asleep again?'	B2. Assess appetite and weight. 'Has your appetite changed recently? If yes, do you tend to eat less or more?' 'Has your weight changed recently? If so, have you lost weight or gained weight. If yes, how many kilograms were involved?'	B3.Assess sexual functions. I hope you would not mind if I ask you some sensitive questions such as sexual problems as depression may affect sexual function. Is it OK with you? 'Have there been any changes in your sexual function recently? If yes, can you tell me more about the nature of sexual dysfunction?' 'When did the sexual dysfunction start? (Does it coincide with the onset of depression?)'
C. Assess cognitive symptoms.	C1. Assess cognitive impairment. 'What has your concentration been like recently? Can you concentrate when you teach?'	C2. Assess feelings towards self and future. 'How do you see yourself?' 'Do you see yourself a failure?'	C3. Explore common cognitive bias. 'Can you tell me more about your negative thoughts?' Look for selective abstraction,

		'How has your memory been?'	'How do you see your future? Do you feel hopeless?'	overgeneralization or catastrophes thinking. Depending on the patient's response, you may gently challenge patient's belief or provide an alternative explanation to seek his or her view?
D.	Assess risk, psychotic features, insight.	D1. Assess suicide risk. 'Have you felt that life is not worth living?' 'Would you do anything to harm yourself or hurt yourself?' 'Have you done anything of that sort?' Have you made any plans? Have your told anybody about it?'	D2. Assess psychotic features. 'When people are under stress, they complain of hearing voices or believing that other people are doing something to harm him. Do you have such experiences?'	D3. Assess insight 'What is your view of the current problem? Do you think that you may suffer from a depressive illness?'
E.	Explore causes and background.	E1. Explore family history of depression. 'Do you have any biologically related relative suffer from depression?' 'Do you have any biologically related relative attempt or commit suicide in the past?'	E2. Explore past psychiatric history and relevant medical illnesses. 'Did you seek help from a psychiatrist or GP in the past for your low mood?' 'Did you receive any treatment from a psychiatrist? If yes, can you tell me more about the medication and side effects?' 'How anxious do you feel in yourself?' (explore comorbidity) 'Do you drink alcohol on a daily basis to cope with stress or help you to sleep?' 'Do you suffer from any chronic medical illness?'	E3. Assess support system. 'Can you tell me the person who is providing emotional support to you at this moment?' 'Is there a person in the school whom you can talk to?' 'What is your career plan at this moment?' 'Have you sought help from Ministry of Education?'

Differential diagnosis

Differential diagnosis of depressive disorder include:
1. Adjustment disorder, dysthymia, bipolar disorder, eating disorders, schizoaffective disorder, schizophrenia with predominance of negative symptoms.
2. Dementia, Parkinson's disease, post-stroke depression and head injury in old people presenting with depression..
3. Addison's disease, Cushing's disease, hypothyroidism, parathyroid dysfunction, hypopituitarism and menopausal symptoms..
4. Systemic lupus erythematosus.

5. Syphilis and HIV encephalopathy.
6. Medication induced (e.g. beta-blockers, steroids, oral contraceptive pills)
7. Substance misuse (e.g. benzodiazepines, alcohol and opiates).

Investigation

Routine laboratory tests should be ordered (e.g. FBC, ESR, B12/Folate, RFT, LFT, TFT, calcium panel and PTH). Sodium level is important in elderly who are prone to hyponatraemia as a result of SSRI treatment.
Further investigations include urine drug screen, urine FEME and urine culture (for elderly), thyroid antibodies (for people with abnormal TFT), antinuclear antibody (suspected SLE), syphilis Serology, HIV testing, CT/MRI.

Questionnaire

Beck Depression Inventory (BDI)
The BDI is a 21-item self rate instrument to measure the presence and degree of depression in both adolescents (Reading age of approximately 10 years is required) and adults. It is self-rated and was designed to measure attitude and symptoms characteristic of depression. The BDI covers the 2 weeks prior to evaluation. It consists of 21 items, each categorised into various level of severity (with a range of score from 0 to 3). The total score is the sum of items. A total score <9 indicates no or minimal depression. A total score >30 indicates severe depression.

Hamilton depression scale (HAM-D)
The HAM-D scale is a clinician rated semi-structured scale. It is designed to measure the severity of depressive symptoms in patients with primary depressive illness. It has two versions: 17-item scale and 21-items scale. The 17-item version covers mood, suicide, guilt, sleep, appetite, energy, somatic complaints, sexual function and weight. The 21-item consists of addition 4 items on diurnal variation of mood, derealisation / depersonalisation, paranoid idea and obsession / compulsions. The HAM-D scale monitors changes in the severity of symptoms during treatment. The HAM-D scale is not diagnostic and its validity is affected if the person has concurrent physical illness. The total scores range from 0 (no depression), 0-10 (mild depression), 10-23 (moderate depression) and over 23 (severe depression).

Montgomery-Asberg Depression Rating Scale (MADRS)
The **MADRS** is a clinician-rated scale for patients with major depressive disorder. It measures the degree of severity of depressive symptoms and is a particularly sensitive measure of change in symptom severity during treatment. The 10-item checklist measures current mood state. In contrast to HAM-D, the MADRS is useful for people with concurrent physical illness as it puts less emphasis on somatic symptoms.

Management [Mahendran and Yap (2005) and NICE guidelines (UK)]

Goal	The goal of treatment is to achieve symptomatic remission of all signs and symptoms of depression, restore occupational and psychosocial functioning,
Initial treatment	• Counselling and supportive therapy alone may benefit those patients with mild depression. • If sleep is a problem, doctor should offer sleep hygiene advice. • If antidepressants are used as the first line of treatment, SSRI is the first line treatment. Tricyclic antidepressants (TCAs) must be avoided in suicidal patients because of their lethality in overdose. Doctors need to inform patient that the antidepressants will take 4 to 6 weeks to achieve its effect. Doctors should be familiar with side effects and be able to explain to patients about the side effects. Monotherapy with a single antidepressant is recommended. • In patients who are reluctant to start antidepressants, or patients with comorbid medical conditions who may be unable to tolerate the antidepressants, psychotherapy may be considered as a first-line treatment. • Hospitalisation may be required if the patient poses high suicide risk to self.
Acute phase of treatment	• The acute phase of treatment is accepted as lasting 12 weeks. • Efficacy of the treatment is gauged by amelioration of symptoms and the dose should be titrated according to clinical response. • Monitor all patients recently started on antidepressants closely for increased agitation and suicidal behaviour, especially young patients (younger than 25 years). • Some symptoms, such as sleep and appetite, may improve more quickly.

		• If partial response or non-response, increase the dose or switch to another antidepressant. The first line is an alternative SSRI. The second line is an antidepressant from a different class. • Doctors have to consider the half-life of the antidepressant before switching. A washout period is needed when switching from fluoxetine which has a long half-life and moclobemide (reversible MAOI) which requires a three-day washout period. • If there is inadequate response to a single drug treatment, other agents such as another antidepressant (e.g. mirtazapine), mood stabiliser (e.g. lithium) or antipsychotics (e.g. olanzapine) can be added as augmentation therapy. • Combination with psychotherapy such as cognitive behaviour therapy is recommended for patients with moderate depressive episode.
Stabilisation phase		• Antidepressants should be continued for at least six months after the acute phase. • 3 to 4 months of psychological intervention is recommended for mild depressive episode. • 4 to 6 months of psychological intervention is recommended for moderate and severe depressive episode. • If a patient needs to stop antidepressants, stop gradually over a four-week period to avoid discontinuation symptoms. Common discontinuation symptoms include anxiety, giddiness, flu like symptoms, low mood nausea and insomnia. Antidepressants with shorter half live such as paroxetine and venlafaxine need to discontinue over a longer of time.

Selective serotonin reuptake inhibitors (SSRIs)

Indications	Contraindications	Mechanism of action	Side effects
Depressive disorder (first line treatment, less sedative than TCAs). Anxiety disorders Obsessive compulsive disorder Bulimia nervosa (fluoxetine) Premature ejaculation.	Mania	SSRIs selectively block reuptake of serotonin at presynaptic nerve terminals and increase synaptic serotonin concentrations	Nausea, abdominal pain, diarrhoea, constipation, weight loss. Agitation, tremor or insomnia. Sexual dysfunction

Common SSRIs used in Singapore (2012 price in SGD)
1. **Fluoxetine [Prozac]** (Dose range: 20 – 60mg/day; $0.46 for 10mg and $0.24 for 20mg) **Special features** 1) Non linear elimination kinetics 2) safe in overdose. **Other indications:** 1) OCD (>60mg/day); 2) Panic disorder; 3) Bulimia Nervosa; 4) PTSD; 5) Premenstrual dysphoric disorder; 6) Premature ejaculation and 7) Childhood & adolescent depression. **Pharmacokinetics:** 1) Fluoxetine inhibits the P450 3A3/4, 2C9, 2C19 & 2D6 and it also inhibits its own metabolism. 2) Due to non-linear pharmacokinetics, higher doses can result in disproportionately high plasma levels and of some side-effects (e.g. sedation) rather late in the course of treatment with this drug. Its metabolite norfluoxetine is much less potent. Long $t_{1/2}$ = 72 hr. **Pharmacodynamics:** The serotonin system exerts tonic inhibition on the central dopaminergic system. Thus, fluoxetine might diminish dopaminergic transmission leading to EPSE. **Side effects:** 1) <u>Anxiety, agitation</u> 2) Delayed ejaculation/orgasmic impotence 3) Hypersomnolence (high doses) 4) Nausea 5) Dry mouth. **Drug interaction:** 1) The washout period for fluoxetine before taking MAOI is 5 weeks. 2) Through inhibition of P450 2D6, fluoxetine may elevate the concentration of other drugs especially those with narrow therapeutic index such as flecainide, quinidine, carbamazepine and TCAs.

2. Fluvoxamine [Faverin] (Dose range: 50-300mg/day; $0.24 for 50mg)

Special features: 1) Highly selective SSRI 2) FDA approval for OCD 3) Lower volume of distribution, low protein binding and much shorter elimination half-life compared to SSRI.

Other indications: 1) Efficacious for social phobia 2) Panic Disorder 3) PTSD

Pharmacokinetics: 1) Well absorbed 2) $t_{1/2}$ =19 hours 3) metabolized to inactive metabolites. 4) lower volume of distribution and low protein binding 5) maximum plasma concentration is dose dependent. 6) Steady-state levels is 2 to 4-fold higher in children than in adolescents especially females. 7) Well tolerated in old people and in people with mild cardiovascular disease or epilepsy. 8) It offers potent inhibition of P450 1A2.

Pharmacodynamics: The specificity for 5HT re-uptake is greater for other SSRIs. Two neuro-adaptive changes: 1) specific serotonin receptor subtypes that change following presynaptic blockade 2) neurogenesis of hippocampal brain cells occurs and results in changes in behaviour.

Side effect: 1) Nausea – more common than other SSRIs; 2) Sexual side effects with fluvoxamine are similar in frequency to those with other SSRls; 3) It minimal effects on psychomotor and cognitive function in humans.

3. Sertraline [Zoloft] (Dose range: 50-200mg/day; $1.92 for 25mg)

Special features: 1) For young women with mood disorders. 2) For mood and anxiety disorders.

Other indications: 1) Premenstrual dysphoric disorder 2) OCD 3) PTSD

Pharmacokinetics: 1) It inhibits P450 2C9, 2C19, 2D6, 3A4. 2) $t_{1/2}$ is 26 – 32 hours 3) More than 95% protein bound 4) Its metabolite, desmethylsertraline, is one-tenth as active as sertraline in blocking reuptake of serotonin.

Pharmacodynamics: The immediate effect of sertraline is to decrease neuronal firing rates. This is followed by normalization and an increase in firing rates, as autoreceptors are desensitized.

Side effects: 1) GI side effects 27%; 2) headache 26%; 3) Insomnia 22%; 4) Dry mouth 15% ; 5) Ejaculation failure 14%.

4. Paroxetine CR [Seroxat CR] (Dose range: 12,5-50mg/day; $2.23 for 12.5mg; $2.17 for 25mg)

Special features: 1) Most sedative and anticholinergic SSRI; 2 Risk of foetal exposure resulting in pulmonary hypertension

Other indications: 1) Mixed anxiety and depression; 2) Panic disorder; 3) social anxiety disorder; 4) generalised anxiety disorder 5) Post-traumatic stress disorder; 6) Premenstrual disorder

Pharmacokinetics: 1) Paroxetine is well absorbed from the GI tract. 2) it is a highly lipidophilic compound. 3) It has a high volume of distribution. 4) It is 95% bound to serum proteins. 4) It undergoes extensive first pass metabolism. 5) Paroxetine CR slows absorption and delay the release for 5 hours. 6) The Short $t_{1/2}$ of original paroxetine leads to discontinuation syndrome 7) It inhibits its own metabolism.

Side effects: 1) <u>Anticholinergic side effects</u> 2) Nausea 3) Sexual side effects emerge in a dose-dependent fashion 4) Closed angle glaucoma (acute). MOH guidelines state that first-trimester paroxetine use should be avoided, as it is associated with increased risk of serious congenital (particularly cardiac) defects.

Drug interaction: 1) Clinically significant interaction: MAOI, TCA, Type 1C antiarrhythmics, 2) Probably significant interaction: β-adrenergic antagonists, antiepileptic agents, cimetidine, typical antipsychotics, warfarin.

5. Escitalopram [Lexapro] (Dose range: 10-20mg/day; $1.5 for10mg; $3.8 for 20 mg)

Special features: 1) Most selective SSRI 2) Relatively weak inhibition of liver P450 enzymes. 3) Escitalopram has fewer side effects, more potent, shorter $t_{1/2}$, less likely to inhibit P450 system, more selective than citalopram.

Other indications: 1) OCD 2) Panic Disorder 3) CVA 4) Anxiety with major depression 5) Emotional problems associated with dementia

Pharmacokinetics: 1) Escitalopram is well absorbed after oral administration with high bioavailability. 2) Peak plasma concentration is normally observed 2-4 hours following an oral dose. 3) It is subject to very little first-pass metabolism.

Side effects: Nausea and vomiting (20%), increased sweating (18%), dry mouth & headache (17%), anorgasmia and ejaculatory failure, but no significant effect on cardiac conduction and repolarisation.

Others:

6. Trazodone (Dose range: 150-300mg/day; this drug is not available at NUH)

Special features: trazodone is a mixed serotonin antagonist/agonist.

Pharmacokinetics: 1) Trazodone is well absorbed after oral administration, with peak blood levels occurring about 1 hour after dosing. 2) Elimination is biphasic, consisting of an initial phase ($t_{1/2}$ =4 hrs) followed by a slower phase, with $t_{1/2}$ = 7 hours. 3) Its metabolites, mCPP, is a non-selective serotonin receptor agonist with anxiogenic properties.

Pharmacodynamics: Trazodone antagonises both α_1 and α_2 adrenoceptors but has very weak anticholinergic side-effects.

Side effects: 1) Priapism 2) Orthostatic hypotension 3) Increased libido 4) Sedation 5) Bone marrow suppression.

7. Noradenaline Specific Serotonin Antidepressant (NaSSa)

Mirtazapine (Dose range: 15-45mg/day; $ 0.60/15mg)

Special features: 1) Mirtazapine blocks negative feedback of noradrenaline on presynaptic α_2 receptors and activates noradrenaline system; 2) Mirtazapine stimulates serotonin neuron and increases noradrenaline activity; 3) Mirtazapine has no effects on seizure threshold or on cardiovascular system. 4) Suitable for patients who cannot tolerate SSRI induced sexual dysfunction..

Other indications: insomnia or poor appetite, dysthymia (40% reduction), PTSD (50% reduction) and chronic pain.

Pharmacokinetics: 1) The peak plasma level is obtained after approximately 2 hours. 2) Linear pharmacokinetics and a steady-state plasma level is obtained after 5 days. 3) The elimination $t_{1/2}$ is 22 hours 4) Metabolised by P450 1A2, 2D6, and 3A4 4) 75% excreted by the kidney and 15% excreted by GI tract.

Pharmacodynamics: 1) Blockade of release-modulating α_2-adrenoceptors leads to enhanced noradrenaline release 2) the released noradrenaline stimulates serotonin neurons via the activation of α_1 adrenoceptors which in turn results in an enhanced noradrenaline effect, together with the selective activation of $5-HT_{1A}$ receptors, may underlie the antidepressant activity; 3) $5HT_{1A}$ agonism: antidepressant and anxiolytic effects. 4) $5HT_{2A}$ antagonism: anxiolytic, sleep restoring and no sexual restoration 5) $5HT_{2c}$ antagonism: anxiolytic & weight gain 6) $5HT_3$ antagonism: no nausea, no gastrointestinal side effects. It also blocks histaminergic receptors and results in drowsiness.

Side effects: 1) drowsiness, 2) weight gain, 3) increased appetite, 4) dry mouth, 5) postural hypotension.

8. Serotonin noradrenaline reuptake inhibitors (SNRIs)

Venlafaxine XR [Efexor XR] (Dose range: 75mg – 375mg/day; $ 0.96/ 75mg tablet)

Special features: Low doses of venlafaxine blocks serotonin reuptake. Moderate doses of venlafaxine block noradrenaline reuptake. High dose of venlafaxine block noradrenaline, dopamine and serotonin reuptake. 2) Metabolised by P450 3A4 to inactive metabolites while P450 2D6 to active metabolites 3) More rapid onset action and enhanced efficacy in severe depression

Other indications: generalised anxiety disorder.

Pharmacokinetics: 1) minimally protein bound (<30%) 2) renal elimination is the primary route of excretion 3) the original version (venlafaxine) has relatively short $t_{1/2}$ = 5-7 hours; 4) prominent discontinuation syndrome (dizziness, dry mouth, insomnia, nausea, sweating, anorexia, diarrhoea, somnolence and sensory disturbance) and hence venlafaxine extended release (XR) is available.

Side effects: 1) nausea (35%); 2) sustained hypertension is dose related and 50% remitted spontaneously 4) dry mouth, constipation, 5) sexual dysfunction

Drug interaction: The toxic interaction with MAOIs, leading to a serotonin syndrome, is the most severe drug interaction involving venlafaxine.

---

9. **Duloxetine [Cymbalta]** (Dose range: 30-120mg/day; $3.8/ 30mg tablet)

Other indications: 1) Depression and chronic pain; 2) Fibromyalgia

Pharmacokinetics: Blood levels of duloxetine are most likely to be increased when it is co-administered with drugs that potently inhibit cytochrome P450 1A2.

Pharmacodynamics: Duloxetine exerts a more marked influence on noradrenaline reuptake than on serotonin reuptake.

Side effects 1) Nausea, dry mouth, dizziness, headache, somnolence, constipation and fatigue are common. 2) A small but significant increase in heart rate was observed 3) Rate of sexual dysfunction is low.

Tricyclic antidepressants (TCA)

Indications	Contraindications	Mechanism of action	Side effects	Examples
TCA is an old antidepressant and it is not first-line antidepressant treatment due to potential cardiotoxicity if patient takes an overdose. Depression Anxiety disorder Severe OCD (Clomipramine) Neuropathic pain Migraine prophylaxis Enuresis	Cardiac diseases (e.g. post myocardial infarction, arrhythmias) Epilepsy Severe liver disease Prostate hypertrophy Mania	TCA inhibits the reuptake of both serotonin and noradrenaline and increase the concentration of these neurotransmitters. TCA also blocks histaminergic H_1, α-adrenergic and cholinergic muscarinic receptors on the postsynaptic membrane.	Anticholinergic (e.g. constipation, blurred vision, urinary retention, dry mouth dizziness, syncope, postural hypotension, sedation). Histaminergic and dopaminergic blockade: nausea, vomiting, weight gain, sedation, Other side effects: sexual dysfunction, hyponatraemia Cardiac: arrhythmias, ECG changes (QTc prolongation), tachycardia, heart block TCA overdoses may lead to delayed ventricular conduction time, dilated pupils and acidaemia due to	**Amitriptyline** (25mg to 150mg daily) has the most potent anticholinergic effect. **Clomipramine** (100 – 225mg daily): Most potent TCA at D_2 receptors; More selective inhibitor of serotonin reuptake.

			central respiratory depression and a fall in pH reducing protein binding.	

Monoamine oxidase inhibitors (MAOIs)

Reversible MAOI – Moclobemide (Dose range: 75mg to 225mg daily; $0.45 per 150mg tablet)

Indications	Contraindications	Mechanism of action	Side effects
Atypical depression. Depression with predominantly anxiety symptoms (e.g. social anxiety). Hypochondriasis.	Acute confusional state. Phaeochromocytoma.	Monoamine oxidase A acts on • Noradrenaline • Serotonin • Dopamine • Tyramine •	Visual changes. Headache. Dry mouth. Dizziness. GI symptoms.

The old irreversible MAOIs may lead to hypertensive crisis with food containing tyramine. Irreversible MAOIs are seldom used nowadays. The following food should be avoided :
1) Alcohol: avoid Chianti wine and vermouth but red wine <120 ml has little risk.
2) Banana skin.
3) Bean curds especially fermented bean curds.
4) Cheeses (e.g. Mature Stilton) should be avoided but cream cheese and cottage cheese have low risk.
5) Caviar.
6) Extracts from meats & yeasts should be avoided but fresh meat and yeast.

Other antidepressants

Bupropion [Wellbutrin] (Dose range: 150-300mg/day; $1.6 / 150mg tablet)

Similar efficacy as SSRI but voluntary withdrawal in the US due to induction of seizure at doses if the daily dose is higher than 450mg/day.

Other indications include patients encountering SSRI induced sexual dysfunction, female depressed patients do not want weight gain from medication and smoking cessation.

Pharmacodynamics: blocking dopamine reuptake.

Side effects: agitation, tremor, insomnia, weight loss and seizure. Bupoprion is not associated with sexual side effect.

Electroconvulsive therapy (ECT)

Indications:
1) Severe depressive disorder which does not respond to an adequate trial of antidepressants.
2) Life threatening depressive illness (e.g. high suicide risk).

3) Stupor or catatonia
4) Marked psychomotor retardation
5) Psychotic depression
6) Treatment resistant mania
7) Treatment resistant schizophrenia.

Relative contraindications:
1) Raised intracranial pressure
2) Myocardial infarction
3) Valvular heart diseases
4) Aneurysm
5) Recent stroke
6) Peptic ulcer.

Mechanism of actions:
1) Release of noradrenaline, serotonin, dopamine but reduction of acetylcholine release.
2) Increase in permeability of the blood-brain barrier.
3) Modulation of neurotransmitter receptors such as GABA or acetylcholine.

Administration:
1. Usually bilateral temporal ECTs for adults for 6 treatments (3 times a week).
2. Unilateral ECT is reserved for old patients with the risk of cognitive impairment.
3. Bilateral ECT is more effective than unilateral ECT.
4. An informed consent is required prior to the ECT. A second opinion from another consultant psychiatrist is required for patient who lacks capacity or in cases of patient under Mental Disorder and Treatment Act who refuses treatment.
5. ECT is given under general anaesthesia. Muscle relaxant is given to prevent muscular spasms.
6. Electric current generates a seizure for less than one minute.
7. Before ECT, avoid long acting benzodiazepine which will affect the duration of seizure.
8. After ECT, patient is recommended to continue antidepressant for at least 6 months.

Side effects:

Common side effects include headache, muscle pain, jaw pain, drowsiness, loss of recent memories (retrograde amnesia), anterograde amnesia (less common than retrograde amnesia), prolonged seizures (longer than 1 minute) and confusion after ECTs.

Other side effects include anaesthesia complications arrhythmia, pulmonary embolism and aspiration pneumonia.

Factors increase seizure threshold: old age, male gender, baldness, Paget's disease, dehydration, previous ECT and benzodiazepine treatment.
Factors decrease seizure threshold: caffeine, low CO_2 saturation of blood, hyperventilation and theophylline

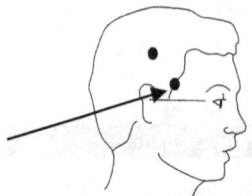

Electrode positioning in unilateral ECT

First electrode: 4cm above the midpoint of lateral angle of eye & external auditory meatus

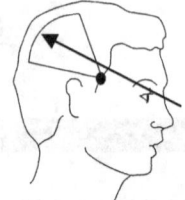

Second electrode or d'Elia positioning: second electrode is placed in the midpoint of the arc. The radius of arc is around 18 cm.

OSCE video – Explain ECT (Refer to Clinical OSCE Videos)

> You are the resident and you have admitted an elderly woman suffering from severe depressive episode with delusion of guilt. She does not respond to the antidepressant and antipsychotic drug. Your consultant has recommended ECT and her daughter is very concerned and wants to speak to you.
>
> Task: Talk to her daughter and address her concerns.

- **Approach**: Express empathy. (e.g. I can imagine the idea of ECT sounds very scary for you, and it's clear you want the best care for your mother. I would like to discuss what ECT involves, because it is very different than what is portrayed in the media. This way, you can make an informed decision)
- **Core information about ECT:**
    - ECT involves inducing a fit, while the patient is under general anaesthesia.
    - ECT is the most effective treatment for depression, particularly for those who have high risk of suicide, very poor appetite and not responding to oral medication; sometimes in pregnant women because it has no side effects to the foetus.

    It is very safe and has been with us for the past 50 years.
- **Will my mother be awake during ECT?** No, your mother will be given **anaesthesia** to put her into sleep and a medication **that paralyze muscles**, so the risk of breaking bones is rare. The patient is given **oxygen** before the procedure. The patient's blood pressure, heart rhythm, and medical status is monitored throughout the procedure and when she comes out of the anaesthesia.
- **How often will my mother get ECT and for how long?** 3 times per week, Mon, Wed, Fri and for 6 sessions (2 weeks); some patients may need 9 to 12 sessions..
- **How do you know the ECT is successful or not?** We will monitor the duration of her fit. It has to be at least 25 second in duration. We will monitor her muscle movement through electrical recordings (i.e. EEG). If response is poor, we will increase the energy level 5% each time.
- **How do you decide on the dose of ECT?** By age-based dosing: Energy level = patient's age divided by 2.
- **What tests do you include in your pre-ECT work-up?** Physical exam, FBC, RFT, ECG, CXR. Assess patient's dentition, especially for elderly or those who have inadequate dental care.
- **What is the preparation for the night before the ECT?** Fasting is required after 12:00 midnight and she should avoid sleeping pills if possible.
- **What is the risk involved?** ECT itself is safe. Risk is associated with anaesthesia.
- **How does ECT affect memory?**
    - Anterograde and retrograde amnesia can occur, though in the majority of patients this does not last more than a few months following the last ECT treatment.
    - Amnesia of events immediately preceding and following ECT treatments may be permanent (reassure the relative those memory is not important).
    - Anterograde amnesia is always transient. In a very small number of patients, the symptoms of retrograde amnesia may be permanent.
- **What are other common side effects?** Memory problems, confusion, nausea, muscle aches and headache are the most common in the morning after the ECT.
- **What are the risk factors associated with confusion after ECT?** Old age; prior cognitive impairment; lithium; anticholinergic and bilateral placement.
- **How would you reduce confusion after ECT?** Unilateral treatment on right – side of the brain, lower electrical energy, increasing the time between ECT treatments and holding off lithium or sleeping pills.
- **What is the mortality rate associated with ECT?** The mortality rate is very low, and is the same as that for general anaesthesia, which is 1 in every 20 000 people.

Psychotherapy

1. **Cognitive behaviour therapy (CBT).** The frequency of CBT is usually weekly or fortnightly. It requires 12 to 16 sessions. The cognitive therapy involves identifying negative automatic thoughts and use dysfunctional thought diary to identify pattern between the time, events, negative thoughts and resulted emotions and behaviours. The psychologist will read the diary and help patients to gently challenge the negative automatic thoughts. Behaviour therapy involves activity scheduling (for those depressed patients with psychomotor retardation), relaxation techniques (for those patients with mixed anxiety and depression).

2. **Interpersonal therapy (IPT)** IPT is held weekly or fortnightly. It involves 12 to 20 sessions. IPT is indicated for depressed patients whom precipitating factor is interpersonal problems. The psychologist closely examines interpersonal relationship and works with the patient to look at interpersonal relationship from another angle to minimise impact on the mood and use role-play to improve communication skills.

    CBT and IPT have the strongest evidence in treating depressive disorder.

3. **Brief dynamic therapy** Brief dynamic therapy originates from psychoanalysis. Brief dynamic therapy is suitable for depressed patients whose predisposing factor is related to past experiences (e.g. unpleasant childhood experience with one of the parents) and these experiences have lead to the use of maladaptive defence mechanisms and affect current mood and personality development. Brief dynamic psychotherapy is contraindicated in psychotic patients.

4. Other psychotherapies include supportive psychotherapy, problem solving therapy or marital therapy depending on clinical history and case formulation.

Course and prognosis

- Depressive episodes may last from 4-30 weeks for mild or moderate depressive disorder to an average of around 6 months for severe depressive disorder.
- 10-20% of patients would have depression as a chronic disorder, with signs and symptoms lasting for around 2 years.
- The rate of recurrence is around 30% at 10 years and around 60% at 20 years.
- Suicide rates for depressive individuals is 20% higher when compared to the general population.
- Good prognostic factors include acute onset of depressive illness, endogenous depression and earlier age of onset. Poor prognostic factors include insidious onset, old age of onset, neurotic depression, low self-esteem and residual symptoms.

Bipolar disorders

Epidemiology

	Prevalence	Mean age of onset	Gender ratio
International	0.3 – 1.5% (overall) 0.2 – 4% (Bipolar I disorder) 0.3 – 4.8% (Bipolar II disorder).	20 years.	Male: female = 1:1.

Aetiology

Genetics:
- Family studies have demonstrated that children of parents suffering from bipolar disorder have a 9-fold increase in lifetime risk compared to the general population.
- The heritability of bipolar disorder is 79-93%.
- Twin studies indicate that monozygotic twins have 70% concordance rate and dizygotic twins have 20% concordance rate.
- Genes related to ion channels are implicated in the aetiology of bipolar disorder (e.g. calcium channels on chromosome 12). (Ferreira et al, 2008).

Monoamine theory states that increased levels of noradrenaline, serotonin and dopamine have been linked with manic symptoms. Excitatory neurotransmitter glutamate is also implicated.

The onset and first manic episode: The diagnosis is commonly delayed until early adulthood (median age of onset is in mid-20s and mean age of first hospitalisation is at 26 years) because there is abnormal programmed cell death or apoptosis in neural networks responsible for emotional regulation. The first manic episodes are often precipitated by life events such as bereavement, personal separation, work-related problems or loss of role. High expressed emotion and sleep deprivation are important precipitating factors.

The relationship between depressive and manic episodes: 1 in 10 patients who suffer from a depressive episode will subsequently develop a manic episode. Monotherapy of antidepressant is a recognised precipitant of the first manic episode in patients who are predisposed to suffer from bipolar disorder. In general, depressed patients with early age of onset, family history of bipolar disorder, depressive episode occurring during postnatal period, hypersomnia and psychotic symptoms are more likely to switch to mania.

Sleep deprivation and flying overnight from west to east may trigger relapse of mania.

Kindling hypothesis: The persistence of neuronal damage leads to recurrence of mania without precipitating factors. This is known as kindling and subsequent manic episodes become more frequent. The episode duration remains stable throughout the course of bipolar illness

Organic causes of mania

Cerebrovascular accident

Mania is associated with right-sided cerebral vascular lesions and it is commonly associated with lesions in the frontal and temporal lobes.

Endocrine causes

1. Thyrotoxicosis
2. Thyroid hormone replacement
3. Cushing's syndrome

Head injury

Mania is associated with right-sided hemispheric damage. Family history of mania is uncommon and patients are more irritable than euphoric.

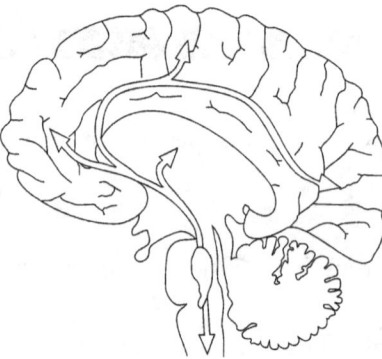

Lesions in right cerebral hemisphere are associated with mania

Other CNS disorders

1. Cerebral tumour
2. Dementia
3. Epilepsy
4. AIDS
5. Multiple sclerosis

Illicit substances:

1. Amphetamine
2. Cannabis
3. Cocaine

Medications:

1. Anticholinergic drugs
2. Dopamine agonists (e.g. bromocriptine and levodopa)
3. Corticosteroids or anabolic steroids
4. Withdrawal from baclofen, clonidine and fenfluramine.

Diagnostic criteria for bipolar disorder

Appearance

Increased sociability or over-familiarity

Thoughts
1. Difficulty in concentration with distractibility
2. Flight of ideas or racing thought (In ICD-10, this only occurs in mania but not hypomania)
3. Inflated self esteem and grandiosity (In ICD-10, this only occurs in mania but not hypomania)
4. Constant change in plans (In ICD-10, this only occurs in mania but not hypomania)

Affect:

Elevation of mood and irritability

Hallucination

Mood congruent: voices telling the patient that he has superhuman powers.

Mood incongruent: voices speaking to the patients about affectively neutral subjects.

Appearance - Behaviour
Thought-Speech axis
Affect-Interest axis
Reality axis

Mania

Delusion

Mood congruent: grandiose delusions

Mood incongruent: delusion of reference and persecution

Interest

increase in goal – directed activity (either socially, at work or school or sexually) or excessive involvement in pleasurable activity that have a high potential for painful consequences (e.g. unrestrained buying sprees, sexual indiscretion or foolish business)

Speech

Increased talkativeness

Behaviour
7. Increased activity and physical restlessness
8. Decreased need for sleep
9. Increased sexual energy (hypomania) / Sexual indiscretions (mania)
10. Mild overspending or other types of reckless or irresponsible behaviour (hypomania)/Foolhardy and reckless behaviour with lack of awareness (mania)
11. Loss of social inhibition, resulting inappropriate behaviour

DSM 5 criteria for bipolar disorder

Diagnosis	DSM-5
Bipolar I and II	**Bipolar 1 Disorder** **Manic Episode** DSM-5 specified that an individual needs to have at least 1 manic episode in order to fulfill the diagnostic criteria of Bipolar I disorder. A manic episode is characterized by a period of time, of at least 1 week, during which the individual has persistent elevated or irritable mood and present for most of the days. In addition, the individual needs to have at least 3 (4 if mood is only irritable) of the following symptoms: a. Increased self confidence b. Reduction in the need for sleep c. More chatty than usual, with increased pressure to talk d. Racing thoughts e. Easily distractible f. Increase in number of activities engaged g. Involvement in activities that might have a potential for serious consequences There must be marked impairments in terms of functioning with the onset of the above symptomatology. Clinicians should note that for manic episode triggered by antidepressants usage, or even electroconvulsive therapy, the diagnosis of Bipolar I disorder could be made if symptoms persist even

		upon the discontinuation of the existing treatment. **Bipolar II Disorder** The DSM-5 diagnostic criteria states that for an individual to be diagnosed with bipolar II disorder, there must be a past or current hypomanic episode, in addition to a current or past major depressive episode. In addition, the symptoms affect functioning but not severe enough to cause marked impairment. Clinicians should not make this diagnosis if there has been previous manic episode.
Hypomania		A hypomanic episode is characterized by a period of time of at least 4 days, during which the individual has persistent elevated or irritable mood, present for most of the days. In addition, the individual needs to have at least 3 (4 if the mood is only irritable) of the following symptoms: a. Increased self confidence b. Reduction in the need for sleep c. More chatty than usual, with increased pressure to talk d. Racing thoughts e. Easily distractible f. Increase in number of activities engaged g. Involvement in activities that might have a potential for serious consequences Clinicians to note that for individuals with hypomanic episodes, their level of functioning will not be markedly impaired.
Mania		A manic episode is characterized as a period of time, of at least 1 week, during which the individual has persistent elevated or irritable mood, present for most of the days. In addition, the individual needs to have at least 3 (4 if the mood is only irritable) of the following symptoms: a. Increased self confidence b. Reduction in the need for sleep c. More chatty than usual, with increased pressure to talk d. Racing thoughts e. Easily distractible f. Increase in number of activities engaged g. Involvement in activities that might have a potential for serious consequences There must be marked impairments in terms of functioning with the onset of the above symptomatology
Mixed episode and rapid cycling disorders		For mixed episodes, patients fulfil both manic and major depressive symptoms for at least 1 week. Rapid cycling as a course of bipolar disorder which consists of at least 4 episodes of mood disturbance (manic, hypomanic and major depressive episode) in one year. Ultra-rapid cycling describes 4 or more episodes in a month and it is a rare condition. Note: Rapid cycling is more common in women, occurring later in the course of bipolar illness and antidepressants can increase the frequency of rapid cycling episodes.
Other specified Bipolar and Related Disorder		Other specified Bipolar and related disorder includes: a. Hypomanic episodes of short duration (2-3 days) and major depressive episodes b. Hypomanic episodes with lack of symptoms and major depressive episodes c. Hypomanic episodes without previous major depressive episode d. Cyclothymia of less than 2 years duration

OSCE grid: Assess bipolar disorder

You have been asked to see a 28-year-old unemployed man who has not slept for 5 day and claims to have full energy. He claims to be the President of Singapore and his plan is to unite all the world leaders to fight for poverty in developing countries.

Task: Take a history to establish the diagnosis of bipolar disorder..

A. Assess mood symptoms.	A1. Assess mood. How's your mood today? If I ask you to rate your mood from 1 to 10, 1 means very depressed and 10 means very happy, how would you rate your mood today? How long have you been feeling high? Do you have mood swings? How about feeling low? If so, roughly how many low or high episodes you would experience in a year?	A2. Assess irritability. How do you get on with people recently? Do you feel that they annoy you? Do you lose your temper easily? What would you do if these people irritate you?	A3. Assess grandiosity How would you compare yourself with other people? Are you special? If yes, please tell me more. Could your special ability be a misunderstanding? Can you provide more evidence about it? Do you feel that you are at the top of the world (i.e. above all the other people)?
B. Assess biological symptoms.	B1. Assess sleep and energy. How has your sleep been lately? What is your energy level like? Do you feel that you need much less sleep but full of energy?	B2. Assess appetite and weight. How has your appetite been lately? Have you lost weight recently?	B3. Assess sexual function and contraceptive method. I am going to ask you some sensitive questions. How has your interest in sex been lately? Do you have sex with new partners? Do you take any precaution to protect yourself (e.g. condom)? If your patient is a woman, you need to ask LMP and chance of pregnancy.
C. Assess cognitive and psychotic symptoms.	C1. Assess interests and plans. Could you tell me about your interests? Have you developed any new interests lately? Do you have any new plan or commitment at this moment? (for example, starting a new business or investment)	C2. Assess thought and speech. Has there been any change in your thinking lately? Have you noticed that your thoughts speed up? Do you find your thoughts racing in your mind? Do your family members say that the topics in your speech change so fast and they cannot follow.	C3. Assess psychotic features. When people are under stress, they have unusual experiences such as hearing a voice talking to them but cannot see the person. Do you encounter such experiences? If so, what did the voices say? How many voices spoke at one time? Do you believe that you have special power or status which other people do not have? If yes, can you tell me about your special power or status? Are you very certain that you have such ability or status?
D. Assess risk	D1. Assess risk.	D2. Explore comorbidity.	D3. Assess insight.

and insight	Have you been buying a lot of things? Have you incurred a lot of debts (e.g. credit card debts?)	Do you take recreational drugs on a regular basis to get the high feelings?	Is there any reason why you encounter those experiences?
	Do you drive? Have you been involved in speeding or traffic offences?	How about alcohol? Do you drink on a regular basis?	Do you think there is a illness in your mind? For example, this illness affects your mood?
	Have you been in trouble with the police lately? (e.g. due to violence).		If so, do you think you need treatment?
	When you feel sad, have you thought of harming yourself?		

Investigations and questionnaire

- FBC, ESR.
- LFT, RFT, TFT, fasting lipid, glucose and body weight measurement (as mood stabilisers are associated with metabolic syndrome).
- VDRL.
- Urine drug screen
- Pregnancy test (for female patients who may be pregnant).
- CT/MRI to rule out space occupying lesion, infarction, haemorrhage.
- ECG to rule out prolonged QTc.
- EEG to rule out epilepsy.

The Young mania rating scale (YMRS)

The **YMRS** is an 11-item questionnaire which helps clinicians to measure the severity of manic episodes in children and adolescents between the ages of 5 and 17 and adults. Its structure is similar to the Hamilton depression scale.

Management

Summary of MOH guidelines (Mok et al 2011) and NICE guidelines (UK)

Acute treatment of mania	Hospitalisation may be necessary in patients present with severe manic symptoms or pose serious risk e.g. violence, sexual indiscretions). Some manic patients who refuse treatment may require admission under the Mental Disorder and Treatment Act.
	Haloperidol may be used for the treatment of acute mania.
	Aripiprazole, olanzapine, quetiapine, risperidone or ziprasidone may be used for the treatment of acute mania.

	Sodium valproate or carbamazepine monotherapy may be used for the treatment of acute mania.

Lamotrigine should not be used for the treatment of acute mania, as it lacks efficacy in this area.

Combination pharmacotherapy with an antipsychotic and a mood stabiliser may be used for patients showing inadequate response to mood stabiliser monotherapy.

Clonazepam or lorazepam (IM or oral) may be used in the acute treatment of agitation in mania.

In the emergency setting, Haloperidol (IM or oral) or olanzapine (oral) may be used in the acute treatment of agitation in mania. |
| Acute treatment of bipolar depression. | For mild depressive symptoms, it is recommended to review patients in 1 to 2 weeks without giving an antidepressant.

If depressive symptoms are moderate to severe, consider adding antidepressant to a mood stabiliser.

If antidepressants are to be used in combination with mood stabilisers as treatment for bipolar depression, they should be used cautiously due to conflicting evidence of efficacy and risk of inducing a manic episode. SSRI such as fluoxetine is the first line of treatment.

Lithium may be used in the treatment of bipolar depression.

Quetiapine monotherapy, olanzapine monotherapy or olanzapine-fluoxetine combination may be used in the treatment of bipolar depression.

Monotherapy with sodium valproate or carbamazepine is not recommended in the treatment of bipolar depression due to conflicting evidence regarding efficacy.

There is insufficient evidence to recommend lamotrigine monotherapy in the treatment of bipolar depression. However, it is recommended as an add-on for patients already on lithium for treatment of bipolar depression.

Psychotherapy (e.g. CBT) is recommended for patients with bipolar depression. |
| Rapid cycling and mixed state. | Mood stabilisers may be used when treating mixed states. Of these, valproate and carbamazepine should be preferred over lithium, as there is more evidence for the efficacy of valproate and carbamazepine than for lithium.

Non-pharmacological treatments should not be used for patients with rapid cycling bipolar disorder due to insufficient evidence. |
| Maintenance treatment | Lithium, valproate or olanzapine may be used as maintenance therapy in preventing relapse to either pole of illness in patients with bipolar disorder.

Aripiprazole may be used as maintenance therapy in bipolar patients with recent manic or mixed episode.

Quetiapine, in combination with lithium or valproate, may be used as maintenance therapy in patients with bipolar I disorder.

In very severe patients, the combination of lithium and valproate is possible and this should be a specialist's decision.

A patient is advised to continue treatment for at least 2 years after an episode of bipolar disorder and up to 5 years if there is a significant risk of relapse. |

Bipolar disorder and pregnancy	Typical antipsychotics, such as chlorpromazine and haloperidol, may be considered for treatment of pregnant women with bipolar disorder, as they appear less likely than atypical antipsychotics to cause metabolic complications and other serious adverse effects.
	Electroconvulsive therapy may be considered as a treatment option in pregnancy for the same indications as in nonpregnant patients.
	Women on combined oral contraceptive pills who are concurrently taking carbamazepine and/ or lamotrigine should be advised of the risk of decreased contraceptive effect as a result of drug interactions.
	Lithium may cause Ebstein's abnormality in the foetus' hearts.
	Use of sodium valproate in women of childbearing age should be balanced against the risk of decreased fertility, foetal malformations and perinatal complications. Sodium valproate may cause polycystic ovary syndrome and reduce the chance of pregnancy. Sodium valproate may cause neural tube defects in foetuses. Periconceptional folate supplementation should be prescribed to protect against neural tube defects.
	For pregnant women with bipolar disorder, consider switching to antipsychotic treatment, or gradual decrement (over 15–30 days) of monotherapy mood stabiliser to the lowest effective amount in divided doses during pregnancy. Concurrent careful foetal monitoring is recommended.
	Sodium valproate and carbamazepine are preferable to lithium and lamotrigine during Breastfeeding. Mothers who require the latter two medications may be advised to consider abstinence from breastfeeding for the infant's safety.
	In the event of breastfeeding while taking mood stabilisers, consider administering feeds before taking medication and discarding the first post-dose batch of expressed milk, so as to minimise the infant's consumption of medication via breast milk.

Mood stabilisers

Compare and contrast mood stabilisers:

Lithium	Sodium valproate CR [Epilim Chrono]	Carbamazepine CR [Tegretol CR]	Lamotrigine [Lacmatil]
Dose: Oral, start at 400mg. Maximum 1200mg per day. Lithium carbonate CR: $ 0.38 / 400mg tablet.	Dose: Oral, starts with 500mg daily, Maximum 1300mg per day. $0.5 per 300 mg tablet.	Dose: 400mg – 800mg Carbamazepine CR $0.16 / 400mg tablet.	A slow titration is required and this will avoid a serious skin rash: 25 mg/day for 2 weeks doubling the dose every two weeks to a maximum of 400 mg/day. $ 1.09 / 50mg tablet $1.72 / 100mg tablet
Monitoring: Lithium level should be checked every 3 months. (0.4 – 0.8 mmol/L: maintenance range.	LFT is a must before starting a patient on sodium valproate.		

RFT is checked every 6 months. TFT is checked every year.			
Properties: • Remains a first line treatment in bipolar disorder. Lithium increases Na/K/ATP-ase activity in patients. It affects serotonin, noradrenaline, dopamine and acetylcholine. Lithium also interferes cAMP (second messenger system). • Onset of action: 5-14 days. • Anti-mania effect is proportional to plasma levels. The level is set to be between 0.4 – 0.8 mEq/L for Asian patients. •Lithium is often used with antidepressants and other mood stabilizers in bipolar depression. • Lithium reduces suicidal ideation in bipolar patients Lithium is contraindicated to people with renal failure, thyroid diseases and rapid cycling disorder.	•Efficacy is superior to placebo but equal to lithium, haloperidol, and olanzapine. Valproate enhances GABA function and produces neuroinhibitory effects on mania. • Valproate can be combined with antipsychotics in treatment of mania and lower dose of antipsychotic drug is required. • Valproate is effective in maintenance treatment to prevent mood episodes. • Effective plasma levels: 50-99 mg/L but clinical response is more important. Valproate is indicated in patients with renal failure and rapid cycling disorder but contraindicated in people with liver failure.	• • The application of carbamazepine is limited by its properties as an enzyme inducers and side effects such as diplopia, blurred vision, ataxia, somnolence, fatigue, nausea, and blood dyscrasia. • Generally effective in maintenance treatment to prevent mood episodes. • No data on plasma levels and response. Carbamazepine is indicated to patients with bipolar disorder who are concerned about weight gain caused by lithium and valproate because carbamazepine does not cause significant weight gain.	• Doubtful efficacy in mania • Effective in bipolar depression in bipolar I patients with clear efficacy at 200 mg/day.
Side effects: Common side effects: Metallic taste Nausea Polydipsia Polyuria Oedema Weight gain Fine tremors. Long term complications: Hypothyroidism Renal failure	Common side effects: Weight gain Nausea Gastric irritation Diarrhoea Serious side effect: Thrombocytopenia.	Common side effects: Dizziness Somnolence Nausea Dry mouth Oedema Hyponatraemia (due to potentiation of ADH) ↑ALP and ↑GGT. Uncommon side effects: Ataxia Diplopia Nystagmus Serious exfoliative dermatological reactions (3% of patients and requires	Common side effects: Dizziness Headache Diplopia Nausea Ataxia. Uncommon side effects If patients develop for lamotrigine-associated rash (10%), hold the next dose and seek immediate medical attention.

		cessation of carbamazepine. Agranulocytosis Leucopenia Aplastic anaemia.	

Lithium toxicity

Lithium level	Signs of lithium toxicity
1.5 – 2.0 mmol/L (Mild toxicity)	CNS: Drowsiness, malaise and poor concentration. Toxic signs in CNS does not closely follow changes in lithium blood levels. PNS: Muscle weakness and severe fine tremor. GIT: Anorexia and diarrhoea which resemble gastroenteritis.
2.0 -3.0 mmol/L (Moderate toxicity)	CNS: Disorientation and dysarthria. CVS: Cardiac arrhythmia. PNS: Coarse tremor, restlessness and ataxia. GIT: Vomiting.
> 3.0 mmol/L (Severe toxicity)	CNS: Confusion and convulsion and coma. CVS: Cardiovascular collapse. Respiratory: Severe viscosity of respiratory secretions. Haemodialysis may be necessary when serum levels exceed 3mmol/L.
> 5.0 mmol/L	It may lead to permanent physical damage and mortality.

Treatment of lithium toxicity involves cessation of lithium and dialysis.

OSCE video – Explain lithium and side effects

A patient was admitted to the psychiatric ward after a manic episode. The consultant psychiatrist has advised him to consider taking lithium as a maintenance treatment. The patient is very concerned about bipolar disorder and lithium after reading the information from internet.

Task: address his concerns about lithium treatment.

1. **Why do you want to prescribe lithium?** Lithium is used to stabilise your mood. After my assessment, your mood seems to be elevated and you suffer from a condition called mania in the context of bipolar disorder.
2. **What is mania?** Feeling high, irritable, full of energy, very good appetite, no need for sleep, high sexual drive, racing thoughts, grandiose ideas, overspending, poor judgement, dangerous behaviour and unusual experiences such as hearing voices.
3. **Why do I sometimes feel depressed?** Periods of depression occur in bipolar disorder. Your mood will go up and down.
4. **What exactly is lithium?** It is a type of salt and can be found naturally.
5. **How long have psychiatrists been using lithium?** 50 years already.
6. **What is the usual dose of lithium?** Starting dose 400mg a day, increase slowly to 800mg to 1200mg per day.
7. **How do you decide the right dose for me?** Based on serum levels 0.4 – 0.8 mmol/L; clinical response.
8. **What time of the day should I take lithium?** Usually at night. The modern lithium has long release version and can last for whole day.
9. **What should I do if I miss a dose?** If you forget a dose, take it ASAP as you remember.
10. **Can I take lithium now?** No, we need to do some blood tests for you.
11. **What do you need those blood tests?** To check it is safe for you to take lithium. Your kidney and thyroid have to be in good condition.
12. **Do I only need to have those blood tests once?** The lithium may affect the function of kidney and thyroid. We have to check every 6 months.
13. **Lithium sounds scary. How do you know it is safe for me to take?** It is usually safe if your kidney and thyroid

are in good condition. Extra care if you take pain killer, medication containing sodium.
14. **How do I know lithium works for me?** Your highs and lows become less extreme. It will reduce thoughts of harming oneself. It may take weeks or months to appreciate the beneficial effects of lithium..
15. **Can I mix alcohol with lithium?** No, it will lead to drowsiness if lithium combines with alcohol, ↑ fall risk & accidents. Avoid alcohol in 1st & 2nd months; if you need to drink socially, try a small amount & see how you feel. Don't drink and take lithium when you drive.
16. **When I feel better, can I stop taking lithium?** You should not stop suddenly. Need to consult your doctor. Lithium is usually a long-term treatment.
17. **Is lithium addictive?** No, it is because you do not need to take more and more lithium to achieve the same effect.
18. **Do I need to know anything else as I stay in Singapore?** Drink enough water in hot weather. Lack of water in body may cause more side effects.
19. **My younger brother likes to steal my medicine. What would happen to him if he swallows a large amount of lithium?** Lithium is toxic if a person takes an overdose. A person will first present with loose stool/vomiting, then very shaky hands, unsteady walking, confusion and may die. You need to send the person to the Emergency Department immediately.
20. **What are the other alternatives besides lithium?** There are other medications which can stabilise patient's mood which are anti-fit / epilepsy medication.

Non-pharmacological treatment

- **Cognitive therapy** to challenge grandiose thoughts.
- **Behaviour therapy** to maintain regular pattern of daily activities.
- **Psychoeducation** on aetiology, signs and symptoms, management and relapse prevention of bipolar disorder.
- **Family therapy:** To work on impact of manic symptoms on family and resolve interpersonal problems.
- **Relapse drills**: to identify symptoms and to formulate a plan to seek help in early manic phase..

Course and prognosis

- Manic episodes usually last between 2 weeks to 4 months. Depressive episodes usually last for 6 months.
- Length of time between subsequent episodes may begin to narrow and remission time decreases with increasing age.
- Lithium can bring 60-70% remission rate.
- Good prognostic factors: female gender, short duration of manic episode, later age of onset, no suicidal thoughts, less psychotic symptoms, few comorbid physical conditions and good compliance.
- Poor prognostic factors include male gender, long duration of manic episode, early age of onset, suicidal thought, depressive symptoms, psychotic symptoms, comorbidity (e.g. alcohol or drug misuse) and poor compliance.

Suicide and deliberate self harm

	Suicide	Deliberate self harm
Epidemiology [Chia BH, 2010]	• Male : Female = 3:1. • More common in older people. • Suicide rates in Singapore remained stable between 9.8-13.0/100,000 from 1955 to 2004. Rates remain highest in elderly men. • Rates in ethnic Chinese and Indians were consistently higher than in Malays. • The rates among female Indians and Chinese have declined significantly between 1995 and 2004, some increase was noted in female Malays. •	• More common in women • Most common in adolescents • It is estimated that 7-14% of adolescents have self-harmed (UK).
Aetiology	**Demographics:**	An expression of emotional distress

• Male gender • Older age. • Single/Divorced. • Professions (policemen or guards with access to firearms, bartender, medical professionals).(based on UK findings) **Past psychiatric history:** • Previous suicide attempt. • Past history of depression or psychosis. • Alcohol/drug misuse. **Past medical history:** • Chronic painful illness (e.g. terminal cancer). **Social factors:** • Isolation/ lack of social network. • Significant life event e.g. death, losing job, relationship breakdown, abuse.	(e.g. sadness, loneliness, emptiness, grief) An expression of social problems and trauma (e.g. difficult relationships, bullying) Problems with identity for example borderline personality disorder, sexuality problem. Secondary to a psychiatric disorder (e.g. command hallucinations in schizophrenia).

Common methods used in Singapore between 2000 and 2004 were jumping (72.4%), hanging (16.6%), and poisoning (5.9%) [Chia et al 2011].

Types of self harm include: cutting usually of the wrists or forearms, scratching, burning skin or banging the head against the wall.

Questionnaire to assess suicide risk – The SAD PERSONS assessment tool [Patternson et al 1983].

Item	Score	Item	Score
Sex = Male.	1	Psychosis.	2
Age < 19 or > 45.	1	Separated/widowed/divorced.	1
Depression or hopelessness.	1	Serious attempt (e.g. hanging, stabbing).	2
Previous suicide attempts.	1	No social support.	1
Excessive alcohol or drug use.	1	Stated future intent.	2

Total score < 6 (may be safe to discharge); 6-8 (refer to psychiatric assessment) and > 8 = urgent admission.

OSCE grid – Assess suicide risk

A 24-year-old woman took an overdose of 20 tablets of paracetamol. She is brought in by her partner to the Accident and Emergency Department and you are the resident on duty at the Accident and Emergency Department.

Task: Assess her suicide risk.

A. Assess her suicide plan and intent	A1. Introduction. I am Dr. XXX. I can imagine that you have gone through some difficult experiences. Can you tell me more about it?	A2. Assess her plan. Was the overdose planned? If yes, how long have you thought	A3. Assess intent. Have you thought about taking your own life by the overdose?

	Can you tell me why you took the 20 tablets of paracetamol tonight? Was there any life event leading to this suicide attempt?	about it? How did you collect the paracetamol tablets? What did you think would happen when you took the paracetamol?	
B. Assess circumstances of suicide attempt.	B1. Assess location of the suicide attempt. Where did you take the medication? Was anyone else there/were you likely to be found? Did you lock the door or take precaution to avoid discovery?	B2. Assess severity of overdose and other self-harm. Besides the paracetamol, did you take other tablets? Did you mixed the paracetamol with alcohol? Did you harm yourself by other means? (e.g. cutting yourself)	B3. Suicide note or good-bye message. Was a suicidal note left? Did you send a SMS or email to say 'good-bye' to your partner or family members?
C. Assess events after suicide attempts.	C1. The discovery How did you come to be in A&E? Were you discovered by other people? If yes, how did they discover you?	C2. Assess physical complications. Did the overdose lead to any discomfort? E.g. severe vomiting. Did you have a period of black out?	C3. Assess current suicide risk. How do you feel about it now? Are you regretful of your suicide attempt? Would you do it again?
D. Assess other risk factors or protective factors.	D1. Past history of suicide. Have you attempted suicide previously? If yes, how many times? What are the usual causes of suicide attempts? Did you try other methods	D2. Past psychiatric / medical history. Do you have a history of mental illness? (e.g. depression) and take a brief mood history and past treatment if depression is present. Are you suffering from any other illnesses? (e.g. chronic pain)	D3. Assess protective factors. We have discussed quite a lot on the overdose and some of the unhappy events. Are there things in life you are looking forward to? Who are the people supporting you at this moment? How about religion?

	like hanging, stabbing yourself, jumping from heights or drowning?		

Bereavement and abnormal grief

Grief and depression

Bereavement	Intense anniversary reaction	Pathological Grief	Depressive episode
- Phase I: Shock and protest includes numbness, disbelief and acute dysphoria. **- Phase II: Preoccupation** includes yearning, searching and anger. **- Phase III Disorganization** includes despair and acceptance of loss. **- Phase IV: Resolution**	e.g. On Christmas day, a 64-year-old woman was brought to the emergency department as she suddenly becomes tearful and starts hitting herself. On further enquiry, her daughter was killed in a road traffic accident on the Christmas eve 2 years ago.	**- Inhibited grief**: absence of expected grief symptoms at any stage. **- Delayed grief**: avoidance of painful symptoms within 2 weeks of loss. **- Chronic grief**: continued significant grief related symptoms 6 months after loss.	A person may undergo depression after bereavement. The following features suggest depression rather than bereavement: - Guilt, suicidal thoughts & hallucinations not related to the deceased. - Feelings of worthlessness. - Psychomotor retardation. - Prolonged and marked functional impairment.

Management for abnormal grief: Grief therapy which focuses on talking about the deceased, prepare for future life without the deceased, plan to discard items related to the deceased and have a closure of unresolved issues related to the deceased.

Seasonal affective disorder (SAD)

Definition: SAD is a form of recurrent depressive disorder, in which sufferers consistently experience low mood in winter months. Symptoms include increase appetite, craving for sugar or rice, low energy, increased sleep and weight gain.

Worst months: November & December in Europe; January & February in US.

Aetiology: 1) melatonin/pineal gland abnormalities, replaced by theories on disordered brain 5HT regulation, phase-advanced circadian rhythms. **2)** Biologically vulnerable individuals is affected by the actual effect of the changes in the seasons and specific anniversary or environmental factors in winter.

Epidemiology: 3% in Europe

Clinical features: SAD presents with features of atypical depression – hypersomnia, hyperphagia, tiredness and low mood in winter. ICD-10 criteria specifies 3 or more episodes of mood disorder must occur with onset within the same 90-day period of the year for 3 or more consecutive days. Remission also occurs within a particular 90-day period of the year. Seasonal episodes substantially outnumber any non-seasonal episodes that may occur..

Treatment: Light therapy involves a special light box which emits 2500 lux and mimics the effect of sunlight for at least 2 hours every morning (or 10000 lux for 30 minutes). Exposure to eyes is important as it alters circadian rhythm. Effects are seen within a few days but it takes 2 weeks for the full effect. The person should use the light box half an hour a day starting in autumn & throughout winter months to prevent relapse. Melatonin levels are lowered with light therapy. 50% of people with SAD show clinically significant response to light therapy. Untreated episodes resolve by spring time. Side effects include jumpiness, headache & nausea (15% of patients) Other treatment includes sleep deprivation.

Revision MCQs

1. A mother worries that her daughter will develop depression because of the family history of depressive disorder. Which of the following genes is associated with increased risk?
A. APO E4 gene on chromosome 21
B. COMT gene on chromosome 21
C. Presenilin-2 gene on chromosome 1
D. Presenilin-1 gene on chromosome 14
E. Serotonin transporter gene

Answer: E
The serotonin transporter gene is implicated in the aetiology of depressive disorder. Short allele (SS) variation in the promoter region of the 5-hidroxytryptamine transporter gene (5-HTTLPR) decreases the transcriptional efficacy of serotonin and cause major depressive disorder in response to stressful life events.

2. A 32-year-old woman suffers from a severe depressive episode. She has three young children studying in primary school. She is unemployed with no confiding relationship. Which of the following works provides an explanation in her case?
A. Brown and Harris: Social Origins of Depression
B. Durkheim E: Anomie
C. Habermas J: The Theory of Communicative Action
D. Parsons T: The Social System
E. Sullivan HS: The Interpersonal Theory of Psychiatry

Answer: A.
G.W. Brown and T. Harris published the *Social origins of depression: A study of psychiatric disorder in women* in 1978. In this book, Brown and Harris stated that women with three young children under the age of 14, unemployed and with no confiding relationship are more likely to develop depression.

3. You are teaching depressive disorder to a group of medical students. They want to know what percentage of patients admitted to the university hospital will have recurrence and require further admission in long run without committing suicide. Your answer is:
A. 20%
B. 30%
C. 40%
D. 60%
E. 80%

Answer: D
An old British study showed that approximately 60% of patients had been re-admitted at least once. Only 20% had recovered fully with no further episodes and 20% were incapacitated throughout or died of suicide.

4. A 30-year-old woman suffers from depression with melancholic features. When compared with depressed patients without melancholia, which of the following statements is incorrect?
A. Cortisol is less likely to be suppressed when this patient is administered with dexamethasone suppression test
B. She is more likely to develop psychomotor retardation
C. She has greater symptom severity
D. She has increased REM latency
E. She has lower placebo response

Answer: D
Depressed patients with melancholic features have decreased REM latency.

5. A 40-year-old woman suffers from severe depressive episode with psychotic features. Which of the following statements is incorrect?
A. Mood incongruent psychotic features predict a better outcome
B. Psychotic symptoms must occur after manifestations of depressive symptoms
C. She has more biological abnormalities compared with depressed patients without psychotic features
D. She has poorer long-term outcome
E. She may be benefitted by receiving ECT

Answer: A
Mood incongruent psychotic features predict a poorer course and outcome.

6. A 23-year-old woman complains of hearing voices. A core trainee is not certain whether this patient suffers from schizophrenia or bipolar disorder. Which of the following features suggest the diagnosis of bipolar disorder rather than schizophrenia?
A. Bizarre delusions
B. Persecutory delusions
C. Prominent affective symptoms and mood congruent delusions
D. Systematized delusions
E. Thought broadcasting

Answer: C
Prominent affective symptoms and mood congruent delusions support the diagnosis of bipolar disorder.

7. A 30-year-old woman suffers from severe depressive episodes, but she tends to forget to take her medication at least twice a week. She finds it very difficult to take medication on a daily basis. She requests that you should prescribe an antidepressant which suits her needs. Which of the following antidepressants would you recommend?
A. Duloxetine

B. Fluoxetine
C. Paroxetine
D. Sertraline
E. Venlafaxine

Answer: B
The half-lives of the antidepressants are listed in descending order: fluoxetine (1–3 days), sertraline (26 hours), paroxetine (24 hours), duloxetine (12 hours) and venlafaxine (10 hours).

8. A 60-year-old woman complained of depression and was started by her GP on escitalopram. After two weeks of treatment, she complains of lethargy, muscle weakness and nausea. The GP wants to know the most likely cause for her symptoms. Your answer is:
A. Acute confusional state
B. Generalised anxiety disorder
C. Hyponatraemia
D. Serotonin syndrome
E. Somatisation disorder

Answer: C
Hyponatraemia is common in old people receiving SSRI treatment. They present with lethargy, muscle ache and nausea. More severe cases present with cardiac failure, confusion and seizure.

9. Which of the following is least likely to be found in patients taking lithium when the lithium level is within therapeutic range?
A. Changes in ECG
B. Endocrine abnormalities
C. Nystagmus
D. Peripheral oedema
E. Weight gain

Answer: C
Nystagmus occurs in lithium toxicity.

10. A 50-year-old woman with bipolar disorder is admitted to the medical ward and the medical consultant discovers that she has thrombocytopenia. The consultant wants to find out which of the following psychotropic medications is most likely to be responsible for thrombocytopenia. Your answer is:
A. Lithium
B. Olanzapine
C. Quetiapine
D. Sodium valproate
E. Zopiclone

Answer D.
Sodium valproate is associated with thrombocytopenia although it is an uncommon side effect.

11. A 30-year-old woman suffers from bipolar disorder and she is very concerned that she became pregnant although she takes oral contraceptive pills. Which of the following medications have led to the contraceptive failure?

A. Lithium
B. Lamotrigine
C. Carbamazepine
D. Valproate
E. Topiramate

Answer: C
Carbamazepine is an inducer of cytochrome P450 and it has led to the contraceptive failure in this woman.

Revision MEQs

A 70-year-old man is brought by his wife to the Accident and Emergency Department because he wanted to jump from his HDB flat. He has history of prostate cancer and the oncologist has started a new chemotherapy which results in side effects. He is concerned with somatic complaints and appears to be anxious during the interview. He has history of depression 10 years ago. His GP started fluoxetine 20mg OM two weeks ago but his symptoms have not improved.

1. Is his suicide risk high or low?
This man has high suicide risk.

2. List the tell-tale signs in the history which support your risk assessment.
1. Dangerous methods of suicide attempt – i.e. jumping.
2. Old age, male gender.
3. History of depression.
4. Neurotic depression and preoccupation with somatic complaints.

3. His wife is ambivalent about admission to psychiatric ward. State 4 reasons why this man should be admitted.
1. Prevention of suicide attempt.
2. To find a right antidepressant or to adjust the dose of current antidepressant.
3. To liaise with the oncologist about the side effects of chemotherapy.
4. Refer him to see a psychologist for psychotherapy.

4. If this man is admitted to the ward, what would you suggest the nurses to do?
To put this man under closing monitoring for suicide attempts [i.e. suicide precaution].

5. He is concerned side effects of fluoxetine. Please list 5 common side effects associated with fluoxetine.
1. Anxiety
2. Insomnia
3. Nausea
4. Headache
5. Diarrhoea.

EMIS:

Affective disorders

1. Bipolar disorder 1
2. Bipolar disorder 2
3. Cyclothymia
4. Hypomania
5. Rapid cycling Bipolar disorder
6. Ultra rapid cycling bipolar disorder

Please choose the most appropriate diagnosis from the above list for each of the above clinical vignettes. Each option may be used once, more than one or not at all.

1. John, over the past 2 years, has been having many episodes of feeling elated, needing less sleep than usual. This is then accompanied by a couple of weeks of feeling extremely depressed and lethargic. He is still able to function during these episodes (Cyclothymia – In this case, there is the presence of hypomania / depressive episodes for a total duration of 2 years. There is clearly no experience of normal mood for longer than 2 months. There is no functional impairment as well.)
2. David, is a patient with affective disorder and have been on lithium treatment. He has 5 episodes of either mania or depression each year and now wishes to change his medications to be able to control the frequency of these episodes. (Rapid cycling bipolar disorder)
3. Charles, has been originally diagnosed with depression and have been started on medications. He now presents with a history of feeling elated, needing less sleep and feeling very energetic – (Bipolar Type 1 Disorder)

References:

Beck AT, Ward CH, Mendelson M, Mock J, Erbaugh (1961) An inventory for Measuring Depression. *Archives of General Psychiatry.* **4**: 53-63.

Hamilton M. (1960) A rating scale for depression. *Journal of Neurology, Neurosurgery and Psychiatry* **23**: 56-62.

Montgomery SA, Asberg M. (1979) A new depression scale designed to be sensitive to change. *British Journal of Psychiatry* **134**: 382-389.

R Mahendran, H L Yap (2005) Clinical practice guidelines for depression *Singapore Med J* 2005 Vol **46(11)** : 610

NICE guidelines on depression in adults
http://guidance.nice.org.uk/CG90 Accessed on 30th December, 2011.

Ferreira MAR et al (2008) Collaborative genome wide association analysis supports a role for ANK3 and CACNA1C in bipolar disorder. Nature Genetics, 40 (9): 1056-58.

Young RC, Biggs JT, Ziegler VE, Meyer DA (1978). A rating scale for mania: reliability, validity and sensitivity. *The British Journal of Psychiatry* **133**: 429–35.

Mok YM, Chan HN, Chee KS et al (2011) Ministry of health clinical practice guidelines: bipolar disorder. *Singapore Med J.* 2011 Dec;52(12):914-9.

MOH guidelines on bipolar disorder: http://www.moh.gov.sg/content/moh_web/home/Publications/guidelines/clinical_practiceguidelines/2011/bipolar_disorder.html Accessed on 30th December, 2011.

NICE guidelines for bipolar disorder
http://guidance.nice.org.uk/CG38 Accessed on 30th December, 2011.

Chia BH, Chia A, Yee NW, Choo TB. Suicide trends in Singapore: 1955-2004. *Arch Suicide Res.* 2010 Jul;14(3):276-83

Chia BH, Chia A, Ng WY, Tai BC. Suicide methods in singapore (2000-2004): types and associations. *Suicide Life Threat Behav.* 2011 Oct;41(5):574-83.

Patterson WM, Dohn HH, Patterson J, Patterson GA et al (1983). Evaluation of suicidal patients: the SAD PERSONS scale. *Psychosomatics* 24 (4) 343-9.

Lange K & Farmer A (2007) The causes of depression in Stein G & Wilkinson G (2007) *Seminars in general adult psychiatry*. London: Gaskell.

Brown, G. W & Harris, T. (1978) *Social Origins of Depression: A Study of Psychiatric Disorder in Women*. London: Tavistock.

World Health Organisation (1992) *ICD-10: The ICD-10 Classification of Mental and Behavioural Disorders: Clinical Descriptions and Diagnostic Guidelines*. Geneva: World Health Organisation.

American Psychiatric Association (1994) *DSM-IV-TR: Diagnostic and Statistical Manual of Mental Disorders (Diagnostic & Statistical Manual of Mental Disorders) (4th edition)*. Washington: American Psychiatric Association Publishing Inc.

American Psychiatric Association (2013) DSM-5: Diagnostic and Statistical Manual of Mental Disorders (Diagnostic & Statistical Manual of Mental Disorders) (5th edition). American Psychiatric Association Publishing Inc: Washington.

Gelder M, Mayou R, Cowen P (2001) *Shorter Oxford Textbook of Psychiatry*. Fourth Edition. Oxford: Oxford University Press.

Shapiro DA (1998) Efficacy of psychodynamic, interpersonal and experiential treatments of depression In Checkley S. (1998) *The Management of depression*. London: Blackwell Science.

Watkins E & Williams R (1998) The efficacy of cognitive behavioural therapy In Checkley S. (1998) *The Management of depression*. London: Blackwell Science.

Lader MH (1998) The efficacy of antidepressant medication In Checkley S. (1998) *The Management of depression*. London: Blackwell Science.

Cowen PJ (1998) Efficacy of treatments for resistant depression

In Checkley S. (1998) *The Management of depression*. London: Blackwell science

Taylor D, Paton C, Kapur S (2009) *The Maudsley prescribing guideline*. London: Informa healthcare.

Checkley S. (1998) The management of resistant depression in Checkley S. (1998) *The Management of depression*. London: Blackwell Science.

Salloum IM, Daley DC, Thase ME (2000) *Male depression, alcoholism, and violence*. London: Martin Dunitz.

King DJ (2004) *Seminars in Clinical Psychopharmacology*. 2nd edition London: Gaskell.

Schatzberg AF & Nemeroff CB (2004) *Textbook of Psychopharmacology*. 3rd edition. Washington: American Psychiatric Publishing.

Johnstone EC, Freeman CPL, Zealley AK editors (1998) *Companion to Psychiatric Studies*. 6th edition. London: Churchill Livingstone.

Sadock BJ, Sadock VA. (2003) *Kaplan and Sadock's Comprehensive Textbook of Psychiatry*. 9th ed. Philadelphia: Lippincott Williams and Wilkins.

Lee A (2007) Clinical features of depressive disorder in Stein G & Wilkinson G (2007) *Seminars in general adult psychiatry*. London: Gaskell.

Jones S & Roberts K (2007) Key topics in psychiatry. Edinburgh: Churchill Livingstone Elsevier.

Kolvin I DJ (1994) Child psychiatry In Paykel ES, Jenkins R (1994) *Prevention in Psychiatry*. Gaskell: London.

Chapter 5 Anxiety, phobia and stress-related disorders

Principles of learning theory	OSCE video: assessing anxiety and phobia
Overview of epidemiology and aetiology	Obsessive compulsive disorder
Overview of management (MOH guidelines)	OSCE grid: assess obsessive compulsive disorder
Generalized Anxiety disorder	Acute stress reaction and adjustment disorder
Panic disorder	Post traumatic stress disorder
Agoraphobia	OSCE grid: assess post traumatic stress disorder
Social phobia	Depersonalization and derealisation
Specific phobia	Revision MCQs and MEQ

Principles of learning theory

Classical conditioning and the Pavlov's experiment

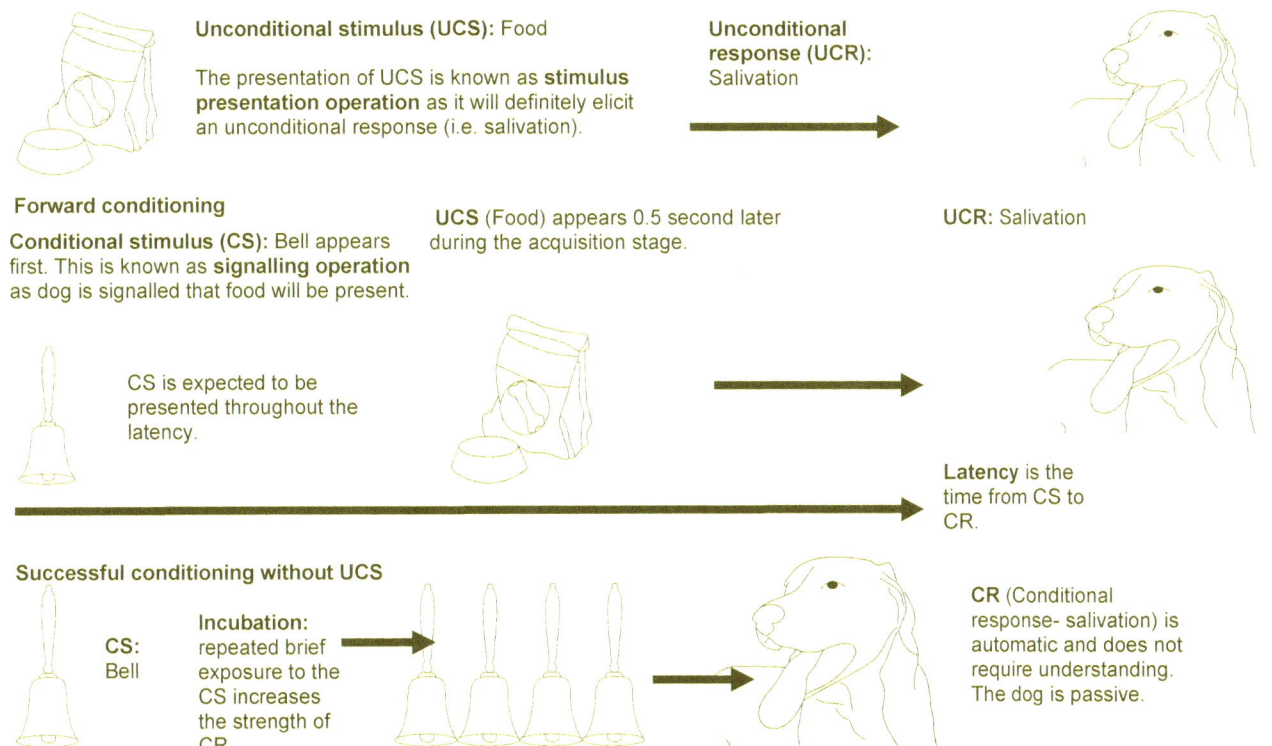

Unconditional stimulus (UCS): Food

The presentation of UCS is known as **stimulus presentation operation** as it will definitely elicit an unconditional response (i.e. salivation).

Unconditional response (UCR): Salivation

Forward conditioning

Conditional stimulus (CS): Bell appears first. This is known as **signalling operation** as dog is signalled that food will be present.

UCS (Food) appears 0.5 second later during the acquisition stage.

UCR: Salivation

CS is expected to be presented throughout the latency.

Latency is the time from CS to CR.

Successful conditioning without UCS

CS: Bell

Incubation: repeated brief exposure to the CS increases the strength of CR.

CR (Conditional response- salivation) is automatic and does not require understanding. The dog is passive.

Clinical example: A cancer patient complains of nausea whenever he sees the hospital building. He comes to the hospital to receive chemotherapy and he feels nausea every time after he receives the chemotherapy. The chemotherapy is UCS and the hospital building is CS. The CR and UCR is nausea. Classical conditioning may explain why this patient always feels nausea when he sees the hospital building.

Operant conditioning and the Skinner's experiments

The organism learns by operating on the environment and this is known as **operant conditioning**. The **operant** is a response that has an effect on the environment.

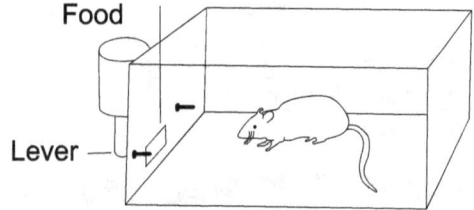

Positive reinforcement: Food is presented by pressing the lever.

The rat discovers that after pressing the lever (the response), food will be released (positive reinforcer).

In this example, food is a **positive reinforcer** which strengthens the response (i.e. pressing the lever). If there is an absence of positive reinforcer, the response will decrease. If there is no release of food by pressing the lever, the rat will stop pressing the lever and extinction occurs. The speed of conditioning is proportional to the size or impact of the reinforcer.

Primary reinforcers: affects biological process and they are naturally reinforcing (e.g. food or sex).

Secondary reinforcers: associated with primary reinforcers and based on previous learning (e.g. money or being praised).

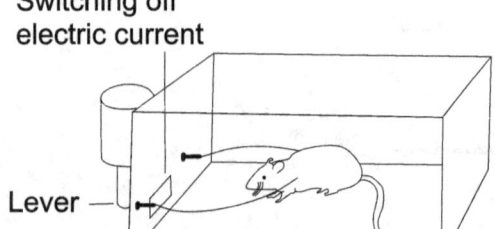

Negative reinforcement: Electric current is switched off (escape learning) or electric shock is avoided altogether (avoidance learning) by pressing the lever.

The rat receives continuous electric current. Then it discovers that the electric current (aversive stimulus) will be switched off by pressing the lever (the response).

The success in switching off of electric current is a **negative reinforcer** which strengthens the response (i.e. pressing the lever) in removing or avoiding of the aversive stimulus.

Clinical example: A 40-year-old man has developed pathological gambling and he cannot stop gambling despite multiple complaints from his family. When the doctor asks him why he still wants to gamble, he mentions that he won $10,000 in one day two years ago and this always encourages him to gamble. In this example, money is a secondary reinforcer and he continues to gamble as a result of positive reinforcement.

Clinical example: A 30-year-old woman develops agoraphobia and she tries to avoid leaving her house. When the doctor asks her why she refuses to leave the house, she replies that her home is a safer place and she can prevent panic attacks by staying at home. Hence, she develops avoidance of going out as a result of negative reinforcement.

Overview of epidemiology

Summary of epidemiology [Lim, 2005, Bienvenu et al 2006, Beesdo et al 2010, Ham 2005, Kendler 1992]

	Generalised anxiety disorder (GAD)	Panic disorders	Social phobia	Agoraphobia
Incidence	4.3%	**General population:** 1 year incidence: 10% Patients presenting to emergency departments with 'chest pain' and 25% have panic disorder.	**General population:** 1-year incidence: 6%	Large numbers of people in the community who have agoraphobia without panic disorder may not seek help. In the US, about 2/1000 person-years.
Prevalence and lifetime risk	**General adults:** The lifetime prevalence of GAD in Singapore is 3.3% The prevalence of GAD in Singapore is 3%. **Children and adolescents** The prevalence is 2%. Girls have somewhat higher rates of anxiety disorders than boys. **Elderly** 5-15% of elderly over 65 years old. Prevalence increases in older people in Singapore	**General adults** Lifetime prevalence is 8.6% in the UK. **First degree relatives of panic disorder:** 8-31%. **Children and adolescents:** <1%.	**Lifetime prevalence:** 6% **Children and adolescents:** Social phobia: 1% Simple phobias: 2-9% Specific phobia: 3%	Lifetime prevalence : 4% 6 month prevalence 3-6%.
Person	**Gender:** Women > Men (3.6:1 in Singapore) **Age:** The age of onset is usually in the 20s. **Life events:** stressful or traumatic life events are important precipitant for GAD and it may lead to alcohol misuse. **Comorbidity:** GAD is probably one of the most common psychiatric disorder which coexists with other psychiatric disorders (e.g. major depressive disorder, dysthymia, panic	**Gender:** Women > Men **Age:** Bimodal peak: 15-24 years and 45-54 years. **Life events:** Recent history of divorce or separation. Panic disorder is associated with the following conditions: mitral valve prolapse, hypertension, cardiomyopathy, COPD and irritable bowel syndrome. **Comorbidity:** Alcohol misuse (30%), agoraphobia (40%), social phobia (50%) and depressive	**Gender:** the gender ratio may be closer in social phobia. Some studies find equal ratio. **Age:** 2 peaks: 5 years and between 11-15 years. **Life events:** being criticised or scrutinised and resulted in humiliation. **Psychiatric comorbidity:** depressive disorder and alcohol misuse. **Inheritance:** Social phobia is more common	**Gender:** Women > Men (2:1) **Age:** 25-35 years Agoraphobia may occur before the onset of panic disorder. **SE status:** more common among housewives **Life events:** Onset usually follows a traumatic event **Disease:** Half of patients have panic disorder. **Comorbidity:** Panic disorder

	disorder, agoraphobia and social phobia based on Singapore study) **Inheritance:** genetic factors play a role.	disorder (70%) **Inheritance:** Heritability: 44% Risk increased by 5 times for the presence of panic disorder in the first-degree relatives of probands with panic disorder.	among relatives of people with social phobia than in the general population.	Depression Other anxiety disorders.

Overview of aetiology

Noradrenaline (NA)
GAD: 1) downregulation of α_2 receptors and ↑ in autonomic arousal; 2) Electrical stimulation of locus coeruleus releases noradrenaline and generates anxiety.

Panic disorder: hypersensitivity of presynaptic α_2 receptor and ↑ in adrenergic activity Yohimbine has high affinity for the α_2-adrenergic receptors and it can induce panic attacks.

Serotonin (5HT)
GAD: dysregulation of 5-HT system

Panic disorder: subsensitivity of $5HT_{1A}$ receptors & exaggerated post-synaptic receptor response.

OCD: dysregulation of 5HT system

GABA
GAD: ↓ in GABA activity.

Panic disorder: ↓ inhibitory receptor sensitivity and causes panic attack.

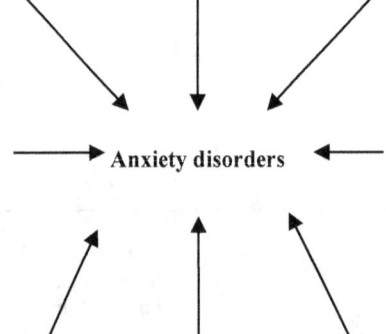

Cognitive theories
GAD: selective attention to negative details, distortions in information processing & negative views on coping.

Panic disorder: classical conditioning and negative catastrophic thoughts during attacks.

Agoraphobia and specific phobias: conditioned fear responses lead to learned avoidance

OCD: compulsions are learned and reinforced

Psychodynamic theories

GAD: symptoms of unresolved unconscious conflicts, early loss of parents, separation in childhood, overprotective parenting, anxious parent or parenting lacking warmth and responsiveness.

Panic disorder: arise from unsuccessful attempts to defend against anxiety provoking impulses.

Agoraphobia and specific phobias: unconscious conflicts are repressed and may be transformed by displacement in phobic symptoms.

Genetics

Heritability
GAD: 30%
Panic disorder: 30%
Agoraphobia relatives: ↑ social phobia, other neurotic disorders, alcoholism & depressive disorders.
OCD: MZ: DZ = 50-80%: 25%; First degree relatives: 10% risk; Heritability: 30%.
Social phobia: 50% MZ:DZ = 24%: 15%
Animal phobia MZ:DZ = 26%: 11%

Endocrine causes

GAD: 30% of patients have reduced suppression to dexamethasone suppression test.
Panic disorder: hypothalamus, amygdala and brainstem are involved.
PTSD: Low cortisol levels after trauma leads to PTSD
(↑ glucocorticoid receptors in hypothalamus and leads to decreased peripheral cortisol) while high cortisol levels lead to depression. Enhanced response to dexamethasone suppression test.

Organic causes

GAD: cardiac, thyroid, medication such as thyroxine.

Panic disorder: hypoglycaemia, thyrotoxicosis, phaeochromocytoma

CO_2 act as a panic stimulant as an indicator for lack of O_2 in the brain. Hence, breathing in –out of the paper bag makes panic attack worse. CCK and sodium lactate induce symptoms of panic disorder. There is increase in nocturnal melatonin production.

OCD: cell-mediated autoimmune factors against basal ganglia are involved.

Neuroimaging findings

In OCD, there is an increase in resting blood flow and glucose metabolism in the orbital cortex and caudate nucleus. Dysfunction of the cortico-striatal-thalamic-cortical circuitry is found in patient with OCD.

Overview of management

Investigations:
1. Thyroid function test: thyrotoxicosis.
2. Blood glucose: hypoglycaemia.
3. ECG or cardiac echocardiogram: atrial fibrillation, arrhythmias and other cardiac problems.
4. Urine drug screen in cases of suspected stimulant use.
5. Lung function test: suspected COPD.
6. 24 hour urine catecholamine (to rule out phaeochromocytoma especially hypertension and panic attacks coexist)

Recommendation from the MOH guidelines:

Pharmacological treatment is indicated when symptoms are severe, there is significant impairment of social, occupational and role functioning, or there is concurrent moderate or severe depressive disorder.

Antidepressants are recommended as effective agents for the treatment of panic disorders, social phobia, obsessive compulsive disorders, generalized anxiety disorder and post-traumatic stress disorder. Selective serotonin reuptake inhibitors (SSRIs) are recommended as the first line drug treatment for anxiety disorder.

For benzodiazepine, the lowest effective dose to achieve symptom relief should be used over a limited period. The dose should be gradually tapered off. Long term use should be closely supervised for adverse effects, abuse, tolerance, dependency and withdrawal symptoms. Cognitive behaviour therapy (CBT) may facilitate the tapering of benzodiazepines.

Antidepressants have good anti-anxiety properties and should be the medication of choice in comorbid depression and anxiety. Some selective serotonin reuptake inhibitors (SSRIs) such as paroxetine and venlafaxine have demonstrated efficacy for treatment of co-morbid depression and anxiety.

Relapse is common after discontinuation of medication for most anxiety disorders. Maintenance therapy may be indicated for individuals who frequently relapse.

Generalized anxiety disorder

Generalised anxiety is commonly described as a sensation of persistent worry and apprehension about common day problems and events, associated with symptoms involving the chest / abdomen, mental state symptoms, general symptoms and other symptoms.

Common signs and symptoms of generalised anxiety disorder

Autonomic arousal symptoms	Symptoms involving chest/ abdomen	Mental symptoms	General symptoms	Other symptoms
- Palpitation/ ↑ HR. - Sweating. - Trembling/ Shaking. - Dry mouth.	- Difficulty breathing. - Choking sensation. - Chest pain. - Nausea/ stomach churning.	- Giddiness / fainting. - Derealisation or depersonalisation. - Fear of losing control. - Fear of dying or "going crazy".	- Hot flushes/cold chills. - Numbness / tingling. - Muscle tension/aches. - Restlessness. - Feelings of keyed up, on the edge. - Lump in the throat.	- Exaggerated responses to minor surprises. - Easily being startled - Persistent irritability. - Poor sleep (initial insomnia, night terrors, waking and feeling unrefreshed). - Poor concentration.

- Mind goes blank.

The ICD-10 criteria specify that the patient must have symptoms for several weeks or more. It involves free-floating anxiety/ apprehension, motor tension and autonomic over-activity.

Generalized anxiety disorder

The DSM-5 specified that individuals would fulfill the diagnostic criteria if they have been experiencing excessive anxiety and worries for most everyday events for at least 6 months in duration. The DSM-5 further specified that these excessive worries are difficult to control, and that these worries are associated with at least 3 of the following symptoms:

a. Restlessness
b. Easily tired
c. Attentional and concentration difficulties
d. Feeling irritable
e. Muscle tension
f. Sleep difficulties

These worries must have caused significant impairments in an individual's level of functioning.

Common questionnaires

Hamilton Anxiety Rating Scale (HAM-A) (Hamilton, 1959)

The HAM-A scale is a clinician-rated scale which quantifies the severity of anxiety symptoms in a total score. HAM-A assess symptoms over the past week and it contains 14 items which assess anxiety, tension, fear, poor concentration, somatic complaints associated with anxiety and mood. The score of each item ranges from 0 to 4. HAM-A can assess baseline anxiety score the response to therapeutic interventions after a period of time.

Differential diagnosis
1. Panic disorder, stress-related disorder, phobia, mixed anxiety and depression.
2. Arrhythmia, ischaemic heart disease, mitral valve prolapse, congestive heart failure
3. Asthma, COPD
4. Hyperthyroidism, hypoparathyroidism, hypoglycaemia, phaeochromocytoma and anaemia.
5. Medication: antihypertensives, antiarrhythmics, bronchodilators, anticholinergics, anticonvulsants, thyroxine, and NSAIDS.

Treatment (including recommendations from the MOH guidelines)

1. **Psychotherapy**: For **CBT**, it should be offered in weekly or fortnightly sessions of 1-2 hours and be completed within 4- 6 months. The optimal range is 16-21 hours in total. If the psychologist decides to offer briefer CBT, it should be about 8-10 hours with integration of structured self-help material. CBT in generalised anxiety disorder delivered by an experienced therapist shows good evidence of efficacy. Two-thirds of patients show clinically significant improvement at 6 months follow-up.

2. For **pharmacological therapy**, psychiatrist should consider the patient's age, previous treatment response, risk of deliberate self harm, cost and patient's preference. Antidepressants can be considered as first-line agents over benzodiazepines in the treatment of generalised anxiety disorder over the long term. Hydroxyzine 50 mg/day has shown efficacy for treatment of generalized anxiety disorder.

3. Inform the patient on the potential side effects, possible discontinuation withdrawal and the time course of treatment. If one SSRI is not suitable, consider another SSRI. The psychiatrist should review the patient within 2 weeks of starting treatment and again at 4, 6 and 12 weeks. Then the psychiatrist can review the patient at 8-12 week intervals.

4. For GAD not responding to at least two types of intervention, consider venlafaxine. Before prescribing, the psychiatrist should consider the presence of pre-existing hypertension.

Comorbidity: concurrent panic disorder (25%) and depression (80%).

Prognosis: 70% of patients have mild or no impairment and 9% have severe impairment. Poor prognostic factors include severe anxiety symptoms, frequent syncope, and derealisation and suicide attempts.

Panic disorder

Common signs and symptoms of panic disorder

Autonomic arousal symptoms	Symptoms involving chest/ and abdomen	Mental symptoms	General symptoms
- Palpitation. - Increase in heart rate. - Sweating. - Trembling. - Shaking. - Dry mouth.	- Difficulty breathing. - Choking sensation. - Chest pain. - Nausea. - Stomach churning.	- Giddiness. - Derealisation. - Depersonalisation. - Sudden fear of losing control. - Sudden fear of dying or 'going crazy'.	- Hot flushes. - Cold chills. - Numbness. - Tingling.

Panic Disorder

The DSM-5 characterized panic attack as the sudden onset of intense fear that usually peaks within minutes and during which the following symptoms might occur:

Physical symptoms:
a. Palpitations
b. Sweating
c. Tremors
d. Difficulties breathing
e. Choking sensations
f. Chest pain or discomfort
g. Abdominal discomfort
h. Dizziness
i. Feeling hot or cold

Mental Symptoms:
a. Derealization
b. Depersonalization
c. Feelings of losing control and going crazy
d. Feelings of death

The DSM-5 specified that at least one of the attacks has been followed by at least 1 month of either a. Persistent concerns about having additional attacks or b. Marked changes in behaviour in relation to the attacks.

Agoraphobia

The DSM-5 diagnostic criteria states that an individual would fulfill the diagnostic criteria if there has been significant anxiety and fear in at least 2 of the following situations:
a. Being alone outside of home
b. Being in a crowd
c. Being in enclosed places

d. Being in open spaces
e. Using public transport modalities

During which, the individual has preoccupation of worries that escape might be difficult or help might not be available when needed.

These anxieties and worries must have affected an individual's level of functioning for at least 6 months in duration.

Clinicians are advised to take note that agoraphobia could be diagnosed in the presence of absence of panic disorder.

ICD-10 also stresses that people suffering from panic disorder should be free of anxiety symptoms between attacks.

Differential diagnosis

Compare and contrast panic disorder and hyperventilation syndrome

	Panic disorder	Hyperventilation syndrome
ICD – 10	Listed under the ICD-10 criteria	Not listed under the ICD – 10 criteria
DSM-5	Codable disorder, to specify with or without agoraphobia	Not a codable disorder
Overlap between the two disorders	50-60% of patients with panic disorder or agoraphobia have HVS symptoms.	25% of HVS patients have symptoms of panic disorder.
Aetiology	Biological and psychological causes are well defined	Less well defined. Lactate, CCK, caffeine and psychological stressors also play a role.
Clinical features	Both disorders share autonomic arousal symptoms and symptoms involving chest and abdomen. Panic disorder has more mental symptoms.	High thoracic breathing or excessive use of accessory muscles to breathe results in hyperinflated lungs.
Metabolic disturbances	- Less well established	- Acute hypocalcaemia (positive Chvostek and Trosseau signs and prolonged QT interval) - Hypokalaemia with generalised weakness - Respiratory alkalosis - Acute hypophosphatemia leading to paresthesias & generalised weakness
Investigations	Thyroid function test, ECG 24 hour urine catecholamine if patients have hypertension.	Investigation may include d-dimer and possible V/Q scan to rule out pulmonary embolism.
Management	Pharmacological agents and CBT may play an important role.	Relaxation and deep breathing exercise play an important role to reduce arousal during hyperventilation.

Other DDX for panic disorder include hypoparathyroidism, phaeochromocytoma, COPD, asthma, mitral valve prolapse, DM, hypoglycaemia, thyrotoxicosis and anaemia.

Treatment (MOH guidelines)

Evidence supports the use of combined cognitive behaviour therapy (CBT) with medication as superior to either therapy alone in the longer term maintenance phase.

Medications: Almost all the selective serotonin reuptake inhibitors (SSRIs) (fluoxetine, sertraline, fluvoxamine, escitalopram, paroxetine) have documented efficacy in the treatment of panic disorder. High potency agents like alprazolam and clonazepam are effective in providing rapid relief. With discontinuation of these agents, however, patients should be closely monitored for recurrence of symptoms, as the rates of relapse are very high, especially for shorter-acting agents. After improvement with medication, antidepressant treatment for panic disorders should be continued for at least 6 months.

Psychotherapy: CBT is the psychotherapy of choice for panic disorder. Possible treatment components for panic disorder, with or without agoraphobia include psychoeducation, exposure to symptoms or situations, cognitive restructuring, breathing exercise and monitoring for panic attacks.

Comorbidity: 30% of patients experience major depressive episode and 40% of them meet the criteria for social phobia.

Prognosis: Recurrence is common especially when new stressors emerge.

Social phobia / Social anxiety disorder

Patients have marked fear which is brought in by social situations (e.g. being the focus of attention or fear of behaving in a way that will be embarrassing). This has led to marked avoidance of being the focus of attention.

Society Anxiety Disorder (Social Phobia)
The DSM-5 specified that individuals must have significant anxiety about one or more social situations, for which individuals worry about being evaluated negatively by others. Consequently, these social situations are been avoided.
The DSM-5 specified a time duration of at least 6 months and there must be significant impairments in terms of functioning.
Subtypes - Performance only: Characterized as when social anxiety disorder is restricted to public performances

Treatment (MOH guidelines):

Medications: Selective serotonin reuptake inhibitor (SSRI) antidepressants are effective for the treatment of social phobia, and their favourable side-effect profile make them the recommended first-line treatment for social phobia. Paroxetine has been the most extensively studied SSRI for social phobia. After improvement with medication, antidepressant treatment for panic disorders and social phobias should be continued for at least 6 months.

Psychotherapy: Cognitive behaviour therapy (CBT) is recommended as effective treatment for social anxiety disorder. Exposure to feared situations is a crucial component. Group approaches are useful and often include elements of social skills training.

Specific phobia

Specific phobia is considered to be one of the most common type of anxiety disorder and occurs in 10% of the population. It usually starts in childhood and may persist into adult life. There is an anticipatory fear of an object (e.g. needles, spiders) or situations (e.g. flying) that cannot be explained or reasoned away. Furthermore, this fear is beyond voluntary control.

Specific Phobia
The DSM-5 specified that for an individual to fulfill this diagnostic criteria, there must be:
1. Significant anxiety about a particular object or situation
2. Encounters with the object or situation always cause marked anxiety
3. The specified object or situation is avoided
4. The anxieties and worries are excessively out of proportion in consideration of the actual threat posed.
The DSM-5 specified a time duration of at least 6 months and there must be significant impairments in terms of functioning.
Subtypes include:

a. Animal
b. Natural environment
c. Blood injection injury type
d. Situational
e. Others

Treatment (MOH guidelines)

Beta-blockers are effective for specific and circumscribed anxiety, especially for patients with prominent sympathetic hyperarousal such as palpitations and tremor. Propranolol 10-40 mg taken 45-60 minutes before the performance is sufficient for most patients.

OSCE video: Assess anxiety, panic attacks and phobia
You are the resident working at the Accident and Emergency Department. A 26-year-old married man is referred by her GP because of his fear that he is going to lose control with hyperventilation in his office. He seems to be very stressed.
Task: Assess anxiety, panic attack and phobia.

OSCE grid: assess anxiety, panic attack and phobia.

| A. Introduce and assess generalised anxiety | A1. Introduction

I am Dr. XXX, a resident of the Accident and Emergency Department. I understand that your GP has referred you because you are afraid that you are losing control.

I can imagine that it is a terrible experience. In the next 7 minutes, I want to find out more about your experiences. Is it ok with you?

Can you tell me more about your stress? | A2. Assess generalised anxiety

Do you tend to worry a lot? If yes, how many days in the last month?

Do you worry about anything in particular? | A3) Assess physical symptoms

What sort of symptoms do you get when you feel worried?

- Do you feel shaky?
- Do you sweat a lot?
- Do you have difficulties with breathing?
- Do you feel that your heart is beating very fast?
- Do you have loose stools?
- Do you feel dizzy or light-headed? |
|---|---|---|---|
| B. Assess panic attacks and agoraphobia | B1. Assess panic attacks

Have you ever had the experience that you felt as if you might have a heart attack or that you might even lose control? If yes, can you describe the symptoms to me.

How frequent have these attacks been for you?

Do you always anticipate about another attack? (anticipatory anxiety) | B2. Assess triggers

Is there anything that trigger the attacks?

Tell me how you felt when you knew the attack was coming along?

Are you very concerned and worried about these attacks? | B3. Assess agoraphobia.

Do you have tend to feel anxious in crowded places or public transport?

Do you have fear when away from home?

Can you tell me what happens when you have this fear?

Do you avoid those places? |
| C. Assess social phobia, specific phobia, comorbidity, past history | C1. Assess social phobia.

Do you worry about social situations where you are being the focus of attention?

Do you feel that other people are observing you and you feel very uncomfortable?

Can you tell more about your concern? | C2. Assess specific phobia.

Are you scared of other situations or object?

Can you tell more about the situation or object? | C3. Assess comorbidity and past psychiatric history

I am sorry to hear that you are affected by the above signs and symptoms. How does this condition affect your life?

How do you cope? Did you seek help from your GP or psychiatrist? Did they offer you any treatment? If yes, what is the effect on your condition?

How is your mood?

How is your sleep and appetite? |

			Do you need to drink alcohol or take sleeping pills to overcome those symptoms?
			Can you tell me more about your medical history? So you suffer from thyroid or heart disorders?

Obsessive compulsive disorder (OCD)

Epidemiology [Nestadt et al, 1998]

Incidence	Prevalence	Gender ratio	Mean age of onset
0.55 per 1000 person-years	**Prevalence:** 1% **Lifetime prevalence**: 0.8%	F:M = 1.5:1	Mean age of onset is around 20 years (70% before 25 years; 15% after 35 years).

Obsessions:

Obsessions are persistent and recurrent Doubts, Impulses, Ruminations and Thoughts (mnemonics: DIRT). This phenomenon is not simply excessive worries about real-life problems. The person attempts to ignore or suppress them and recognise that this phenomenon is the product of his or her own mind.

- **Doubts:** repeating themes expressing uncertainty about previous actions. E.g. Have I turned off the tap?
- **Impulses or images**: repeated urges to carry out actions that are usually embarrassing or undesirable e.g. shout obscenities in church or mentally seeing a disturbed images e.g. seeing one stabs oneself.
- **Ruminations**: repeated worrying themes of more complex thought e.g. worrying about the end of the world.
- **Thoughts**: repeated and intrusive words or phrases.

Compulsion is a repetitive behaviour or mental act which. is usually associated with an obsession as if it has the function of reducing distress caused by obsession. E.g. cleaning, checking and counting. Carrying out the compulsive act should not be pleasurable.

The most common obsessions (in descending order)	The most common compulsions (in descending order)
1. Fear of contamination (45%). 2. Doubting (42%). 3. Fear of illness, germs or bodily fear (36%). 4. Symmetry (31%). 5. Sexual or aggressive thoughts (28%).	1, Checking (63%). 2. Washing (50%). 3. Counting (36%).

DSM-5: (Please note that OCD is now classified under obsessive-compulsive and related disorders in DSM-5 and is no longer part of the anxiety disorders chapter)
Obsessive-Compulsive disorder
The DSM-5 specified that individuals would fulfill the diagnosis only if there is the presence of (a) Obsessions and (b) Compulsions that have caused much impairments in terms of functioning.
The DSM-5 defined Obsessions as:
a. Repetitive thoughts, urges or images that are experienced recurrently, which individuals find them to be intrusive and have resulted in significant anxieties
b. Efforts made by individuals to try to suppress these thoughts, urges or images with other thoughts or actions
The DSM-5 defined Compulsions as:

a. Repetitive behaviours or even mental acts that individual feels obliged to perform as a response to the underlying obsessive thoughts
b. These repetitive behaviours or even mental acts are being performed by individuals in order to reduce the anxiety experienced, or to prevent some dreadful event from happening.

It is important for Clinicians to distinguish the 3 subtypes of OCD, which are with good or fair insight, with poor insight and with absent insight or even delusional beliefs.

Questionnaire:

Yale-Brown Obsessive Compulsive Scale (Y-BOCS) (Goodman et al, 1989)

The **Y-BOCS** is a clinician-rated semi-structured questionnaire and rates the severity of OCD symptoms. It covers the week prior to the interview. The questionnaire is divided into obsession and compulsion subsets. The questionnaire takes about 15 to 30 minutes to complete. It is often used to monitor changes over the course of treatment.

Differential diagnosis:
1. Recurrent thoughts and worries in a normal person.
2. Anankastic or obsessive compulsive personality disorder
3. Generalised anxiety disorder.
4. Schizophrenia.
5. Delusional disorder.
6. Depressive disorder.
7. Organic causes (e.g. Paediatric Autoimmune Neuropsychiatric Disorders Associated with Streptococcal infections - PANDAS).

Management (including recommendations from the MOH guidelines):

Inpatient treatment is indicated when patients 1) pose severe Risk to self or others, 2) severe self-neglect (e.g. poor hygiene or eating) 3) extreme distress or functional impairment, 4) poor response to treatment and the need to monitor compliance.

Pharmacotherapy: The recommended first line of pharmacotherapy for OCD is a 10-12 week trial with a selective serotonin reuptake inhibitor (SSRI) at adequate doses. Fluvoxamine, fluoxetine, citalopram, sertraline and paroxetine, have all been shown to be effective in adults with OCD. The usual dose of SSRIs to treat OCD is 2-3 times higher than the dose for treating depression. The minimum mean daily dosage of one of the SSRI is listed as follows:
fluvoxamine 150 mg.
fluoxetine 40 mg.
sertraline 150 mg.
paroxetine 40 mg.
paroxetine CR 50mg.

Clomipramine is used when 1) there is an adequate trial of at least one SSRI which was found to be ineffective, 2) SSRI is poorly tolerated, 3) the patient prefers clomipramine and 4) there has been a previous good response to clomipramine. You need to carry out an ECG and a blood pressure measurement before prescribing clomipramine. Clomipramine is a TCA (derivative of imipramine) and minimum daily dose is 150 mg.

For patients who do not respond to SSRI, clomipramine is an effective treatment for obsessive compulsive disorder (OCD) in the dose range of between 150-300 mg/day.

CBT will begin with anxiety management and asking the patient to keep a diary. Then it will move onto response prevention in excessive washing with cognitive re-structuring and coping strategies. Behaviour therapy using Exposure-Response Prevention (ERP) is the treatment of choice for limiting the dysfunction resulting from obsessions and compulsions.

OSCE grid: Assess obsessive compulsive disorder

A GP has referred a 26-year-old woman to you who has severely chapped hands due to repeated hand washing. She is very concerned about contamination.

Task: Take a history to establish the diagnosis of OCD.

| A) Assess obsessions | A1) Introduction and assess the reasons for excessive hand washing.

I am Dr. XXX. The GP has referred your case to me due to excessive hand washing. Can you tell me why you need to wash your hands so many times a day?

Can you tell me more about your concerns?

Do you come up with this thought? (Assume the patient tells you that she is concerned about contamination). | A2) Assess the nature of obsessions and resistance.

Do you feel that your thoughts are excessive?

Are those ideas reasonable?

Do you feel unpleasant about those thoughts?

Do you want to stop those thoughts? | A3) Assess obsessional doubts.

Do you ask yourself the same question over and over again? For example, you cannot be certain whether you have closed the door even though you have checked a few times. | A4) Assess obsessional impulses.

Do you have impulses which you cannot control? (e.g. impulse to do inappropriate thing)

Do you have recurrent thoughts of harming yourself or others? | A5) Explore other obsessions:

Do you like things to be in a special order?

Do you feel upset if someone changes it? |
|---|---|---|---|---|---|
| B) Assess compulsions | B1. Assess compulsive washing.

Can you tell me how many times you need to wash your hands per day?

Why do you need to wash your hands so many times a day?

How long does it take for you to take a bath?

Why does it take so long? What do you do inside the bathroom? | B2. Assess compulsive checking.

Do you need to check things over and over again?

What kinds of items do you check? (e.g. windows, doors).

How long does it take for you to finish checking all items before leaving your house? | B3. Assess compulsive counting.

Do you count things over and over again? If yes, why?

Is there a number you like or do not like? | B4) Assess other rituals

Do you perform a regular ritual or ceremony to prevent something bad from happening? | B5) Assess nature of compulsions

How do you find the repetitive behaviours?

Are they excessive?

Are they pleasurable?

What would happen if you do not clean your hands?

How long have you been washing your hands excessively? |
| C). Assess impact, comorbidity, risk and insight. | C1) Assess psychosocial impact.

Since you wash your hands very frequent, does it affect your work?

Does it affect your relationship with other people? | C2) Assess comorbidity.

Do you feel stressed or nervous?

How's your mood?

How's your sleep and appetite? | C3) Assess biological complications

Since you wash your hands many times a day, do you have any complication on your skin? Do you consult a dermatologist? | C4) Assess insight.

What is your view of the current problem?

Do you think you have an illness in your mind?

Do you think you | Assess her insight and expectations

Does she think that she has a psychiatric illness? If not, what are her views and explanations?

Has she read any information on OCD? |

				need check to reduce the hand washing behaviour?	What are her expectations on treatment? Are they realistic or achievable?
	Are you slow at work? Can you tell me your water bill? Is it very high?	How you thought of ending your life? Can you tell me more about your character? Are you a perfectionistic person? Do you have abnormal twitching movement in your face? (Assess tics)			What type of treatment does she prefer? (Medication, psychotherapy or both).

Poor prognostic factors include a strong conviction about the rationality of obsession, prominent depression, comorbid tic disorder and underlying medical condition.

Other related clinical conditions:

Hoarding disorder
The DSM-5 specified that individuals afflicted with this condition has marked difficulties with disposing of items, regardless of their actual value. The difficulties with disposing of items has been attributed to a preoccupation with regards to needing to save the items, and to the distress associated with disposing. This behavioural difficulties would have resulted in the accumulation of items that clutter up personal living spaces. The disorder must have resulted in significant impairments in terms of functioning, and not be attributed to another medical disorder such as due to underlying brain injury, cerebrovascular disease or Prader-willi syndrome.

The DSM subtypes include:
a. With good or fair insight
b. With poor insight
c. With absent insight or delusional beliefs

Acute stress reaction and adjustment disorder

Compare and contrast acute stress reaction and adjustment disorder

	Acute stress reaction	Adjustment disorder
Aetiology	Severe acute stress such as rape, assault, natural catastrophe, sudden unemployment or loss of status.	Major change in a life situation e.g. migration, entering university, entering national service, or a newly diagnosed chronic illness.
Duration of illness	ICD-10: Exposure to the stressor is followed by an immediate onset of symptoms within 1 hour and begin to diminish after not more than 48 hours. DSM-5: The DSM-5 Diagnostic criteria specified that the following symptoms must be fulfilled within a duration of 3 days to 1 month after experiencing the traumatic event. The symptoms are: 1. Exposure to a severe or threatened death,	ICD-10: The duration is within 1 month of exposure to stressor. DSM 5: The DSM-5 diagnostic criteria specified that there must be behavioural or emotional symptoms that have occurred within 3 months from the onset of the stressor. The behavioural and emotional symptoms experienced must have

	serious injury or even sexual violence. 2. Repetitive, intrusive, and distressing memories of the traumatic events (intrusion symptoms) 3. Marked inability to experience positive emotions (negative mood) 4. Dissociative symptoms characterized as feelings of an altered sense of reality or selective dissociative amnesia for the events (Dissociative symptoms) 5. Marked efforts to avoid distressing memories and external reminders (avoidance symptoms) 6. Hyperarousal symptoms: Sleep difficulties, irritable mood, keyed up and always on edge, issues with concentration and easily getting startled.	caused impairments in terms of functioning, and must be considered to be out of proportion in terms of severity and intensity to what is normally expected of the stressors. The DSM-5 also states that the behavioural and emotional symptoms should resolve within 6 months after the stressors have been removed. The various subtypes include: a. With depressed mood b. With anxiety c. With mixed anxiety and depressed mood d. With disturbances in conduct e. With mixed disturbances of emotions and conduct
Clinical features	Physical: palpitations, chest pain Psychological: withdrawal, inattention, anger, aggression, despair, purposeless overactivity, numbness, derealisation, depersonalisation and amnesia.	ICD-10 criteria specify 1) Brief depressive reaction (< 1 month). 2) Prolonged depressive reaction (< 2 years). 3) Mixed anxiety and depressive episode. 4) With predominant disturbance of other emotions. 5) With predominant disturbance of conduct (adolescent grief reaction). 6) With mixed disturbance of emotions and conduct.
Management	Symptomatic relief: short term anxiolytic agents. Crisis intervention and reassurance.	Counselling. Problem solving therapy. Short duration of anxiolytic and antidepressant.

Post traumatic stress disorder (PTSD)

Epidemiology [Kessler et al, 1995]

Incidence	Prevalence	Gender ratio	Mean age of onset
On average, about 10% of people experiencing a significant traumatic event actually go on to develop PTSD.	Lifetime prevalence: 1in 10 in the general population. 1 in 5 fire-fighters. 1 in 3 teenager survivors of car crashes. 1 in 2 female rape victims. 2 in 3 prisoners of war.	Women > Men (F:M = 2:1)	Most prevalent in young adults.

Clinical features

PTSD is a prolonged response to a traumatic event such as abuse, serious road traffic accident, disaster, violent crime, torture and terrorism. The event should be extraordinary and most people find it traumatic. For example, one cannot suffer from PTSD after failure in an examination because it is not as traumatic as a serious road traffic accident. The development of PTSD symptoms is usually within 6 months after the traumatic event.

Main symptoms include re-experiencing (e.g. flashbacks in the day time, nightmare at night), avoidance (e.g. place and objects associated with the event), hyperarousal (e.g. increased vigilance, irritability, poor concentration, exaggerated startle response) and emotional numbing (e.g. detachment, lack of interest).

DSM-5 Diagnostic Criteria (Please note that PTSD is no longer part of anxiety disorders chapter in DSM-5 but classified together within the trauma and stress related disorders chapter).
The DSM-5 Diagnostic criteria specified that individuals diagnosed with this condition must have had exposure to a severe or threatened death, serious injury or even sexual violence.
To fulfill the diagnosis, the following symptoms must be present:
1. Repetitive, intrusive, and distressing memories of the traumatic events
2. Marked efforts to avoid distressing memories and external reminders
3. Dissociative amnesia towards important aspects of the traumatic event
These symptoms must have resulted in marked impairments in terms of psychosocial functioning.
At times, it is important for clinicians to specify whether individual experience persistent or recurrent symptoms of either (a) depersonalization or (b) Derealization.
DSM-5 has also a delayed expression criteria, for which the typical criteria are not fulfilled until at least 6 months after the experience of the traumatic event.

OSCE grid: Assess post traumatic stress disorder
The GP has referred a 35-year-old driver to you for assessment. He was almost killed in a road traffic accident 6 months ago and, he is suing the other party for compensation. Task: take a history to establish the diagnosis of PTSD.

A) Exploration on his trauma	A1) Explore the accident.	A2) Explore immediate outcome of the accident.	A3) Assess the extent of injury and suffering.	A4) Assess outcome of other people involved in the accident (if any)
	I am Dr. XXX. I am sorry to hear that you were involved in a road traffic accident. In the next 7 minutes,, I would like to find more about the recent event. Is it ok with you? Can you tell me what happens on that night? Were you driving the car alone or with someone? Can you describe the severity of the accident? Was it life threatening?	How long did you wait for the rescue to come? Do you remember what happened next? Were you brought to the accident and emergency department? Were you admitted to the hospital? What kind of treatment did they offer? Did you undergo operation?	Can you tell me some of the complications after the accident? Do you lose any ability or function? For example, memory, mobility or sensation. Are you in pain at this moment? If yes, for how long?	Were the other passengers injured? If so how many? What happened to them? What is your relationship with them? Do you feel sorry towards them?
B) Assess PTSD symptoms	B1) Re-experiencing Identity the latency period between the incident and the onset of PTSD symptoms. How does the memory relives itself? How vivid is it? Does the memory come in the form of repetitive distressing images? How often do those mental images come in a day? Do you have nightmares at night? Can you tell me more about it?	B2) Avoidance Do you try to avoid driving a car? How about sitting in a car? Do you try to avoid the place where the accident occurred?	B3) Hyperarousal "Are you always on the edge?" How about excessive sweating, fast heart beats and difficulty in breathing? How do you find your concentration?	B4) Emotional detachment: Are you able to describe your emotion? Do you feel blunted?
C) Assess comorbidity, vulnerability, compensatio	C1) Assess comorbidity. How is your mood?	C2) Assess vulnerability. Did you encounter any traumatic event when you were young? (e.g. abuse,	C3) Explore compensation and legal procedure. What is the status of	C4) Assess current support system and past treatment. I am very sorry to hear the road traffic accident and the

n issues.	How do you see your future?	past accident)	the legal procedure?	complications you have gone through.
		Did you stay with your family when you were young? (explore social isolation)	Is your case due to be heard in court soon?	Do you get any support from your partner or family members?
	Do you have thought of harming yourself?			
	How do you cope? Do you turn to alcohol or recreational drugs?	Can you tell me your education level? (low education is associated with PTSD).		Did you see a doctor for the anxiety symptoms? If yes, did the doctor offer treatment to you? How do you find the treatment?

Questionnaire (Horowitz, 1979)

The **Impact of Events Scale (IES)** is a 15-item questionnaire assessing symptoms of intrusion and avoidance. It assesses self-reported levels of distress with regard to a specific life event.

Management (MOH guidelines):

When symptoms are mild and have been present less than 4 weeks after the trauma, doctors can offer reassurance and close monitoring. If PTSD symptoms persist after 1 month, the following treatment can be offered.

Pharmacotherapy: SSRIs are generally the most appropriate medication of choice for post-traumatic stress disorder (PTSD), and effective therapy should be continued for 12 months or longer. Paroxetine, sertraline and fluoxetine all have well documented evidence of efficacy.

Psychological treatment: Studies of trauma-focused cognitive behaviour therapy (tf-CBT) have shown the most effective results in the treatment of post-traumatic stress disorder (PTSD). Tf-CBT involves exposure therapy to overcome avoidance associated with accident. Eye movement desensitization and reprogramming (EMDR) involves recalling the traumatic event when the patient performs a dual attention movement such as bilateral hand tapping. The objective of EMDR is to give new insight to the accident and reduce sensitivity to negative re-experiences.

Prognosis: 50% will recover within the first year and 30% will run in a chronic course. An initial rapid resolution of symptoms predicts a good prognosis. Bad prognostic factors include completed rape, physical injury, and perception of life threat during the accident or assault.

Depersonalization and derealisation

Depersonalisation and derealisation syndrome refers to the phenomenon when the patient complains spontaneously that the quality of mental activity, body and surroundings are changed to appear to be unreal and remote. This syndrome is a subjective and unpleasant experience with insight retained. This syndrome can be a primary phenomenon or secondary to sensory deprivation, temporal lobe epilepsy, phobic anxiety disorders and generalised anxiety disorders.

DSM-5 Criteria:
The presence of persistent or recurrent depersonalisation, derealisation, or both:

Depersonalisation: experiences of unreality, detachment, or being an outside observer with respect to one's thoughts, feelings, sensations, body, or actions(e.g., unreal or absent self, perceptual alterations, emotional and/or physical numbing, distorted sense of time).

Derealisation: experiences of unreality or detachment with respect to surroundings (e.g., individuals or objects are experienced as unreal, dreamlike, foggy, lifeless, or visually distorted).

During the depersonalisation and/or derealisation experiences, reality testing remains intact.

Depersonalisation	Derealisation
The patient complains of a feeling of being distant or not really there.	The patient complains of feeling that the environment is unreal.
e.g. A stressful resident complains of his emotions and feelings are detached after the night duty. He feels that his emotions and movements belong to someone else.	e.g. A stressful resident complains of feeling of unreality after the night duty. The ward looks strange as the colour of the wall is less vivid and the staff look soulless. He feels that the time passes very slowly on that morning.

Treatment: For short term depersonalisation and derealisation, reassurance and relaxation training are recommended. For long term depersonalisation and derealisation, CBT and/or SSRI may be useful.

Revision MCQS

1. A 50-year-old woman meets the diagnostic criteria for panic disorder with agoraphobia. Each time she leaves the house, she experiences high levels of anxiety. When she goes back home, her anxiety level goes down. After some time, she learns that by staying at home, she can avoid any possibility of a panic attack in public. This contributes to the maintenance of her disorder. Which of the following statements about the above phenomenon is incorrect?

A. The reinforcement is contingent upon the behaviour
B. The behaviour is voluntary
C. A negative reinforcer positively affects the frequency of response
D. The reinforcement can occur before the behaviour
E. An alteration in frequency of behaviour is possible after reinforcement

Answer: D
This phenomenon is known as operant conditioning, specifically negative reinforcement. Escape and avoidance learning are two examples of negative reinforcement. The removal of the unpleasant stimulus leads to reinforcement of the behaviour. In operant conditioning, the reinforcer (reduction in anxiety levels) is presented only after the behaviour (going home) is executed, which is why Option D is incorrect.

2. A 25-year-old woman is referred by her lawyer after a road traffic accident which occurred one month ago. Her lawyer wants you to certify that she suffers from post-traumatic stress disorder (PTSD). Which of the following clinical features is not a predisposing factor in PTSD?

A. Childhood trauma
B. Inadequate family support
C. Low premorbid intelligence
D. Lack of control of the accident
E. Recent stressful life events

Answer: D
The above options are predisposing factor for PTSD except option D. Other risk factors for PTSD include female gender, previous exposure to trauma including childhood abuse and pre-existing anxiety or major depression.

3. Which of the following statements regarding diagnostic criteria for panic disorder is false?

A. Based on the DSM-IV-TR, panic disorder is classified into panic disorder with agoraphobia and panic disorder without agoraphobia

B. Based on the DSM-IV-TR, agoraphobia is classified into agoraphobia without history of panic disorder and panic disorder with agoraphobia

C. Based on the ICD-10, agoraphobia is classified into agoraphobia without history of panic disorder and panic disorder with agoraphobia

D. Based on the ICD-10, panic disorder is classified into panic disorder with agoraphobia and panic disorder without agoraphobia

E. Based on the ICD-10 criteria further define severe panic disorder of having at least 4 panic attacks per week over a 4-week period

Answer: D
Only the DSM-IV-TR but not the ICD-10 classifies panic disorder into panic disorder with agoraphobia and panic disorder without agoraphobia.

4. In clinical practice, it is often difficult to differentiate obsession from delusion. Which of the following strongly indicate that a patient suffers from obsessive-compulsive disorder rather than delusional disorder?

A. Better occupational functioning
B. No other psychotic phenomenon such as hallucinations
C. The thought content is less bizarre
D. The patient believes that the origin of thoughts is from his or her own mind
E. The patient tries to resist his thoughts

Answer: E
Resistance is seen in people with obsessive-compulsive disorder but not delusional disorder.

5. You are a GP and a 30-year-old man with obsessive compulsive disorder (OCD) wants to know more about psychological treatment for his condition. He read the information from the internet and provided a list of psychological interventions. Which of the following therapies is specially indicated for OCD?

A. Biofeedback
B. Exposure and response prevention
C. Eye movement desensitisation and reprocessing
D. Psychoanalysis
E. Relaxation exercise

Answer: B
Exposure to dirt and response prevention (e.g. no hand

washing for 3 hours after contact of dirt) is a form of psychological treatment for an OCD patient with obsessions about contamination by dirt.

Revision MEQ

You are the resident working at the Accident and Emergency Department (AED). A 45-year-old woman presents with episodes of hyperventilation, fear of losing control, palpitations and tremor. She works in the bank and being stressed at work. Few days ago, she consulted her GP who gave her 'piriton' which did not work. She was very concerned and called an ambulance to send her to the nearest hospital for treatment.

1. **List five possible diagnoses.**
Panic disorder
Panic disorder with agoraphobia
Hyperventilation syndrome
Mixed anxiety and depression.
Medical causes: hyperthyroidism, hypoglycaemia.

2. She requests to have investigations done at the AED. Name three investigations you would order?
1. ECG
2. Thyroid function test
3. CXR.

3. You have discussed with the consultant psychiatrist. The diagnosis is panic disorder. The patient wants a medication to stop her panic attack in the emergency department. She is scared of injection. Which oral medication would you order?

Oral alprazolam 0.25mg stat or clonazepam 0.5mg stat or lorazepam 1mg stat.

4. The patient wants to get regular medication to treat her panic disorder. Which oral medication would you recommend?
1. Regular SSRI (e.g. paroxetine, fluvoxamine).
2) Anxiolytics for 2 weeks (e.g. alprazolam or clonazepam TDS prn for 2 weeks)

5. The nurse in the AED has asked the patient to breath in and out of a page bag to control her panic attacks. The patient wants to seek your view on this method. What is your advice?
Breathing in and out of a paper bag is not a good method because it may cause retention of carbon dioxide and makes her hyperventilation worse. Deep breathing exercise is a better option.

EMIs:
Anxiety Disorders

A – Acute stress reaction
B – Adjustment reaction
C – Agoraphobia
D – Bipolar disorder
E – Cyclothymia disorder
F – Dysthymic disorder
G – Generalized anxiety disorder
H – Major depressive disorder
I – Post traumatic disorder

Please select one diagnosis from the above list for each of the clinical vignettes below. Each option could be used once, more than once or not at all.

1. Jane, presents to the emergency services with the complaints of feeling lethargic, associated with inability to concentrate in her daily work. However, she denies having any issues with her sleep, appetite or ever feeling suicidal. She then claimed that she has had a 1-month period during which she felt extremely elated and full of energy. She wishes to get back to how she was 1-month ago – Bipolar disorder
2. Victor met with a car accident 4 months ago and has not driven since. He still experiences flashbacks of the accident occasionally – PTSD
3. Christine has just started with a new marketing job. She has been having frequent episodes of feeling on edge, associated with palpitations and also dizziness. She worries a lot about everyday events. She claimed that she has seen a doctor previously and was told that she has no medical issues – GAD
4. Daniel has been finding it increasingly difficult to leave home by himself as he is preoccupied with thoughts that something nasty would actually happen to him if he goes out of his house – Social Phobia

References

Lim L, Ng TP, Chua HC, Chiam PC, Won, V, Lee T, Fones C, Kua EH (2005) Generalised anxiety disorder in Singapore: prevalence, co-morbidity and risk factors in a multi-ethnic population. *Soc Psychiatry Psychiatr Epidemiol* **40**: 972–979

Hamilton M. (1959) The assessment of anxiety states by rating. *British Journal of Medical Psychology* **32**: 50-55.

Bienvenu OJ, Onyike CU, Stein MB, Chen LS, Samuels J, Nestadt G, Eaton WW. (2006) Agoraphobia in adults: incidence and longitudinal relationship with panic. Br J Psychiatry;188:432-8.

Beesdo K, Pine DS, Lieb R, Wittchen HU. (2010) Incidence and risk patterns of anxiety and depressive disorders and categorization of generalized anxiety disorder. Arch Gen Psychiatry.;**67(1)**:47-57.

Ham P et al ;(2005) Treatment of Panic Disorder. *Am Fam Phys* ;**71(4)**:733-739.

] Kendler KS, Neale MC, Kessler Re, et al. (1992) A population-based twin study of major depression in women. The impact of varying definitions of illness. *Arch Gen Psychiatry*; **49**: 257-266.

Anxiety disorder guideline: http://www.gov.sg/moh/pub/cpg/cpg.htm [Accessed on 5th January, 2012]

Nestadt G, Bienvenu OJ, Cai G, Samuels J, Eaton WW. (1998) Incidence of obsessive-compulsive disorder in adults. J Nerv Ment Dis.;**186(7)**:401-6.

Goodman WK, Price LH, Rasmussen SA et al. (1989) The Yale-Brown Obsessive Compulsive Scale, I: development, use, and reliability. *Archives of General Psychiatry* **46**: 1006-1011.

Kessler RC, Sonnega A, Bromet E,et al. (1995) Posttraumatic stress disorder in the National Comorbidity Survey. *Archives of General Psychiatry*. **52**: 1048-1060

Horowitz M, Wilner N, Alvarez W. (1979) Impact of Events Scale: a measure of subjective stress. *Psychosomatic Medicine* **41**(3): 209-218

Puri BK, Treasaden I (eds) *Psychiatry: An Evidence-Based Text*. London: Hodder Arnold, 2010.

American Psychiatric Association. Diagnostic Criteria from DSM-IV-TR. Washington: American Psychiatric Association, 2000.

World Health Organization. ICD-10 Classification of Mental and Behavioural Disorders. Edinburgh: Churchill Livingstone, 1994.

Argyropoulos S, Campbell A and Stein G (2007) Anxiety disorders In Stein G & Wilkinson G (2007) *Seminars in general adult psychiatry*. London: Gaskell.

Semple D, Smyth R, Burns J, Darjee R, McIntosh A (2005) *Oxford Handbook of Psychiatry*. Oxford: Oxford University Press.

Sadock BJ, Sadock VA. (2003) *Kaplan and Sadock's Comprehensive Textbook of Psychiatry*. (9th ed.) Philadelphia (PA): Lippincott Williams and Wilkins.

Gelder M, Mayou R and Cowen P (2006) *Shorter Oxford Textbook of Psychiatry (5th ed)*. Oxford: Oxford University Press.

Taylor D, Paton C, Kapur S (2009) *The Maudsley prescribing guideline*.10th edition London: Informa healthcare.

Bluglass R, Bowden P. (1990) *Principles and practice of forensic psychiatry*. Edinburgh: Churchill Livingstone.

Johnstone EC, Cunningham ODG, Lawrie SM, Sharpe M, Freeman CPL. (2004) *Companion to Psychiatric studies*. (7th edition). London: Churchill Livingstone.

Andrews G, Creamer M, Crino R, Hunt C, Lampe L, Page A. (2003) *The treatment of anxiety disorders. Clinician guides and patient manuals*. Cambridge: Cambridge University Press.

Tyrer P (1994) Anxiety disorders In Paykel ES, Jenkins R (1994) *Prevention in Psychiatry*. Gaskell: London.

American Psychiatric Association (2013) DSM-5: Diagnostic and Statistical Manual of Mental Disorders (Diagnostic & Statistical Manual of Mental Disorders) (5th edition). American Psychiatric Association Publishing Inc: Washington.

Notes:

Chapter 6 Personality and impulse control disorders

Introduction	OSCE grid: assess borderline personality disorder
Epidemiology and aetiology	Management of borderline personality disorder (NICE guidelines)
Differential diagnosis and general management strategies	Cluster C personality disorders (dependent, avoidant and obsessive compulsive personality disorders)
Course of personality disorders	Impulse control disorders
Cluster A personality disorders (schizoid, schizotypal and paranoid personality disorders)	
Cluster B personality disorders (borderline, antisocial and narcissistic personality disorders)	Revision MCQs, MEQ, EMIs

Introduction

Personality is a dynamic organisation and configuration of trait within a person and this trait determines the characteristic behaviour and thoughts of the person. The personality of a person allows the others to predict how this person would react, feel or think in a particular situation.

Personality disorder represents an extreme variant of a normal personality trait.
DSM5 states the following:
- Impairments in personality (self and interpersonal) functioning.
- The presence of pathological personality traits.
- The impairments in personality functioning and personality trait expression are not better understood as normative for the individual's developmental stage or socio-cultural environment.
- The impairments in personality functioning and the personality trait expression are stable across time and consistent across situations.
- Significant impairments in self (identity or self-direction) and interpersonal (empathy or intimacy) functioning.

Not attributed to:
- Direct physiological effects of a substance (e.g., a drug of abuse, medication).
- General medical condition (e.g., severe head trauma).

The following personality disorders are included within DSM-5:
a. Paranoid personality disorder
b. Schizoid personality disorder
c. Schizotypal personality disorder
d. Antisocial personality disorder
e. Borderline personality disorder
f. Histronic personality disorder
g. Narcissistic personality disorder
h. Avoidant personality disorder
i. Dependent personality disorder
j. Obsessive compulsive personality disorder
k. Personality change due to medical condition
l. Other specified personality disorder and unspecified personality disorder

DSM-5 Classification of Personality Disorder:
Cluster A: Paranoid, Schizoid, Schizotypal
Cluster B: Antisocial, borderline, histronic, narcissistic
Cluster C: Avoidant, dependent, obsessive compulsive disorder

Epidemiology

Prevalence in the general population	Prevalence in patients with a concurrent psychiatric disorder
4 to 13%	50%

Aetiology

Genetics	- Heritability is 40 to 60%. - Higher concordance rates in monozygotic twins over dizygotic twins. - Cluster A personality disorders are more common in first degree relative of patients with schizophrenia.
Neurochemistry	- High levels of testosterone, oestrone and 17-oestradiol are associated with impulsivity seen in patient with borderline and antisocial personality disorders. - Low level of serotonin metabolite 5-hydroxyindoleacetic acid (5-HIAA) is associated with low mood, self-harm, suicidal behaviour in patients with borderline personality disorder.
Environmental factors	- Low socio-economic status. - Social isolation.
Parenting styles	- Low parental affection or lack of care is associated with borderline personality disorder and dependent personality disorder. - Aversive parenting is associated with antisocial personality disorder.
Childhood abuse	- People with personality disorder are more likely to have childhood maltreatment and trauma. For example, sexual and emotional abuse may lead to personality disorder.

Differential diagnosis

Differences between personality disorder and other psychiatric disorders:
1. People with personality disorder may present with psychotic features. Personality disorder is different from schizophrenia because people with personality disorder have relatively intact capacity for reality testing, expression of emotion and the ability to distinguish between thoughts of their own and others.
2. People with personality disorder may complain of mood swings. Personality disorder is different from bipolar disorder because people with personality disorder should not have hypomanic or manic episodes. The mood swings they refer to is from normal mood to irritability.
3. Anxiety disorder is different from personality disorder because people with personality disorder use immature, defences such as projection or denial.

Overall management strategies

The objective of management is to offer support to people with personality disorder and allow them to express their concerns and emotions in a safe environment. This supportive approach would allow the treatment team and patient to develop trust and lead to stabilisation. The specific treatment strategies include crisis intervention, short term hospitalisation, psychotherapy and pharmacotherapy.

Prognosis

1. People with personality disorder have higher chance to stay alone and they are often rejected by family members.
2. People with antisocial, schizotypal and borderline personality disorder are relatively more impaired at work, relationships and leisure.
3. People with obsessive compulsive, histrionic and narcissistic personality disorder have relatively less functional impairment.

4. People with personality disorder and a comorbidity of Axis 1 disorders are at great risk of further functional impairment, more chronic course and poor response to treatment.

*ICD-10 criteria are listed as blueprint on the left hand side and the additional DSM 5 criteria are listed on the right hand side.

Borderline personality disorder

Epidemiology

Prevalence	Gender ratio	Age of onset	Suicide rate	Comorbidity
1-2%.	M: F = 1:2.	Adolescence or early adulthood.	9%.	Depression, PTSD, substance misuse, bulimia nervosa.

Aetiology

Early development	<ul><li>Early insecure attachment results in fear of abandonment.</li><li>Family environment is characterised high conflict and unpredictability.</li><li>Mother is not emotionally available.</li><li>Emotionally vulnerable temperament interacts with an invalidating environment.</li><li>Increased reports of family dysfunction.</li><li>Early separation or loss.</li><li>Parent alcohol or drug misuse.</li><li>Forensic history of parents.</li></ul>
Past trauma and abuse	<ul><li>Childhood trauma.</li><li>Physical abuse.</li><li>Neglect.</li><li>Sexual abuse.</li></ul>
Attachment	<ul><li>All infants possess basic instinct towards attachment. If attachment is not formed, it will impair the child's ability to develop a stable and realistic concept of self. The child will have limited capacity to depict feelings and thoughts in self and other people. This capacity is known as mentalisation.</li></ul>
Defence mechanisms	<ul><li>**Splitting** – adopting a polarised or extreme view of the world where people are either all good or bad and fail to see that each person has good and bad aspects. For example, a patient with borderline personality disorder tries to classify the doctors of the ward into two groups, good doctors and bad doctors. The patient fails to see the strengths and weaknesses of each doctor.</li><li>**Projective identification** – the patient unconsciously project a figure onto the other people. For example, a man does not like his father and projects a bad father figure onto the male doctor (projection) and accuses the doctor as a non-caring individual. Due to the counter-transference, the male doctor tries to avoid the patient as if he does not care about the patient (identification).</li></ul>
Biological factors	<ul><li>Family studies show the risk of relatives of borderline personality disorder to develop such disorder is 5 times higher as compared to the general population.</li><li>More common in first degree relatives of patients with depression.</li><li>Some studies demonstrate abnormal dexamethasone suppression test rest result, decreased REM latency, decreased thyrotropin response, abnormal sensitivity to amphetamine in patients with borderline personality disorder.</li><li>Specific marker for impulsivity – decreased CSF 5HIAA – a metabolite of serotonin.</li><li>Chronic trauma leads to decreased hippocampal volume, decreased hemispheric integration and hypearactive HPA axis.</li><li>Increased bilateral activity in amygdala with emotionally aversive stimuli.</li></ul>

	• Orbitofrontal cortex abnormalities lead to reduction in cortical modulation of amygdala. Such abnormalities lead to impulsivity, affect dysregulation, chronic feeling of emptiness and decreased mentalisation.

Clinical features of emotionally unstable personality disorder and related OSCE interview questions

ICD-10 classifies emotionally unstable personality disorder into impulsive and borderline type. DSM-5 does not have the concept of emotionally unstable personality disorder and only has borderline personality disorder.

Met general criteria for personality disorder, ≥3 symptoms of impulsive type (one of which must be the symptom indicated by *)/

Impulsive type (ICD-10)
Affect:
1. Liability to outbursts of anger or violence, with inability to control the resulting behavioural explosions. *(Do you have a problem with your anger control?)*
2. Unstable and capricious mood. *(How often does your mood change?)*

Behaviour:
1. Marked tendency to act unexpectedly and without consideration of consequences.
2. * Marked tendency of quarrelsome behaviour and conflicts with others, especially when impulsive acts are thwarted or criticized. *(How often do you get into quarrel or fights?)*
3. Difficulty in maintaining any course of action that offers no immediate reward.

Borderline type (ICD-10)
The borderline type requires the person to meet the diagnostic criteria of impulsive type and an additional 2 symptoms in the following:

Affect:
1. Chronic feelings of emptiness *(How often do you feel empty inside yourself?)*.

Behaviour:
1. Liability to become involved in intense and unstable relationships, often leading to emotional crises *(How have your relationship been?)*.
2. Excessive efforts to avoid abandonment. *(How often do you feel abandoned?)*
3. Recurrent threats or acts of self-harm *(How often do you harm yourself?)*.

Cognition:
1. Disturbances in and uncertainty about self-image, aims and internal preferences (including sexual) *(Do you feel that your views about yourself often change?)*

People with borderline personality disorder present with recurrent acts of self harm usually in the form of self laceration. They often report chronic feelings of emptiness and unstable emotion.

DSM-5:
1. Negative affectivity:
   a. Emotional lability: unstable emotional experiences and frequent mood changes.
   b. Anxiousness: intense feelings of nervousness or panic in reaction to interpersonal stresses and fears of losing control.
   c. Separation insecurity: fears of rejection by or separation from significant others and associated with fears of excessive dependency and complete loss of autonomy.
   d. Depression: frequent feelings of being down, miserable or hopelessness.

2. Disinhibition:
   a. Impulsivity: acting on the spur of the moment in response to immediate stimuli; acting on a momentary basis without a plan or consideration of outcomes; difficulty establishing or following plans; a sense of urgency and self-harming behaviour under emotional distress.
   b. Risk taking: Engagement in dangerous, risky, and potentially self-damaging activities.

3. Antagonism:
   a. Hostility: Persistent or frequent angry feelings; anger or irritability.

Additional cognitive symptom includes transient, stress-related paranoid ideation or severe dissociative symptoms. *(Have you ever seen things or heard voices that are not really there?)*

Differential diagnosis
1. **Dependent personality disorder:** people with dependent personality disorder are less chaotic in affect regulation and be less impulsive.
2. **Antisocial personality disorder:** people with antisocial personality disorder have history of forensic problems and do not care the safety of other people.
3. **Histrionic personality disorder:** people with histrionic personality disorder want to be the centre of attention but usually demonstrate less self-harm, emptiness and affective instability.
4. **Narcissistic personality disorder:** people with narcissistic personality disorder see rejection as humiliating. People with borderline personality disorder see rejection as abandonment. People with borderline personality disorder feel that they are entitled to special treatment because they suffered in the past. People with narcissistic personality disorder feel that they are entitled to special treatment because of their special status.
5. **Post traumatic stress disorder (PTSD):** people with PTSD usually present with flashbacks, hypervigilance and nightmares after a recent traumatic event.
6. **Depressive disorder:** people with borderline personality disorder may present with depression but people with depressive disorder do not demonstrate primitive defence mechanisms such as splitting or projective identification.
7. **Schizophrenia:** both borderline personality disorder and schizophrenia may present with psychotic symptoms. People with borderline personality disorder present with transient psychosis, lack of first rank symptoms, visual illusions and lack of negative symptoms.
8. **Bipolar disorder:** both borderline personality disorder and bipolar disorder may present with mood swings. People with borderline personality disorder refer mood swings to fluctuations in mood from normal to irritability without hypomanic or manic features.
9. **Identity confusion in normal adolescence development:** identity confusion is part of a normal adolescence development.

Treatment (NICE guidelines – UK)

Psychotherapy
Long term outpatient psychotherapy is recommended because patients can handle challenges in daily life with the support from psychotherapists. Dialectical Behaviour Therapy (DBT) and mentalisation based therapy are recommended treatment for people with borderline personality disorder. The NICE guidelines (UK) recommend that therapists should use an explicit and integrated theoretical approach and share this with their clients. The guidelines also recommend that the therapists should set therapy at twice per week and should not offer brief psychological interventions (less than 3 month's duration).

Dialectical behaviour therapy (DBT)

Aims of DBT:
a. To reduce life threatening behaviour such as self-harm.
b. To reduce behaviour which interfere therapy and quality of life.
c. To enhance emotion regulation.

Techniques: DBT involves 4 modes of treatment:
a. Weekly individual psychotherapy.
b. Group skills training with skill acquisition focusing on mindfulness, interpersonal effectiveness, emotional regulation and distress tolerance.
c. Skill coaching phone calls.
d. Therapist-consult team meetings.

All the above components require at least one year of commitment.

Mentalisation based therapy

Mentalisation based therapy was developed by Anthony Bateman and Peter Fonagy.

Aims: To help people with borderline personality disorder to develop the capacity to know that one has an agentive mind and to recognise the importance of mental states in others as there was a failure in parental responsiveness during their childhood. Hence, people with borderline personality disorder are unable to mentalise and form a stable and coherent self.

Indications: Borderline personality disorder. It can be applied as individual therapy (in Singapore and western countries) or group therapy (in western countries).

Techniques: 1) Appropriate affect expression to deal with impulse control, reduction of self harm, passive aggression, idealisation, hate and love. 2) Establishment of a stable representational system. 3) Formation of a coherent sense of self. 4) Develop capacity to form secure relationships.

Psychotropic medications:

For mood lability, rejection sensitivity and anger, one can prescribe a SSRI (e.g. fluoxetine) or switch to another SSRI if the first SSRI is not effective. If the second SSRI is not effective, one can add a low dose antipsychotic (e.g. quetiapine) for anger, anxiolytic (e.g. clonazepam) for anxiety. Mood stabilisers such as sodium valproate and carbamazepine can be added if the above medications are not effective. It is recommended not to use psychotropic medications that have narrow therapeutic index such as lithium or TCA which are toxic during overdose.

Prognosis
Poor prognosis is associated with early childhood sexual abuse, early first psychiatric contact, chronicity of symptoms, high affective instability, aggression and substance use disorder.

Obsessive compulsive personality disorder (anankastic personality disorder in ICD-10)

Epidemiology

Prevalence	Gender ratio	Heritability	Comorbidity
Prevalence in the community: 1.7-2.2%. More common in eldest children, Caucasians and high socioeconomic status.	M > F	0.78.	1. Depressive disorder. 2. Anxiety disorders. 3. Somatoform disorders. 4. Hypochondriasis. 5. 30% of people with OCPD have OCD but not same in reverse.

Aetiology
1. Early development: excessive parental control and criticism causes insecurity. This insecurity is defended by perfectionism, orderliness and control.
2. More common in first degree relatives of patients of obsessive compulsive personality disorder.

Clinical features
ICD-10

Affect:
1. Feelings of excessive doubt and caution.

Behaviour:
1. Perfectionism that interferes with task completion.
2. Excessive conscientiousness and scrupulousness (extremely careful).
3. Excessive pedantry (adherence to rules and forms) and adherence to social conventions.

Cognition:
1. Rigidity and stubbornness.
2. Undue preoccupation with productivity to the exclusion of pleasure and interpersonal relationships.

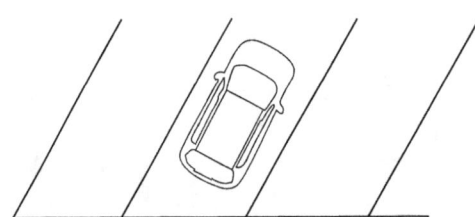

People with anankastic personality disorder are rigid with routine. For example, they may park in the same parking lot everyday and be distressed if the parking lot is taken by others. The person may perform other tasks in a certain way that he/she will not delegate work to other people.

1. **DSM-5**

Compulsivity:

Rigid perfectionism on everything being flawless, perfect, without errors or faults, including one's own and others' performance; sacrificing of timeliness to ensure correctness in every detail; believing that there is only one right way to do things; difficulty changing ideas; preoccupation with details, organisation, and order.

Negative affectivity:

Perseveration at tasks long after the behaviour has ceased to be functional or effective; continuance of the same behaviour despite repeated failures

People with anankastic personality disorder may have excessive adherence to rules.

Differentials diagnosis

Axis I disorders	Axis II disorders
Obsessive compulsive disorder (OCD) – People with OCD present with more clearly defined obsessions and compulsions. People with OCPD are more egosyntonic (in line with ego) with their behaviour and hence, they are less anxious. **Generalised anxiety disorder (GAD)** – People with GAD present with excessive worry.	**Schizoid personality disorder**: People with OCPD may present with constricted affect because they want to maintain control. People with schizoid personality appear to be emotionally cold because there is a fundamental lack of capacity.

Course of OCPD:

Adolescents with strong OCPD traits can grow out of the diagnosis.

Treatment:
1. Psychodynamic psychotherapy is useful in this case. It involves an active therapist who challenges isolation of affect, helps the patient to increase emotional awareness and modifies harsh superego. Patient needs to develop the capacity to accept that he or she is a human being and cannot be perfectionistic in all areas.
2. CBT is useful to enhance tolerance of imperfection and errors. Patients should label the tasks as completed once the result is good enough.

Avoidant personality disorder

Epidemiology

Prevalence	Gender	Heritability estimates	Comorbidity
Prevalence in the community: 0.8-5.0%.	Male: female.	0.28.	Social phobia.

Aetiology

1. **Temperament:** Neuroticism and introversion are vulnerabilities and seem to share with social phobia.
2. **Parenting**: Parents tend to be inconsistent, absent and less demonstration of parental love. Parents are discouraging and rarely show pride in their children. There are higher rates of rejection and isolation. Hence, maladaptive avoidance develops as a defence against shame, embarrassment and failure.

Clinical features

ICD-10
Affect:
1. Persistent, pervasive tension and apprehension.

Behaviour:
2. Unwilling to be involved with people unless certain of being liked.
3. Restricted lifestyle due to need for physical security.
4. Avoidance of social or occupational activities involving significant interpersonal contact

People with anxious (avoidant) personality disorder exhibit persistent, pervasive tension and apprehension. It is characterised by avoidance of interpersonal

DSM-5

Additional behavioural symptoms include:
1. showing restraint intimate relationships because of the fear of being ashamed, ridiculed, or rejected due to severe low self-worth.
2. Inhibition in new interpersonal situations because of feelings of inadequacy.

because of fear of criticism, disapproval or rejection.   contact due to fear of criticism or rejection.

Cognition
5. Belief that one is socially inept, personally unappealing or inferior to others.
6. Excessive preoccupation with being criticised or rejected in social situations.

Differential diagnosis:

Axis I disorders	Axis II disorders
1. **Social phobia**: People with social phobia show more impairment and distress in social situations. Their low self-esteem is lower compared to people with avoidant personality disorder. 2. **Agoraphobia**: people with agoraphobia may have more frequent panic attacks. 3. **Depressive disorder**: negative self-evaluation is related to low mood.	1. **Dependent personality disorder**: People with avoidant personality disorder avoid contact while people with dependent personality disorder focus on being cared for. 2. **Schizoid personality disorder**: isolated but emotionally cold. 3. **Paranoid personality disorder** – People with paranoid personality disorder are isolated due to lack of trust in other people.

Treatment
1. CBT may be more useful and effective as compared to brief dynamic psychotherapy to overcome the avoidance.
2. Assertiveness and social skill training is useful to help patients to make and refuse request.
3. Distress tolerance skill is important to help patients to handle anticipatory anxiety in social situations.

Prognosis
1. People with avoidant personality disorder may do well in familiar environment with known people.
2. Shyness tends to decrease when they get older.
3. People with avoidant personality disorder and comorbid depressive disorder may have high drop-out rate in treatment.

Schizotypal personality disorder

Epidemiology

Prevalence	Gender ratio	Heritability
3%	Male > Female	0.72

Aetiology
- More common in first degree relatives of schizophrenia patients (14% versus 2% in the general population)
- Linked to dopamine dysregulation.

Clinical features

Appearance	Cognitive:	Emotional	Perception	Relationship
Eccentric appearance.	Ideas of reference. Odd beliefs or magical thinking which influences behaviour and inconsistent with subcultural norms.	Inappropriate or constricted affect. Associated with paranoid fears/ mistrust rather than	Unusual perceptions and experiences.	Socially withdrawn. Suspiciousness.

| | Vague or circumstantial thinking. | negative judgements about self. | | |
| | Paranoid ideation. | | | |

DSM-5 Criteria:
Psychoticism:
    a. Eccentricity: odd, unusual, or bizarre behavior or appearance
    b. Cognitive and perceptual dysregulation: Odd or unusual thought processes; odd sensations.
    c. Unusual beliefs and experiences: unusual experiences of reality.

Detachment:
    a. Restricted affectivity: constricted emotional experience and indifference
    b. Withdrawal: avoidance of social contacts and activity.

Negative affectivity:
    a. Suspiciousness: expectations of and heightened sensitivity to signs of interpersonal ill-intent or harm; doubts about loyalty and fidelity of others; feelings of persecution.

Differential diagnosis:
1. Delusional disorder, schizophrenia and severe depressive disorder with psychotic features.
2. Paranoid and schizoid personality disorder: people with paranoid and schizoid personality disorder do not have perceptual or cognitive disturbances.
3. Borderline personality disorder: people with borderline personality disorder may have brief psychotic experiences which are closely related to affective states.
4. Avoidant personality disorder – people with avoidant personality disorder may seek closeness with other people.
5. Pervasive development disorder: autism and Asperger's syndrome.

Management:
1. Supportive therapy and social skill training.
2. Antipsychotic drugs may lead to mild to moderate improvement in psychotic symptoms.

Prognosis:
1. 10-20% of people with schizotypal disorder may develop schizophrenia.
2. Magical thinking, paranoid ideation and social isolation are associated with an increased risk in schizophrenia.

Antisocial personality disorder

Epidemiology

Prevalence	Gender ratio	Comorbidity
Prevalence in the community: 0.6-3.0%. Prevalence in prison: 75%. More common in urban settings.	M:F = 3:1.	Onset usually before 15 years and associated with antisocial personality disorder. Substance misuse.

Aetiology

Developmental causes	<ul><li>Parental deprivation and antisocial behaviour. For example, witnessed abuse when the patients were young, inconsistent or harsh parenting.</li><li>Frequent moves or migration, large family size and poverty.</li><li>Children who go on to develop antisocial personality disorder are innately aggressive, having reactivity levels and diminished ability to be consoled.</li></ul>

Psychological causes	• Temperament: high novelty seeking, low harm avoidance, low reward dependence and uncooperativeness.

Clinical features

ICD-10 criteria:
Affect:
1. Very low tolerance to frustration and a low threshold for discharge of aggression, including violence.
2. Incapacity to experience guilt, or to profit from adverse experience, particularly punishment.

Behaviour:
3. Incapacity to maintain enduring relationships, though no difficulty in establishing them.
4. Marked proneness to blame others, or to offer plausible rationalizations for the behaviour that has brought the individual into conflict with society.

Cognition:
5. Callous unconcern for the feelings of others.
6. Gross and persistent attitude of irresponsibility and disregard for social norms, rules and obligations.

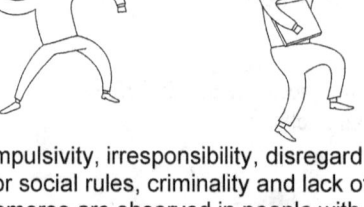

Impulsivity, irresponsibility, disregard for social rules, criminality and lack of remorse are observed in people with dissocial personality disorder. For example, a man with dissocial personality disorder derives his pleasure from the suffering of others.

Salient features of DSM-5

Antagonism:

a. Manipulativeness: frequent use of subterfuge to influence or control others.
b. Deceitfulness: dishonesty and fraudulence or misrepresentation of self.
c. Callousness: lack of concern for feelings or problems of others; lack of guilt or remorse.
d. Hostility: persistent or frequent angry feelings; anger or irritability in response to minor insults.

Disinhibition:

a. Irresponsibility: failure to honour obligations or commitments and lack of follow through on agreements and promises.
b. Impulsivity: acting on the spur of the moment in response to immediate stimuli; acting on a momentary basis without a plan or consideration of outcomes; difficulty establishing and following plans.
c. Risk taking: engagement in dangerous, risky, and potentially self-damaging activities, unnecessarily and without regard for consequences; boredom proneness and thoughtless initiation of activities to counter boredom; lack of concern for one's limitations and denial of the reality of personal danger.

Differential diagnosis:
1. Temporary antisocial behaviour (e.g. vandalism or riot) – focal behaviour, not exploitive and with conscience preserved.
2. Mania / Hypomania – antisocial behaviour (e.g. reckless driving or violence) as a result of impaired judgement and irritability.

Recommendations from the NICE guidelines (UK) on the management of dissocial/ antisocial personality disorder:

1. Consider offering cognitive and behavioural interventions, in order to address problems such as impulsivity, interpersonal difficulties and antisocial behaviour. For people with forensic history, the CBT should focus on reducing offending and other antisocial behaviour.

2. When providing CBT to people with dissocial personality disorder, it is important to assess risk regularly and adjust the duration and intensity of the programme accordingly.

3. Pharmacological interventions should not be routinely used for the treatment of dissocial personality disorder.

4. Pharmacological interventions may be considered in the treatment of comorbid disorders such as depression and anxiety (e.g. SSRIs) and aggression (e.g. low dose antipsychotics or mood stabilisers).

Course and prognosis:

Positive prognostic factors include:
- Show more concerns and guilt of antisocial behaviour.
- Ability to form therapeutic alliance.
- Positive occupational and relationship record.

Narcissistic personality disorder

Epidemiology

Prevalence	Gender ratio	Comorbidity
0.4 – 0.8%	Male > Female	Hypomania. Depression or dysthymia. Substance misuse. Anorexia nervosa.

Aetiology

Psychological causes	<ul><li>Parental deprivation in some cases.</li><li>Pampering and spoiling by parents in some cases.</li><li>Most theories state that people with narcissistic personality disorder develops narcissism in response to low self-esteem. Narcissism has been suggested to be a defence against awareness of low self-esteem.</li><li>People with narcissistic personality disorder have inflated self-esteem and seek information that confirms this illusory bias.</li><li>Temperament: High novelty seeking and reward dependence.</li></ul>

Clinical features
Narcissistic personality disorder is not listed in the ICD-10.

Narcissism refers to the trait of excessive self-love based on the self-image. The term is derived from the Greek myth of Narcissus who falls in love with his own reflection in a pool of water.

Salient features of DSM-5

a. Grandiosity: Feelings of entitlement, either overt or covert; self-centeredness; firmly holding to the belief that one is better than others; condescending toward others.

b. Attention seeking: Excessive attempts to attract and be the focus of the attention of others; admiration seeking.

1. DSM-5 requires a fulfilment of ≥5 symptoms.

Affect:
1. Lacks empathy.
2. Often envious of others or believes that others are envious of him/her.
3. Shows arrogant, haughty behaviours or attitudes.

Behaviour:
1. Tendency to be interpersonally exploitative.

Cognition:
1. Grandiose sense of self-importance.
2. Preoccupation with fantasies of unlimited success, power, brilliance or beauty.
3. Belief that he/she is 'special' or 'unique'.
4. Excessive need for admiration.
5. Sense of entitlement.

Differential diagnosis:
1. Other cluster B personality disorders e.g. histrionic personality disorder.
2. Adjustment disorder with depressive features or depressive disorder with narcissistic defences.
3. Hypomanic episodes.
4. Substance misuse.

Treatment:
1. People with narcissistic personality disorder are often ambivalent about treatment and they tend to feel that it is the others who need to change.
2. People with narcissistic personality disorder may come to seek help when they are depressed after narcissistic injury.
3. Psychotherapies such as dynamic psychotherapy and CBT are useful.

Course and prognosis:
1. Depression is perpetuated by continuing frustration and disappointment and reduced boosters for narcissism.
2. They may encounter difficulty with aging as a result of high value placed on self-image and unrealistic strength. They may not be able to satisfy with life achievements.

Schizoid personality disorder

Epidemiology

Prevalence	Gender ratio	Heritability
0.5 – 1.5%	M: F = 2:1	0.55

Aetiology
- Schizoid personality disorder is more common among first degree relatives of schizophrenia patients.

ICD-10 criteria
Met general criteria for personality disorder and ≥4 symptoms

Affect:
1. Emotional coldness, detachment or flattened affectivity.
2. Limited capacity to express either warm, tender feelings or anger towards others.
3. Appear to be indifferent to either praise or criticism.

Behaviour:
1. Few, if any, activities provide pleasure.
2. Little interest in having sexual experiences with another person (taking into account age).
3. Consistent choice of solitary activities.
4. No desire for, or possession of any close friends or confiding relationships (or only one).

Cognition:
5. Excessive preoccupation with fantasy and introspection
6. Marked insensitivity to prevailing social norms and conventions (which is unintentional).

People with schizoid personality disorder show restricted affective expression, impoverished social relationships and avoidance of social activities.

DSM-5 criteria
Salient features of DSM-5:
1. DSM-5 requires a fulfilment of ≥5 symptoms.
2. Does not occur exclusively during the course of schizophrenia, mood disorder with psychotic features or other psychotic disorder, or a pervasive developmental disorder.

Affective symptoms of DSM-IV-TR Limited capacity to express either warm, tender feelings or anger towards others.

Additional behavioural symptom includes the lack of close friends or confidants.

Cognitive symptom of DSM-5
Marked insensitivity to prevailing social norms and conventions (which is unintentional).

Differentials diagnosis
1. **Other personality disorders** – People with schizotypal personality disorder would be more eccentric with disturbed perception and thought form. People with paranoid personality disorder are easier to be engaged and resentful.
2. **Pervasive developmental disorders**: autism and Asperger's syndrome. People with schizoid personality disorder have better communication and fewer stereotypical behaviours.
3. **Schizophrenia** – People with schizoid personality disorder do not have full-blown first rank symptoms.

Management:
1. Most individuals rarely seek treatment due to low insight into associated problems. They have low capacity for relationships and motivation.
2. **Psychotherapy:** Supportive psychotherapy is useful to establish therapeutic alliance. As trust increases, the therapist may be able to access the fantasies.
3. **Medications:** Low dose antipsychotics and antidepressants have been used with variable outcomes.

Paranoid personality disorder

Epidemiology

Prevalence	Gender ratio	Heritability
0.5 – 2.5%	Male > Female	0.69

Aetiology:
1. Paranoid personality disorder is more common among first degree relatives of schizophrenia patients.
2. Childhood experiences: lack of protective care and support in childhood. Excessive parental rage and humiliation.
3. Temperament: non-adaptability and tendency to hyperactivity and intense emotions.
4. Defence mechanism: projection of negative internal feelings (e.g. hostility and rage) onto other people.
5. Sensory impairments: impaired vision, hearing and victims of traumatic brain injury.

Clinical features:

ICD-10 criteria
Met general criteria for personality disorder and ≥4 symptoms

Behaviour:
1. Tendency to bear grudges (feeling resentful about something) persistently.

Cognition:
1. Excessive sensitivity to setbacks and rebuffs.
2. Suspiciousness and a pervasive tendency to distort experience by misconstruing the neutral or friendly actions of others as hostile or contemptuous.
3. Combative and tenacious sense of personal rights out of keeping with the actual situation.
4. Recurrent suspicious, without justification regarding sexual fidelity of spouse/sexual partner.
5. Persistent self-referential attitude, associated particularly with excessive self-importance.
6. Preoccupation with unsubstantiated 'conspiratorial' explanations of events either immediate to the person or in the world at large.

People with paranoid personality disorder are suspicious and worry about conspiracy against oneself.

DSM-5 criteria
Salient features of DSM-5:
Emphasize that paranoid personality disorder does not occur exclusively during the course of schizophrenia, mood disorder with psychotic features or other psychotic disorder, or a pervasive developmental disorder.

Additional affective symptom includes:
1. Angry reactions to perceived attacks on his/her character or reputation.

Additional behavioural symptom includes:
1. Reluctance to confide in others because of doubts of loyalty or trustworthiness.

Additional cognitive symptoms include:
1. Unjustified doubts about loyalty or trustworthiness of friends or associates.
2. Hidden demeaning or threatening meanings read into benign remarks or events.

Differential diagnosis:

Axis I disorders	Axis II disorders
• **Delusional disorder**: people with delusional disorder have their delusions well encapsulated, systematised but people with paranoid personality disorder do not have well-formed delusions. • **Schizophrenia**: people with schizophrenia have full blown first rank (positive) and negative symptoms. • **Severe depressive episode with psychotic features**: mood congruent delusions such as delusions of guilt or nihilistic delusions. • **Substance misuse**: e.g. amphetamine, cannabis	• **Borderline personality disorder**: transient paranoia but not as persistent as paranoid personality disorder. • **Schizoid personality disorder**: more indifferent • **Narcissistic personality disorder**: paranoia may occur as a result to the threat of imagined success. • **Avoidant personality disorder** – they tend to avoid other people as a result of lack of self-confidence or fear of embarrassment.

Management:
1. **Supportive psychotherapy or problem solving therapy.**
2. **Cognitive Behaviour Therapy** – CBT targets at core beliefs such as others are malicious and deceptive. Patient needs to realise that it will be alright if he or she reduces suspiciousness. In behaviour therapy, they will be trained to expect hostility and personal attacks in daily live. They are advised to record their ideas in the dysfunction though diary and the psychologist will train them to develop ability to gain controls over the sessions.
3. **Psychotropic medications** – Antidepressants and antipsychotics are indicated to treat mood and psychotic symptoms.

Course and prognosis:
1. Patients can be hypersensitive when they are young with poor peer relationships and being eccentric.
2. Paranoid ideas may intensify during stress.
3. Some patients may develop agoraphobia.

Histrionic personality disorder

Epidemiology

Prevalence	Gender ratio	Comorbidity
2-3%	M = F	Somatisation disorder Alcohol misuse

Aetiology

Psychological causes	• Extreme variant of temperamental disposition. • Emotionality (intensity and hypersensitivity), extraversion and reward dependence. • Tendency towards overly generalized cognitive processing.
Developmental causes	• Significant separations in the first 4 years of life. • Association with authoritarian or seductive paternal attitudes during childhood. • Low cohesion in family. • Favourism towards boys or male gender in a family (if patient is a woman) and leads to power imbalance. • Chronic physical illness in childhood. • Absence of meaningful relationships. • Deprived and or traumatic childhood. • May have families high in control, intellectual orientation, low in cohesion.
Defence	• Dissociation.

mechanisms	• Denial.

Clinical features

ICD 10 criteria

Affect:
1. Self-dramatization, theatricality or exaggerated expression of emotions
2. Shallow and labile affectivity.

Behaviour:
3. Suggestibility (he/she is easily influenced by others or circumstances).
4. Continual seeking for excitement and activities in which the individual is the centre of attention.
5. Inappropriate seductiveness in appearance or behaviour.

Cognition:
6. Overconcern with physical attractiveness.

People with histrionic personality disorder feel comfortable in situations where they are the centre of attention.

DSM-5 criteria

Salient features of DSM-5:
DSM-5 requires a fulfilment of ≥5 symptoms.

Additional affective symptom includes discomfort in situation in which he/she is not the centre of attention.

Additional behaviour symptoms include:
1. the consistent use of physical appearance to draw attention to self.
2. Excessively impressionistic style of speech.

Additional cognitive symptom includes the consideration of relationships to be more intimate than they actually are.

Differential diagnosis

Axis I disorders	Axis II disorders
• **Hypomania / mania:** characterised by episodic mood disturbances with grandiosity and elated mood. • **Somatisation / conversion disorder:** people with histrionic personality disorder are more dramatic and attention seeking. • **Substance misuse.**	• **Borderline personality disorder**: more self-harm, chaotic relations and identity diffusion. • **Dependent personality disorder** – more impairment in making important decisions. • **Narcissistic personality disorder** – people with narcissistic personality disorder need attention for being praised and they are very sensitive by humiliation when they are the centre of attention.

Treatment

1. **Dynamic psychotherapy** may be useful for people with histrionic personality disorder. Patients may see power and strength as a male attribute based on their childhood experience and feel inferior about one own gender and need to seek attention. Dynamic therapy can help the patients to analyse their deep seated views and understand how childhood experiences affect perception and personality development. The therapists should help the patient to build self-esteem.
2. **CBT** – challenge cognitive distortions and reduce emotional reasoning.
3. **Psychotropic medications** – SSRIs target at depressive symptoms.

Course and prognosis

1. In general, people with histrionic personality disorder have less functional impairments compared to other personality disorder.

2. Some people with histrionic personality disorder improve with age as a result of maturity.
3. Sensation seeking may lead to substance misuse.

Dependent personality disorder

Epidemiology

Prevalence	Heritability
Prevalence in the community: 1.0-1.7%.	0.57.

Aetiology
- Patients may have indulgent parents who prohibit independent activity.
- Twin studies suggest a biological component to submissiveness.
- Insecure attachment

Clinical features
ICD-10 criteria:

Affect:
1. Uncomfortable or helpless when alone due to exaggerated fears of inability to self-care.

Behaviour:
2. Encourages or allows others to make most of one's important life decisions.
3. Subordination of one's own needs to those of others on whom one is dependent, undue compliance with their wishes.
4. Unwilling to make even reasonable demands on the people one depends on.

Cognition:
5. Preoccupation with fears of being left to care for oneself.
6. Limited capacity to take everyday decisions without an excessive amount of advice and reassurance from others.

People with dependent personality disorder may have self-effacing diffidence, eagerness to please and importuning. They may adopt a helpless or ingratiating posture, escalating demands for medication, distress on separation, and difficulty in termination of psychotherapy.
Background history may reveal childhood separation anxiety, periods of separation from parents or presence of chronic illness.

Salient features of DSM-5:
1. DSM-5 has an additional criterion of disorder beginning in adulthood.
2. DSM-5 requires a fulfilment of ≥5 symptoms.

Additional behavioural symptoms include:
1. Difficulty initiating projects or doing things on his/her own
2. Goes to excessive lengths to obtain nurturance and support from others.
3. Urgently seeks another relationship for care and support when a close relationship ends.

Differential diagnosis:

Axis I disorders	Axis II disorders
1. Depressive disorder. 2. Agoraphobia. Social phobia.	1. **Borderline personality disorder:** both borderline personality disorder and dependent personality disorder share fear of rejection and abandonment. People with borderline personality disorder show more anger, emptiness and dramatic responses as compared to people with dependent personality disorder who are more submissive and clinging. People with dependent personality disorder want to be controlled but people with borderline personality disorder react strongly to the efforts to be controlled. People with borderline personality disorder show more rage and chaotic relationship. 2. **Avoidant personality disorder:** People with avoidant personality disorder show low self esteem, need for reassurance, high sensitivity for rejection. People with avoidant personality disorder react by avoiding while people with dependent personality disorder seek out for relationship. 3. **Histrionic personality disorder:** people with histrionic personality disorder

| | | are more seductive, flamboyant and manipulative to get attention. |

Treatment:
1. CBT and social skill training.
2. Therapy targeted at increasing self-esteem, self-confidence, sense of efficacy and assertiveness.
3. Explores fear of autonomy.
4. Family or couple therapy.

Impulse control disorders

In general, there is an impulse (e.g. gambling or fire-setting) and the person has difficulty to resist. Prior to the act, there is a build up of tension and the person seeks pleasure, gratification, or relief at the time of performing the act.

Impulse control disorders
Disruptive, impulse control and conduct disorder now includes:
1. Pyromania
2. Kleptomania
3. Other specified and unspecified disruptive, impulse control and conduct disorders

Disorders	ICD-10 diagnostic criteria	DSM-5
Pathological fire-setting (pyromania)	1. There are two or more acts of fire-setting without apparent motive. 2. The individual describes an intense urge to set fire to objects, with a feeling of tension before the act and subsequent relief. 3. The individual is preoccupied with thoughts, mental images and related matters such as an abnormal interest in fire-engines, fire stations and the fire service.	The DSM-5 criteria are very similar to the ICD-10. The DSM-5 criteria emphasise that fire setting is not done for monetary gain, expression of political views or anger, riot, to conceal criminal activity, false insurance claim for personal gain, under the influence of delusion or hallucination, or as a result of impaired judgment (e.g. in dementia or intellectual disability). Exclusion: The fire setting is not in the context of conduct disorder, manic episode, or antisocial personality disorder.
Pathological stealing (kleptomania)	1. There are 2 or more thefts in which the individual steals without any apparent motive of personal gain or gain for another person. 2. The individual describes an intense urge to steal, with a feeling of tension before the act and subsequent relief.	The DSM-5 criteria are very similar to the ICD-10 criteria. The DSM-5 emphasize that the stealing is not committed to express anger and the theft is not a response to a delusion or a hallucination. Exclusion: The theft is not in the context of conduct disorder, manic episode or antisocial personality disorder.

Multiple choice questions

1. Which of the following statements regarding diagnostic criteria for multiple personality disorder is false?

A. Based on the ICD-10 criteria, two or more distinct personalities exist within the individual
B. Based on the ICD-10 criteria, two or more distinct personalities are evident at a time
C. Based on the ICD-10 criteria, each personality has its own memories, preferences and behaviour patterns
D. Based on the ICD-10 criteria, the inability to recall important personal information is extensive
E. Multiple personality disorder is known as dissociative identity disorder in the DSM-IV-TR

Answer: B.
Based on the ICD-10 criteria, only one personality is evident at a time.

2. A 25-year-old man always believes that he is socially inept and fears of negative evaluation by others. He seems to be timid and insecure. Which of the following personality disorders best describes this person based on ICD-10 or DSM-IV-TR diagnostic criteria?

A. Anxious (avoidant) personality disorder
B. Anankastic personality disorder
C. Dissocial personality disorder
D. Emotionally unstable personality disorder – borderline type
E. Paranoid personality disorder

Answer: A.
Anxious (avoidant) personality disorder is characterized by persistent apprehension and tension; unwillingness to be involved with people unless certain of being liked; restricted lifestyle as a result of the need for physical security; avoidance of social or occupational activities; feeling socially inept and excessive preoccupation with facing criticism.

3. Which of the following symptoms does not belong to borderline personality disorder?

A. Chronic feelings of emptiness
B. Disturbances in and uncertainty about self-image
C. Excessive efforts to avoid abandonment
D. Impulsivity that are antisocial
E. Liability to become involved in intense and unstable relationships

Answer: D
Option D is not part of the diagnostic criteria of borderline personality disorder.

4. A 40-year-old man wants to transfer from another hospital to your hospital. He is admitted to the psychiatric ward because of low mood. He complains to the ward manager that the inpatient service is not up to standard. After discharge, he has written 15 letters to the CEO about the delay in psychologist appointment. Which of the following personality disorders is the most likely diagnosis?

A. Anankastic
B. Anxious (avoidant)
C. Antisocial
D. Borderline
E. Narcissistic

Answer: E
This man suffers from narcissistic personality disorder which is characterized by sense of entitlement and importance.

5. Which of the following personality disorders has the highest admission rate?

A. Anxious

B. Borderline
C. Dissocial
D. Dependent
E. Histrionic

Answer: B
The rates are as follows: A: 5%; B: 52%; C: 13%; D: 2%; and E: 2%.

EMIS:
A – Obsessive-compulsive personality disorder
B – Paranoid personality disorder
C – Schizotypal personality disorder
D – Hysterical personality disorder
E – Antisocial personality disorder
F – Narcissistic personality disorder
G – Borderline personality disorder
H – Dependent personality disorder
I – Avoidant personality disorder
J – Schizoid personality disorder

Question 1: 40 year old male has always been extremely neat and conscientious. He often stays long after normal working hours to check on errors. – A

Question 2: 40 year old male refused to provide answers to standard questions during an initial clerking and threatens to stop the interview if recording his telephone number is insisted upon – B

Question 3: 30 year old night security at a local hospital prefers to be alone whenever possible. He has no friends and does not socialize. He does not keep update with current affairs and has no sexual interest. He spends most of his time day-dreaming - J

References

Lackwood K (1999) Psychodynamic psychotherapy (Chapter 7) in Stein S, Hiagh R & Stein J (1999) *Essentials of psychotherapy*. Oxford: Butterworth Heinemann.

Lee KM, Chan HN, Cheah B, Gentica GF, Guo S, Lim HK, Lim YC, Noorul F, Tan HS, Teo P, Yeo HN. Ministry of Health clinical practice guidelines: management of gambling disorders. Singapore Med J. 2011 Jun;52(6):456-8

Turner T (2003) *Your questions answered: depression*. London: Churchill Livingstone.

NICE guidelines for borderline personality disorder http://guidance.nice.org.uk/CG78

Palmer RL (2002) Dialectical behaviour therapy for borderline personality disorder. *Advances in Psychiatric Treatment*. **8**: 10-16.

Puri BK, Treasaden I (eds) Psychiatry: An Evidence-Based Text. London: Hodder Arnold, 2010.

American Psychiatric Association. Diagnostic Criteria from DSM-IV-TR. Washington: American Psychiatric Association, 2000.

American Psychiatric Association (2013) DSM-5: Diagnostic and Statistical Manual of Mental Disorders (Diagnostic & Statistical Manual of Mental Disorders) (5th edition). American Psychiatric Association Publishing Inc: Washington.

World Health Organization. ICD-10 Classification of Mental and Behavioural Disorders. Edinburgh: Churchill Livingstone, 1994.

Dasgupta P, Barber J. Admission patterns of patients with personality disorder. *Psychiatric Bulletin* 2004; **28**: 321–323.

Chapter 7 Substance misuse and dependence

Terminology	Solvents, sedatives, caffeine and steroids
Alcohol misuse and dependence	Sample MCQS
OSCE: alcohol dependence	Sample MEQS
Opioid misuse and dependence	Sample EMIs
Stimulants and hallucinogens	
Nicotine	

Terminology used in addiction medicine

Substance related disorders in Psychiatry refer to the inappropriate usage of compounds that would be capable of inducing changes in the normal functioning of the cognitive system. These changes typically include changes in mood, behaviour as well as cognitive capabilities.

Commonly used terminology:
1. **Dependence** – Defined as the maladaptive usage of a chemical or substance that has led to significant impairment or distress.
   It is sub-classified into 2 different types of dependency – physiological dependency and without physiological dependency. Those individuals with physiological dependency would exhibit signs of tolerance or withdrawal.
2. **Abuse** – Typically defined as the maladaptive usage of a chemical or substance in a way that differs markedly from social norms
   (Please note that in the new DSM-5, there is no sub-classification into dependence or abuse.)
3. **Misuse** – Definition similar to that of abuse; but typically applies only to substances that are administered by physicians.
4. **Addiction** – Defined as the repeated and increased usage of a chemical or substance, to an extent that the discontinuation of chemical or substance would lead to significant physiological and psychological disturbances. There is also an inherent urge to continue the usage of the substance in order to avoid the physiological and psychological disturbances.
5. **Intoxication** – Defined as a transient, reversible syndrome during which specific aspects of cognition, behavioural and social functioning are compromised as a result of usage of a particular chemical or substance.
6. **Withdrawal** – A constellation of symptoms that typically is substance specific and occurs when a particular substance has been used regularly and discontinued abruptly. There are usually both psychological and physiological disturbances.
7. **Tolerance** – Defined as either the need for an increased dosage of a drug to achieve the same clinical response achieved previously; or a diminished clinical response achieved with the same routine dosage of the drug.

    Substances that are typically associated with the above-mentioned disorders:
    1. Alcohol
    2. Amphetamine
    3. Caffeine
    4. Hallucinogens
    5. Inhalants
    6. Nicotine
    7. Opioid
    8. Phencyclidine (PCP)
    9. Sedatives, hypnotics and anxiolytics

A more in-depth discussion of the relevant disorders would be further explored.

Alcohol misuse and dependence

Epidemiology

Singapore prevalence of Alcoholic misuse:

Based on the latest Singapore Mental Health Study on Alcohol dependency and abuse, the lifetime prevalence of alcohol abuse and alcohol dependence was 3.1% and 0.5%, while the 12-month prevalence of alcohol abuse and alcohol dependence was 0.5% and 0.3%, respectively.

Aetiology of alcohol misuse:

The aetiology of alcohol misuse is multifactorial and best understood in a biopsychosocial model.

Genetic causes:

Twin studies

	MZ: DZ ratio:
Male alcohol abuse	70%: 45%
Male alcohol dependence	40%: 15%
Female alcohol abuse	50%: 40%
Female alcohol dependence	30%: 25%

Acetaldehyde dehydrogenase (ALDH)
The genotype ALDH2 504Lys are present in 40% of East Asians. High activity isoforms of ALDH2 causes alcohol flushing reactions among Orientals and lead to unpleasant effects after drinking.

The roles of nucleus accumbens and pleasure seeking contribute to the prevalence of misuse and dependency as well.

Cloninger's theory
Type 1: Onset> 25 years, common in both men and women, loss of control, guilt, no forensic history, no family history, greater ability to abstain.

Type 2: Onset < 25 years, common in men but rare in women, inability to abstain, antisocial behaviour and with family history.

Alcohol misuse and dependence

Learning theory
1. Temporarily reduces fear or conflict through classical conditioning
2. Gains from alcoholic behaviour and operant conditioning.
3. Modeling (from people who also misuse alcohol)
4. Social learning

Personality
1. Antisocial personality disorder

Psychodynamic theories:
1. Oral fixation
2. Introjections of anger and slow suicide

Childhood difficulties and psychiatric disorders
1. ADHD
2. Conduct disorder
3. Victims of sexual abuse
4. Family development: disruption to families with- the use of alcohol by parents and siblings

Psychiatric disorders in adulthood:
1. Affective disorders
2. Anxiety disorders

Environmental factors
1. Marital or relationship problems (single or divorced people are more prone)
2. Changing gender roles in females (Men is still 6 times higher in risk)
3. Migration
4. Stress at work and vulnerability of certain occupations (e.g.. sales persons, hoteliers, brewers, bar personnel, entertainment industry, journalists, police, the armed forces and medical practitioners).
5. Alienation and social isolation
6. Poor income
7. Poor education
8. Poor awareness of the dangers of alcohol
9. Consumption during entertainment.
10. Advertising promotion

National and cultural influences

1. **Cultural attitude to drunkenness** (social lubricant, celebratory" use)
2. **Cultural patterns of alcohol consumption.**
3. **Use of alcohol under peer group pressure**
4. **Religion:** (the traditional Catholic "allowance" of the use of alcohol, the Methodist restriction of alcohol, the low consumption among the Jews and the forbiddance of its use in Islamic religions)

Assessment of alcohol misuse – Moderate/Hazardous/ Harmful/ Dependent drinkers

Summary of safe, hazardous and harmful levels of alcohol consumption and recommendations from the UK government on sensible drinking

	Safer levels (there is no such thing as a 'safe' level but levels below indicate 'safer' levels	Hazardous levels	Harmful levels	Dependent levels	Sensible drinking	Against heart disease
Men	21 units per week	21-35 units per week	35-50	> 50 units per week	Max 3-4 units/day	2 units/day for men>40 years
Women	14 units per week	14-21 units per week	21-35	> 35 units per week	Max 2-3 units/day	2 units /day after menopause

The lower limits set for women are due to women's higher ratio of fat to water and hence inability to dilute the alcohol consumed. Men who drink more than 30 units per week are associated with 3.5 times increase in mortality rate at 15 years. Women are more vulnerable to alcohol-related liver disease.

Moderate Drinkers: These could be defined as people drinking at or below the identified limits in a safe pattern.
Hazardous Drinkers: These are people drinking above the safe limits without the occurrence of problems or dependency.
Harmful Drinkers: These are people drinking above the safe limits with the occurrence of problems but without established dependency.
Dependent Drinkers: These are people drinking above the safe limits with problems and established dependency.

Explanation on alcohol withdrawal syndrome (AWS)

Alcohol withdrawal
The DSM-5 diagnostic criteria specified that there must be cessation or reduction in the alcohol usage that has been heavy and prolonged previously.
The criteria specified that at least 2 of the following signs and symptoms must have developed within several hours to few days after stopping the usage of alcohol.
a. Sweating or having tachycardia (autonomic hyperactivity)
b. Increased hand tremor
c. Difficulties falling asleep
d. Nausea or vomiting
e. Transient visual, tactile or auditory hallucinations or illusions
f. Psychomotor agitation
g. Anxiety
h. Generalized tonic-clonic seizures

1. AWS occurs when the hyperactive brain cannot adjust to the sudden drop in blood alcohol concentration.
2. Criteria for AWS include:
    - Tremor of the hands and tongue
    - Sweating
    - Nausea
    - Retching or vomiting
    - Tachycardia
    - Psychomotor agitation
    - Headache
    - Insomnia
    - Malaise

- Transient hallucination
- Grand mal convulsions.

Time of onset: within hours of last drink and peak for DTs is within 24-48 hours

Explanation on delirium tremens (DTs)
1. DTs is a toxic confusion state when AWS are severe.
2. Clinical presentation involves a triad of symptoms:
   1) Clouding of consciousness and confusion; 2) Vivid hallucinations (visual); 3) Marked tremor.
3. In addition to confusion and marked tremor, the above person demonstrates autonomic instability, paranoid delusions, agitation and sleeplessness (with REM sleep rebound) which are common symptoms in DTs.
4. The potential life threatening aspect of DTs: mortality is 5%.

Physical complications of alcohol misuse:

Neurological complications:
Chronic alcohol dependence damages the spinocerebellum and leads to sensory ataxia (failure in the heel-shin test) with depressed deep tendon reflex. This will increase the risk to fall.

Demyelination: optic atrophy, central pontine myelinolysis (bulbar palsy, locked in syndrome, quadriplegia, loss of pain sensations, fatal), Marchiafava-Bignami syndrome (ataxia, dysarthria, epilepsy, impairment of consciousness, spastic paralysis and dementia).

Retrobulbar neuropathy: loss of central vision (e.g. bilateral central scotoma) and preventable with vitamin B replacement.

Upper GIT: Cancer of mouth, pharynx and oesophagus. Mallory-Weiss tear leads to upper GI bleeding.

Lung: Increase risk of pneumonia and tuberculosis, respiratory depression

Liver: Impairments of liver function, ↑GGT and ↑MCV (PPV: 85%), AST:ALT > 2:1 = alcohol hepatitis (AST: ALT <1= viral hepatitis), cirrhosis (more common in female) and liver cancer, hepatic encephalopathy and hepatocerebral degeneration.

Pancreas: Pancreatitis, hypo or hyperglycaemia.

Peripheral: Tremor, peripheral neuropathy, acute proximal myopathy, hypo or hyperthermia and palmar erythema.

Cognition: alcohol damages the mammillary bodies, the floor of the fourth ventricle and cerebral atrophy leading to memory impairments.

Thiamine deficiency will lead to **Wernicke's encephalopathy** (ataxia, ophthalmoplegia, memory disturbance, 10% have classical triad, hypothermia and confusion) 80% of untreated Wernicke's encephalopathy will convert to **Korsakoff's psychosis** (lesions in periventricular and periaqueductal gray matter, with extensive retrograde amnesia and implicit memory is preserved).

Sexual dysfunction: In men, alcohol misuse may lead to erectile dysfunction. In women, consumption of alcohol during pregnancy will give rise to foetal alcohol syndrome

Neuropsychiatric phenomenon:
Change in personality, irritability, depression and anxiety. Morbid jealousy may occur in men secondary to erectile dysfunction.
Alcoholic hallucinosis occurs in clear sensory during abstinence. It persists 6 months after abstinence and occurs in 5-10% of cases. It involves unpleasant sound or threatening voices but absence of thought disorder or mood incoherence. It has good prognosis and shows rapid response to antipsychotic.

Idiosyncratic alcohol intoxication (mania a potu): aggression, psychosis and delirium occur within minutes after intoxication of alcohol.

Head injury results in neuropsychiatric symptoms

Seizure (10%). Tonic clonic seizure and seizure threshold is reduced.

Heart: Coronary artery disease, cardiomyopathy, heart failure, arrhythmia and macrocytic anaemia
Hypertension increases the risk of CVA. Hypotension is possible.

Stomach: inflammation and gastritis.

Intestine: inflammation, diarrhoea, malnutrition and vitamin deficiencies (e.g. Vitamin B_1), malabsorption syndrome,

Urea and electrolytes:
↓Na, ↓ Mg^{2+}, ↓PO^{4+}
↑Ca^{2+}

The top four causes of early death in people who are dependent on alcohol are heart disease, cancer, accident and suicide.

Alcohol decreases sleep latency, REM sleep and stage IV sleep. Alcohol causes more sleep fragmentation and longer episodes of awakening.

The ICD – 10 and DSM-5 Criteria

Categories	ICD – 10 criteria (Aide memories: 2Cs + withdrawal + tolerance +2Ps)
Salient features	Harmful and hazardous drinking or substance abuse which include drinking behavior as the sole criteria for the disorder. This applies to countries with wide range of acceptability and consequences of substance misuse across cultures in the world.
Number of criteria met and duration	Three or more of the following manifestations should have occurred together for at least 1 month. If the manifestations persist for less than 1 month, they should have occurred together repeatedly within a 1 year period.
Compulsion (C)	A strong desire or sense of compulsion to consume alcohol.
Control (C)	Impaired capacity to control drinking in terms of its onset, termination, or levels of use. Alcohol is being often taken in larger amounts or over a longer period than intended. There is persistent desire to or unsuccessful efforts to reduce or control alcohol use.
Withdrawal	Physiological withdrawal state: shaky, restless or excessive perspiration. The severity of withdrawal is related to the amount, duration and pattern of use.
Tolerance	There is a need for significantly increased amounts of alcohol to achieve intoxication or the desired effect, or a markedly diminished effect with continued use of the same amount of alcohol.
Preoccupation (P)	Important alternative pleasures or interests being given up or reduced because of drinking; or a great deal of time being spent in activities necessary to obtain, take, or recover from the effects of alcohol.
Persistence (P)	Persistent alcohol use despite clear evidence of harmful consequences even though the person is actually aware of the nature and extent of harm.

Alcohol use disorder

The DSM-5 diagnostic criteria state that there must be a problematic usage of alcohol that has led to significant impairments occurring over a total of 12 months duration. The problematic usage of alcohol is manifested by at least 2 of the following:
a. Increasing usage of alcohol, or over a longer period than originally intended
b. Repeated unsuccessful efforts to cut down or control usage, despite the desire to do so.
c. Large amount of time is spent on activities to obtain alcohol, use alcohol, or recover from the effects of alcohol.
d. Presence of a strong desire or urge to use alcohol
e. Repeated alcohol usage resulting in a significant failure to fulfill major roles
f. Persistent usage of alcohol despite having recurrent social or interpersonal problems due to the usage of alcohol.
g. Important activities are given up due to the usage of alcohol
h. Repeated usage despite significant impairments in physical health
i. Continued use despite knowing that there have been physical or psychological problems arising from the usage of alcohol
j. Tolerance as defined by either (1) Need for increasing amounts of alcohol to achieve the same or desired effects or (2) Reduced effects with continued use of the same amount of alcohol
k. Withdrawal as defined by either (1) Characteristic withdrawal symptoms (see above section on AWS) or (2) Alcohol is being used to prevent or avoid withdrawal symptoms.

Clinicians might need to specify alcohol usage:
a. In early remission - when none of the criteria for alcohol use has been met for 3 months or more but less than 12 months.
b. In sustained remission - when none of the criteria for alcohol use has been met for the last 12 months or longer.
c. In a controlled environment - when the individual is in an environment where access to alcohol is restricted

Acute intoxication due to use of alcohol [ICD-10 criteria]

Dysfunctional behaviour	Signs
1. Disinhibition 2. Argumentativeness 3. Aggression 4. Lability of mood 5. Impaired attention 6. Impaired judgement 7. Interference with personal functioning	1. Unsteady gait 2. Difficulty in standing 3. Slurred speech 4. Nystagmus 5. Decreases level of consciousness 6. Flushed face 7. Conjunctive injection

Alcohol intoxication

The DSM-5 diagnostic criteria specified that there must be recent usage of alcohol and that clinically significant behavioural or psychological changes have arisen during or shortly after the usage.

This is characterized by at least one of the following signs or symptoms:

a. Slurred speech

b. Incoordination

c. Unsteady gait

d. Nystagmus

e. Impairments in attention or memory

f. Stupor or coma

Assessment Questionnaire for Alcohol addiction: Alcohol Use Disorders Identification Tool (AUDIT)

> **AUDIT questionnaire:**
>
> **Introduction:** Now I am going to ask you some questions about your use of alcoholic beverages during this past year. Please explain what is meant by "alcoholic beverages" by using local examples of beer, wine, vodka.
>
> **Hazardous alcohol misuse:**
> 1. Frequency of drinking: "How often do you have a drink containing alcohol?"
> 2. Typical quantity: "How many drinks containing alcohol do you have on a typical day when you are drinking?"
> 3. Frequency of heavy drinking: "How often do you have six or more drinks on one occasion?"
>
> **Dependence symptoms:**
> 4. Impaired control over drinking: "How often during the last year have you found that you were not able to stop drinking once you had started?"
> 5. Increased salience of drinking: "How often during the last year have you failed to do what was normally expected from you because of drinking?"
> 6. Morning drinking: "How often during the last year have you needed a first drink in the morning to get yourself going after a heavy drinking session?"
>
> **Harmful alcohol use**
> 7. Guilt after drinking: "How often during the last year have you had a feeling of guilt or remorse after drinking?"
> 8. Blackouts: "How often during the last year have you been unable to remember what happened the night before because you had been drinking?"
> 9. Alcohol-related injuries: "Have you or someone else been injured as a result of your drinking?"
> 10. Others concerned about drinking: "Has a relative or friend or a doctor or another health worker been concerned about your drinking or suggested you cut down?"
>
> **AUDIT scores and management:** 0-7: Alcohol education; 8-15: Simple advice; 16-19: Simple advice, brief counseling and continued monitoring; 20-40: Referral to specialist for diagnostic evaluation and treatment. A high AUDIT score is strongly associated with suicidality.

Treatment of Alcohol Dependence / Alcohol Misuse

Non-Pharmacological Methods
1. Cognitive Psychotherapy – Involves assessment of the patient's readiness for behavioural change using the stage of change model and also make use of motivation interviewing to enable patients to be empowered to change.

The stages of change model developed by Prochasaka and DiClemente:

Pre-contemplation	Patient does not recognise or accept the diagnosis yet and does not see the need for change.
Contemplation	Patient now recognize the alcohol misuse and is considering the pros and cons of treatment
Decision	The patient has decided to quit alcohol
Action	The patient has made the change to quit and the change has been integrated into the patient's life.
Maintenance	The changes have been integrated into the patient's life. He or she has been abstinent from alcohol
Relapse	Restart of addictive pattern of use after previous abstinence
Motivational interviewing	Analysis of the pros and cons of continued drinking

Motivational interviewing would be highly effective, if the following key elements are used in the therapeutic session. The therapist must be able to listen effectively during the psychotherapy process and allow the patient to give their inputs without interrupting the patient. The therapist must focus on the usage of open-ended questions during the interview, and reflect upon what the patient has vocalized and focus on collaboration with the patient to set common achievable goals. Appropriate reinforcement must be given and key misconceptions should be clarified during the therapy session. Whenever the patient demonstrates positive intentions, or thoughts or specific behavior, it is crucial that the therapist reinforce the patient for these behaviours and thought processes. The key for a successful motivational interview is also to avoid advice giving. The patient must be empowered to take charge of his issues and problems.

2. Cue exposure
   Cue exposure is a behavioural concept that uses classical conditioning to help patients to avoid the cues that might predispose them to further episodes of drinking.

3. Alcoholic Anonymous movement
   A self-help group, first formed in America, with the sole purpose of enabling patients to quit their drinking habits. Their program is based on 12 core principles:

   12 Steps:
   1. We admitted we are powerless over alcohol - that our lives had become unmanageable
   2. Came to believe that a power greater than ourselves could restore us to sanity
   3. Made a decision to turn our will and our lives over to the care of God as we understood him
   4. Made a searching and fearless moral inventory of ourselves
   5. Admitted to God, to ourselves, and to another human being the exact nature or our wrongs.
   6. Were entirely ready to have God remove all these defects of character
   7. Humbly ask Him to remove our shortcomings
   8. Made a list of persons we had harmed, and became willing to make amends to them all
   9. Made direct amends to such people whenever possible, except when to do so would injure them or others
   10. Continued to take a personal inventory, and when we are wrong, promptly admit to it.
   11. Sought through prayer and mediation to improve our conscious contact with God as we understood Him praying only for knowledge of His will for us and the power to carry that out.
   12. Having had a spiritual awakening as a result of these steps we tried to carry this message to alcoholics and to practice these principles in our affairs.

Pharmacological treatment

(a) Disulfiram (Starting dose 800mg and then reduce to 100-200mg daily)

1. The mechanism of actions of disulfiram is to inhibit aldehyde dehydrogenase and lead to acetaldehyde accumulation after drinking alcohol, resulting in unpleasant effects (such as flushing, tachycardia and hypotension). The aversion effect occurs 10-30 minutes after drinking and is dose dependent. The reaction to alcohol discourages the person from drinking and reduces the number of days spent on drinking. Patient has to be informed that the reaction may last for 1 week. Alcohol should be avoided for 1 week after cessation of alcohol.
2. Addiction specialist should ensure that the person has not consumed any alcohol at least one day before starting disulfiram. It is contraindicated in people with advanced liver diseases. It also increases level of warfarin, diazepam, diazepam and theophylline.
3. Side effects include halitosis (common side effect, bad breath from oral cavity), nausea, reduction in libido, peripheral neuritis and liver damage.
4. Dangerous side effects include arrhythmia, hypotension and collapse and hence, disulfiram is not commonly used.
5. Contraindications include cardiac failure, coronary artery disease and cerebrovascular disease, pregnancy and breastfeeding.

(b) Acamprosate (Starting dose 666mg TDS for adult with body weight ≥ 60 kg)

The mechanism of action of acamprosate

The withdrawal of alcohol (see the diagram on the right) enhances the actions of glutamate at the NMDA receptors and the voltage-sensitive Ca^{2+} channels. This will attenuate the action of GABA at inhibitory $GABA_A$ receptors, resulting in agitation and craving during alcohol withdrawal.

Acamprosate is a GABA agonist and glutamate antagonist. It inactivates the NMDA receptors and prevent the Ca^{2+} influx. This will reverse the GABA and glutamate imbalance when abstaining from alcohol and reduce the long-lasting neuronal hyperexcitability.

Acamprosate can increase the likelihood of abstinence by reducing craving in people with chronic alcohol dependence. It should be started as soon as possible during abstinence. Benefits will continue 1-2 years after stopping acamprosate.

The person is allowed to have only one relapse while they are taking acamprosate. If there is more than one relapse, the psychiatrist should advise the person to stop acamprosate.

Side effects: diarrhoea, nausea, rash, bullous skin reactions and fluctuation in libido.

Acamprosate can be combined with disulfiram. It is not licensed for elderly.

Contraindications include: 1) severe renal or hepatic impairment, 2) pregnancy, 3) breastfeeding.

(c) Naltrexone (Starting dose: 25mg and increase to 50mg per day on weekdays and 100mg per day on weekend)

The mechanism of actions of naltrexone on the opioid receptors

Opioid receptors are responsible for reward, and results in craving.

Naltrexone is an opioid antagonist. Alcohol becomes less rewarding when opioid receptors are blocked and the person does not experience euphoric effects.

It is not licensed to treat alcohol dependence in the UK due to mortality after overdose and potential withdrawal. It is only used in people who are in abstinence and highly motivated.

It is used to treat violent behaviours among people with learning disability.

(d) **Diazepam**

Summary of dosage:

Mild dependence	Moderate dependence	Severe dependence
Small dose of diazepam is enough.	Larger dose of diazepam is required and treatment is over 5-7 days: Day 1-2: 5 mg TDS Day 3-4: 2.5 mg TDS Day 4-5: 2.5 mg BD Day 6-7: 2.5 mg ON	Very high dose of diazepam is required and treatment is over 10 days: Day 1-2 : 10mg QDS Day 3-4: 10 mg TDS Day 5-6: 5 mg TDS Day 7-8: 2.5mg TDS Day 9-10: 2.5mg ON

(e) **Other medications used in the treatment of alcohol withdrawal**:

Conditions	Medications
Liver cirrhosis	Short active benzodiazepine such as alprazolam (available in Singapore).
Prophylaxis or treatment for Wernicke's encephalopathy	IM high potency B – complex vitamins daily for 3-5 days and followed by oral vitamin B compound treatment. Ophthalmoplegia is the first sign to respond.
Seizure prophylaxis	Diazepam
Hallucinations	Haloperidol.

*OSCE: Alcohol dependence

Clinical Vignette: A 40 year-old man was admitted to the medical ward with minor head injury after he was drunk. Routine blood tests showed increased GGT and MCV. The physicians have sent a referral to you because the patient also accuses his wife of having an affair with another man which is not true.

	Introduction	Establishment of drinking habits	Establishment of average alcohol consumption	Establishment of social factors influencing drinking habits
Introduction	Introduction and establishment of therapeutic relationship with the patient. - Hi, I'm Dr. Michael. I understand that you have been referred from your physician as you have had some blood abnormalities. There is also some concern about your relationship with your wife. - Could we spend some time to explore that in further details?	- Could you tell me when you first tasted alcohol? - Could you also tell me when you started drinking occasionally and regularly at weekends, evenings, lunchtimes and in the mornings?	- Can you tell me on average how much do you drink everyday? - What do you usually drink? - What else do you drink?	- Do you usually drink alone or with your friends? - Do you have a tendency to indulge in more alcohol when you are drinking with your friends? - Do you always drink in the same hawker centre? - Do you always drink with the same company?
Tolerance and Withdrawal	Establishment of tolerance - Nowadays, do you need more alcohol to get drunk than what you needed in the past? - What is the maximum you have drunk in a day? - How much can you drink without feeling drunk?	Establishment of withdrawal effects - What happens if you miss your drink? - What would happen if you go without a drink for a day or two?	Establishment of physical effects of alcohol withdrawal - Do you feel shaky or sweat a lot?	Elicit hallucinations - Were there times when you were seeing or hearing things when you could not have your usual amount of alcohol?

Motivation to stop drinking (CAGE Questionnaire)	C – Cut down drinking, requested by family members. - Has anyone in your family advised you that you need to cut down on drinking?	A – Annoyed by family members - Do you feel that your family members are displeased with regards to your current drinking problem? - Have they told you that they are irritated by your drinking issues?	G – Guilty of drinking - Do you feel any guilt with regards to your current drinking issue?	E – Eye opening in morning and then start to drink - Do you find yourself having to resort to alcohol as an eye opener, or to kick start a day?
Complications of drinking	Family and social issues - Have you had issues with your family because of your drinking habits?	Work and financial issues - Has your drinking habit got you into issues with your work? - Do you have any problems currently financing your drinking habit?	Forensic issues - Have you got yourself into trouble because of your drinking? Do you have issues with drunk driving, drunk and disorderly behavior in the public or ended up in fights when you were drunk?	Relationship issues - Have there been any problems with your existing marital relationship?
Treatment & Motivational interviewing	Treatment - Have you undergone any specific treatment previously for your alcohol issues?	Relapse after Treatment - How long have you been successful without relying on alcohol? - Could you tell me more as to why you started drinking again? - When you restarted, how long did it take you before you were back at your normal level of consumption/	Assessment for suitability currently to quit drinking - Do you feel you have a problem with alcohol? - Have you ever thought of giving it up completely? - What do you think will happen if you give up completely?	

Opioid misuse and dependence

Opioid use disorder
The DSM-5 diagnostic criteria state that there must be problematic usage of opioid that has led to significant impairments in terms of functioning over a 12 month period.
The rest of the criteria are similar to that of alcohol use disorder.

Opioid Intoxication
The DSM-5 states that there must be recent usage of opioid, with the presence of pupillary constriction (or pupillary dilation due to anoxia from severe overdose) and at least 1 of the following signs and symptoms:
a. Feeling drowsy or losing consciousness
b. Slurring of speech
c. Impairments in attention or memory
There must also be significant problematic behavioural or psychological changes that have arisen during or shortly after the usage.

Opioid Withdrawal
The DSM-5 diagnostic criteria state that there is either (a) recent cessation of the usage of opioid that was previously heavy and prolonged or (b) recent administration of an opioid antagonist after a period of opioid usage.
This is manifested by at least 3 or more of the following, which develops within minutes to several days:
a. Mood changes - dysphoria
b. Gastrointestinal disturbances which include nausea or vomiting
c. Muscular aches
d. Lacrimation or rhinorrhea
e. Pupillary dilation, piloerection or sweating.

Epidemiology:

In the United States, the estimated prevalence rate of opioid misuse and dependence is 2%. It is most common in individuals between the ages of 30-40s, with males being more predominant (3:1) as compared to females.
In the Asian context (Hong Kong), the estimated prevalence of abuse of opioid has been estimated to be 52.9% in 2011 out of the total proportion of drug abusers.

Classification of opioid:
They could be divided into agonists, antagonists and also partial agonists.
Agonists would bind and directly activate the specific receptors. Examples include morphine, methadone, fentanyl.
Antagonists bind to but do not active the specific receptors. Examples include Naltrexone and Naloxone.
Partial agonists bind to the receptors but only activate them to a limited extend. Common examples include Buprenorphine.

Neurochemistry:
Opioid works via interactions with the mu and delta receptors. There would be resultant increased activity in the mesolimbic system and also increased dopamine released.

Overview of the mechanisms of actions of opioid and the mu receptors

Mu receptors are potassium – channel linked and inhibits adenylate cyclase. Morphine activates mu receptors preferentially.

The binding of morphine to the mu receptors inhibits the release of GABA from the nerve terminal, reducing the inhibitory effect of GABA on the dopaminergic neurones.

The increased activation of dopaminergic neurones in the nucleus accumbens and the ventral tegmental areas which are part of the brain's 'reward pathway" and the release of dopamine into the synaptic results in sustained activation of the post-synaptic membrane.

Continued activation of the dopaminergic reward pathway leads to the feelings of euphoria and the 'high' or disinhibition, impaired attention and judgement, interference with personal functioning associated with heroin use.

Route of administration:
The routes of administration depend on the properties of the individual drug. Opium would be commonly smoked. Heroin is usually injected either via the subcutaneous route or via the intravenous route. In recent years, snorting of heroin has increased in incidences.

Common psychiatric disorders associated with opioid misuse:
Some common psychiatric disorders associated with opioid misuse include opioid intoxication delirium, opioid induced psychotic disorder, and opioid induced mood disorder. Opioid induced sexual dysfunction, opioid induced sexual dysfunction and opioid induced sleep disorder.

Signs of intoxication and tolerance:

1. Typical behaviour or psychological changes following intoxication includes:
a. Initial euphoria followed by feelings of apathy, dysphoria, psychomotor agitation or retardation, and associated with impaired judgment, social or occupational functioning.
b. Pupillary constriction, or at times, pupillary dilation due to anoxia from severe overdose.
c. Drowsiness or coma, slurred speech and impairments in attention and memory, which occurs when a massive overdose of opioid occur.

Overview of physical complications associated with opioid misuse

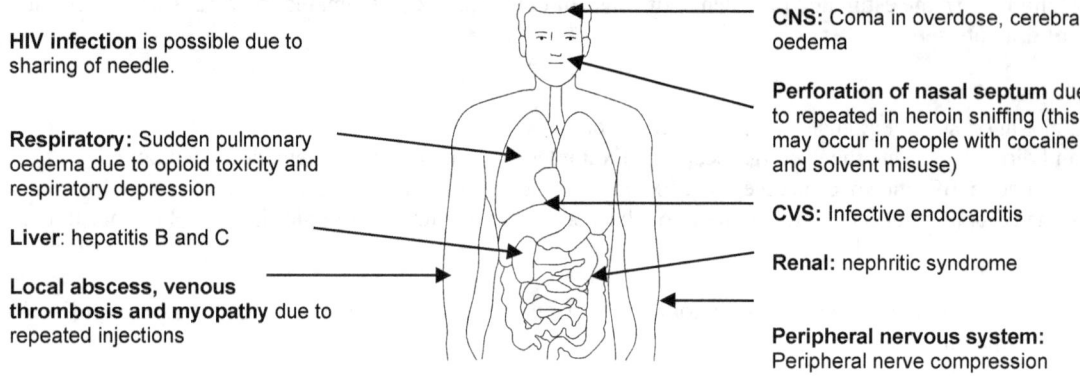

HIV infection is possible due to sharing of needle.

Respiratory: Sudden pulmonary oedema due to opioid toxicity and respiratory depression

Liver: hepatitis B and C

Local abscess, venous thrombosis and myopathy due to repeated injections

CNS: Coma in overdose, cerebral oedema

Perforation of nasal septum due to repeated in heroin sniffing (this may occur in people with cocaine and solvent misuse)

CVS: Infective endocarditis

Renal: nephritic syndrome

Peripheral nervous system: Peripheral nerve compression

2. Typical behaviour or psychological changes following development of tolerance includes:
   a. Development of feelings of euphoria, or even sedation.
   b. Nausea and vomiting.

Withdrawal symptoms:

During the initial stages of withdrawal (which refers to within hours of the last dose), there would be intense craving for the drug, lacrimation (tearing), rhinorrhea (running nose), yawning and diaphoresis (Sweating). Individuals in the latter stages of withdrawal (within 12 hours to 72 hours of the last dose) would experience mild to moderate sleep disturbance, mydriasis, anorexia, piloerection, increased irritability and tremors. Following this stage, individuals might experience even more withdrawal symptoms that would include severe insomnia, violent yawning, weakness, nausea and vomiting, chills and fever, flushing, spontaneous ejaculation and abdominal pain.

Treatment

(a) **Biological treatment**
Opioid substitutes could be used in treatment. Methadone is commonly used and the target is to maintain patients on daily doses of 60mg or less. The starting dose is 10-30mg per day and it would be gradually increased by 5-10mg per week to achieve the dose range that is appropriate. Contraindication to Methadone would include individuals with QTc prolongations.

Buprenorphine, a partial opioid agonist is currently also in use. It can be given just 3 times a week in view of it being much longer acting. The target dose range is between 8-32mg/day for individuals. It is contraindicated in individuals with a history of hypersensitivity and individuals who are consuming alcohol or taking sedative hypnotics such as benzodiazepines.

(b) **Psychological treatment.**
Therapeutic group therapy in a structured environment is helpful for helping individuals with their addiction. Psycho-education with regards to their possibility of HIV transmission when using shared needles would be helpful.

Stimulants (cocaine, amphetamine, cannabis) and hallucinogens

Common stimulants that are abused include cocaine, amphetamine and cannabis.
Stimulants are commonly referred to as substances that would induce cardiovascular stimulation, elevated mood and reduction in the need of sleep.
Hallucinogens are substances that would produce hallucinations, loss of contact with reality, and an experience of expanded or heightened consciousness.

Stimulant use disorder
The DSM-5 diagnostic criteria state that there must be problematic usage of amphetamine-type substance, cocaine, or other stimulant leading to significant impairments in terms of functioning over a 12 month period.
The rest of the criteria are similar to that of alcohol use disorder.

Stimulant intoxication
The DSM-5 diagnostic criteria specified that there must be recent usage of amphetamine-type substance, cocaine or other stimulant that has led to significant impairments in functioning, shortly after usage. This is manifested by at least 2 of the following signs and symptoms:
a. Increase or decrease in heart rate
b. Pupillary dilation
c. Elevated or lowered blood pressure
d. Perspiration or chills
e. Nausea or vomiting
f. Evidence of weight lost
g. Psychomotor agitation or retardation
h. Muscular weakness, respiratory depression, chest pain or cardiac arrhythmias
i. Confusion, seizures and coma

Stimulant withdrawal
The DSM-5 diagnostic criteria state that there must be recent cessation of prolonged usage. Within few hours to several days upon cessation, there must be dysphoric mood and at least 2 of the following:
a. Decreased energy and feelings of fatigue
b. Vivid and unpleasant dreams
c. Sleep changes - either insomnia or hypersomnia
d. Marked increased appetite
e. Psychomotor changes - retardation or agitation

Cannabis Use Disorder
The DSM-5 diagnostic criteria state that there must be problematic pattern of cannabis usage, leading to significant impairments over duration of 12 month. This is manifested by at least 2 of the criteria, which are similar to the criteria set out for alcohol use disorder.

Cannabis intoxication
The DSM-5 diagnostic criteria state that there must be recent usage of cannabis and there have been significant problematic behavioural or psychological changes developed since the commencement of usage.
At least 2 of the following signs and symptoms must develop within 2 hours of usage:
a. Injection of conjunctiva
b. Appetite that is better than normal
c. Dryness of mouth
d. Marked increased in heart rate
The DSM-5 also specified that cannabis intoxication might occur with perceptual hallucinations.

Cannabis Withdrawal

The DSM-5 diagnostic criteria specify that there must be recent cessation of cannabis use that was previously heavy and prolonged. 3 or more of the following signs and symptoms must be present:

a. Mood swings characterized as increased irritability
b. Anxiety features or marked nervousness
c. Sleep problems that present with initiation difficulties or with disturbing dreams
d. Reduction in appetite and associated weight loss
e. Restlessness
f. Physical symptoms including abdominal pain, tremors, sweating, fever, chills and headache.

Phencyclidine Use Disorder

The DSM-5 diagnostic criteria specified that there must be a pattern of continuous usage that has led to impairments in terms of functioning, which has occurred over a 12 month period.
This is manifested by at least 2 of the criteria, which are similar to the criteria set out for alcohol use disorder.

Phencyclidine intoxication

The DSM-5 diagnostic criteria state that there must be recent usage of PCP or other similar substance. The DSM-5 states that within an hour of usage, the individual should experience at least 2 of the following signs and symptoms:

a. Either vertical or horizontal nystagmus
b. Hypertension or tachycardia
c. Reduction in responsiveness to pain
d. Cerebellar signs like ataxia
e. Dysarthria
f. Rigidity of muscles
g. Seizures or even coma
h. Hyperacusis

Hallucinogen Persisting Perception Disorder

The DSM-5 diagnostic criteria state that following the cessation of usage of a particular hallucinogen, there has been re-experiencing of the perceptual disturbances that were experienced while the individual was previously intoxicated with the hallucinogens. This might include:

a. Geometric hallucinations
b. False perception of movement in peripheral visual field
c. Flashes of colour
d. Colours which are intensified
e. Trials of images of moving objects
f. Positive afterimages
g. Halos that appear around objects
h. Macropsia and micropsia

Epidemiology:

1. Amphetamine: In the United states, the abuse of Amphetamine has been estimated to be around 7%, with the highest prevalence amongst those who are between 18-25 years old. There is no gender difference associated with the usage. In Hong Kong, the estimated prevalence is around 12.4% out of the total proportion of drug abusers.
2. Cocaine: In the United states, at least 10% of the population has tried cocaine and the abuse of cocaine has been estimated to be around 2%. The highest prevalence is amongst those who are between 18 to 25 years old with a higher proportion of male abusers as compared to females (2:1). In Hong Kong, the estimated prevalence is around

6.6% out of the total proportion of drug abusers.
3. Cannabis: In the United states, at least 5% of the population are actively abusing the drug. All age groups are similarly affected, but the highest would be amongst those who are from 18 to 21 years old. In Hong Kong, the estimated prevalence is around 3.1% out of the total proportion of drug abusers.
4. Hallucinogens: In the United states, the prevalence of the abuse of MDMA has been estimated to be around 2%, and is especially common amongst young white men between ages of 15-35. In Hong Kong, the estimated prevalence is around 0.9% out of the total proportion of drug abusers.

Neurochemistry:

All of the above-mentioned stimulants and hallucinogens have potential to induce neurochemical changes in both the serotonin and dopamine systems. LSD increases serotonin function. Overdose of LSD may cause serotonin syndrome.

Overview of the neurochemistry and clinical effects of cocaine

Pharmacokinetics: The $t_{1/2}$ is 50 minutes. The route of administration involves the intranasal and intravenous routes.

Pharmacodynamics: There release of dopamine from dopamine containing neurons in nucleus accumbens produces intense feeling of euphoria.

Clinical effects: Cocaine produces a dose – related increase in arousal, improved performance on tasks of vigilance and alertness, and a sense of self-confidence and well-being. It is not uncommon for cocaine users to go on binges with cocaine and lead to a "crash" (depression and exhaustion) after a period of heavy use.

At a higher dose, cocaine produces a brief euphoria, involuntary motor activity, stereotyped behaviour and paranoia.

Sensitization as a result of repeated administration has been linked to paranoid and psychotic manifestations of cocaine.

Cocaine users will develop a desire for more cocaine. Intranasal use results in earlier seeking behaviour (after 10 to 30 minutes) compared to intravenous route (50 minutes). Injection may cause a euphoric rush which lasts for 10-15 minutes.

Overview of the neurochemical effects of amphetamines

Amphetamine causes excessive release of dopamine, this will lead to hyperexcitable state such as tachycardia, arrhythmia, hyperthermia and irritability. It causes pupil dilation.

Psychotic-like state can results from acute or chronic ingestion. It will lead to paranoia, hallucination and sometimes a delirium-like state. The effect may last for 3 to 4 days.

In the withdrawal state (aka crash), the person will develop fatigue, hypersomnia, hyperphagia, depression and nightmare.

Overview of the neurochemical effects of cannabis

1) There are two subtypes of the cannabinoid receptor, CB1 and CB2 receptors.

The CB1 receptors are highly expressed in the hippocampus, cortex, basal ganglia, cerebellum and spinal cord and this accounts for the effects of cannabis on memory, cognition and movement.

Both CB1 and CB2 cannabinoid receptors are coupled to inhibitory G-proteins.

Activation of the cannabinoid receptors causes inhibition of adenylate cyclase and a subsequent decrease in the concentration of cAMP in the cell.

This will ultimately result in the inhibition of neurotransmission.

Route of administration:
1. Cocaine has several routes of administration. The onset of its effect is highly dependent on the mode of administration. The onset would be fastest when it's directly smoked (within 6-8 seconds); as compared to inhalation which would take around 3-5 minutes and achieve peak levels at 30-60 minutes (this is so as nasal absorption is limited by intrinsic vaso-constrictive effects); as compared to Intravenous, which is slower as circulation time in needed.
2. Amphetamines is usually taken orally or being abused via intravenous route, nasal inhalation or via smoking.
3. Cannabis is used either smoked or taken orally.
4. Hallucinogens are usually ingested orally, or could also be sucked out of paper or smoked.

Signs of intoxication and tolerance:
1. Cocaine/amphetamine/cannabis intoxication leads to the following psychological and physiological changes: euphoria or blunted affect, changes in sociability, hypervigilance, impaired judgment, tachycardia or bradycardia, pupillary dilation, elevated or lowered blood pressure, perspiration or chills, nausea or vomiting, weight loss, psychomotor agitation or retardation, muscular weakness, respiratory depression, chest pain or cardiac arrhythmias and lastly, confusion and seizures.
2. Hallucinogen intoxication would usually result in marked anxiety or depression with ideas of references and hallucinations.

Core information to remember with regards to acute intoxication:

Acute intoxication due to usage of cocaine

Dysfunctional behaviour (at least one of the following)	**Signs** (at least one of the following must be present)
1. Euphoria and sensation of increased energy	1. Tachycardia (sometimes bradycardia)
2. Hypervigilance	2. Cardiac arrhythmias
3. Grandiose beliefs or actions	3. Hypertension (sometimes hypotension)
4. Abusiveness or aggression	4. Sweating and chills
5. Argumentativeness	5. Nausea or vomiting
6. Lability of mood	6. Evidence of weight loss
7. Repetitive stereotyped behaviours	7. Pupillary dilatation
8. Auditory, visual, or tactile illusions	8. Psychomotor agitation (sometimes retardation)
9. Hallucinations, usually with intact orientation	9. Muscular weakness
10. Paranoid ideation	10. Chest pain
11. Interference with personal functioning	11. Convulsions

Acute intoxication due to usage of cannabis

Dysfunctional behaviour (at least one of the following)	**Signs** (at least one of the following must be present)
1. Euphoria and disinhibition (e.g. giggles)	1. Increased appetite
2. Anxiety or agitation (20% of patients)	2. Dry mouth
3. Suspiciousness or paranoid ideation	3. Conjunctival injection (reddening of eyes)
4. Temporal slowing (a sense that time is passing very slowly or rapid flow of ideas)	4. Tachycardia
5. Impaired judgment, attention or reaction time	
6. Auditory, visual, or tactile illusions	
7. Hallucinations with preserved orientation	
8. Depersonalization or derealisation	
9. Interference with personal functioning	

Acute intoxication of hallucinogen

Dysfunctional behaviour (at least one of the following)	**Signs** (at least one of the following must be present)
1. Anxiety and fearfulness	1. Tachycardia
2. Auditory, visual, or tactile illusions or hallucinations occurring in a state of full wakefulness and alertness (it may lead to accidents and hallucinations are recurrent)	2. Palpitations
	3. Sweating and chills
	4. Tremor
	5. Blurring of vision
3. Depersonalisation	6. Pupillary dilatation
4. Derealisation	7. Incoordination
5. Paranoid ideation	
6. Ideas of reference	
7. Lability of mood	

8. Hyperactivity 9. Impulsive acts 10. Impaired attention 11. Interference with personal functioning	

Withdrawal symptoms:
1. Cocaine or amphetamine withdrawal would induce the following signs and symptoms: dysphoric mood associated with fatigue, vivid and unpleasant dreams, increased appetite, psychomotor agitation or retardation.
2. Cannabis usually does not have a characteristic withdrawal syndrome.

Treatment:
1. Cocaine: Treatment rendered largely targets the symptoms that individuals are experiencing. To reduce the amount of agitation, chemical and physical restraints could be used. If severe, extremely low dose of appropriate antipsychotics might be indicated for usage. Physical symptoms could be treated with beta blockers
2. Amphetamines: Treatment rendered is also largely symptomatic. Acute treatment makes use of benzodiazepines for agitation. Long term treatment involves the usage of antidepressants to maintain drug-free behaviours after detoxification.
3. Cannabis: Acute treatment for intoxication is usually not needed. If there are persistent hallucinations or delusions, then anxiolytic could be used.
4. Hallucinogens: Benzodiazepines could be used in acute intoxication. In the event that the patient is acutely psychotic and agitated, a consideration should be made to use high potency antipsychotics.

Nicotine and smoking

Tobacco Use Disorder
The DSM-5 diagnostic criteria state that there must be problematic pattern of usage that has led to significant impairments over duration of at least 12 months.
The rest of the criteria are similar to that of alcohol use disorder.

Tobacco Withdrawal
The DSM-5 diagnostic criteria state that there must be daily usage of tobacco for at least several weeks. Upon sudden cessation or even reduction in the amount of usage, the following signs and symptoms develop:

a. Mood changes characterized by irritability or anger

b. Anxiety

c. Difficulties with concentration

d. Marked changes in appetite - increased appetite

e. Feelings of restlessness

f. Depressed mood

h. Difficulties with sleep initiation

Epidemiology:

Worldwide incidence of smoking is around 47%. The mean age of onset of smoking and nicotine addiction and abuse has been estimated to be from 20 years onwards.

Neurochemistry:

Nicotine has the potential to activate the nicotine receptors and also the dopamine system and stimulate the release of multiple neurohormones.

Overview of Neurochemical effects of nicotine

Nicotinic receptors are found on presynaptic dopaminergic neurons. Smoking of tobacco leads to release of dopamine. Effects of nicotine include euphoria, enhancing motivation and sustained vigilance.

Other actions of nicotine: suppression of insulin production from the pancreas (slight hyperglycaemia),

Route of administration: Most commonly, it is inhaled.

Acute intoxication due to use of tobacco

Dysfunctional behaviour (at least one of the following)	Signs (at least one of the following must be present)
1. Insomnia 2. Bizarre dreams 3. Lability of mood 4. Derealisation 5. Interference with personal functioning	1. Nausea or vomiting 2. Sweating 3. Tachycardia 4. Cardiac arrhythmias

Withdrawal symptoms:
This would include a constellation of symptoms that include intense craving, irritability, anxieties, restlessness and difficulties with concentration.

Treatment:
1. Doctor's advice is the strongest factor to motivate patients to quit smoking.
2. Bupropion has been commonly used. It is an antidepressant with noradrenergic activity. It could help in decreasing the effect of nicotine withdrawal. The empirical starting dose is 150mg/day and subsequently titrated to 150mg BD. The common associated side effects include epilepsy (1/1000). It is contraindicated in individuals with a history of epilepsy, eating disorders, CNS tumour and those with a psychiatric history of bipolar disorder. Common side effects include headache (30%), insomnia and rash (01%).
3. Nicotine replacement therapy could be another alternative. The duration of treatment is from 8 to 12 weeks. Common route of administration include sublingual tablets, gum, patch, nasal spray. The common side effects include local irritation, and it might also cause deranged capillary glucose levels in individuals with diabetes. It is important to note that it is not recommended for both bupropion and NRT to be administered together.

Solvents, sedatives, caffeine and steroids

Inhalant Use Disorder
The DSM-5 diagnostic criteria state that there must be a problematic pattern of usage of a hydrocarbon based inhalant substance, which has led to much impairment in functioning occurring within a 12 month period.
The rest of the criteria are similar to that for alcohol use disorder.

Inhalant Intoxication
The DSM-5 diagnostic criteria specified that there must be recent usage of the substance, and there must be at least 2 of the following signs and symptoms being experienced:

a. Dizziness

b. Nystagmus

c. Incoordination

d. Slurred speech

e. Unsteadiness in gait

f. Depressed reflexes

g. Psychomotor changes - retardation

h. Tremor

i. Generalized muscular weakness

j. Blurring of vision or even diplopia

k. Stupor or coma

l. Euphoria

Solvent use
Examples of solvents include glue, butane and toluene.

Route of administration:
It is administered by directly spraying into the mouth or via inhalation from a bag.

Signs of acute intoxication:
The user will have an initial excitatory phase and then follow by acute depression.
Heavy users may have personality change and cognitive impairment as the solvent will lead to white matter changes in the brain.
Serious complications include liver and renal impairment, perforated nasal septum and sudden death.

Acute intoxication due to use of solvent

Dysfunctional behaviour (at least one of the following)	Signs (at least one of the following must be present)
1. Apathy and lethargy 2. Argumentativeness 3. Abusiveness or aggression 4. Lability of mood 5. Impaired judgment 6. Impaired attention and memory 7. Psychomotor retardation 8. Interference with personal functioning	5. Unsteady gait 6. Difficulty in standing 7. Slurred speech 8. Nystagmus 9. Decreased level of consciousness 10. Muscle weakness 11. Blurred vision or diplopia

A withdrawal reaction rarely occurs.

It would give rise to signs and symptoms of increased irritability, insomnia, diaphoresis, nausea and vomiting, tachycardia, and sometimes associated also with hallucinations and delusions.

Treatment:
Largely medical supportive treatment would be adequate.

Sedatives
Refers to drugs that are commonly used to treat insomnia and anxiety.
The prevalence of abuse of sedatives in US is estimated to be around 6%, with the highest prevalence amongst those who are between the ages of 26-35. There is an increased prevalence of usage amongst females as compared to males (3:1).

Neurochemistry: Most of these drugs have direct agonist effects on the GABA(A) receptor complex.

Acute intoxication due to use of benzodiazepine:

Dysfunctional behaviour (at least one of the following)	Signs (at least one of the following must be present)
1. Euphoria and disinhibition 2. Apathy and sedation 3. Abusiveness or aggression 4. Lability of mood 5. Impaired attention 6. Anterograde amnesia 7. Impaired psychomotor performance 8. Interference with personal functioning	1. Unsteady gait 2. Difficulty in standing 3. Slurred speech 4. Nystagmus 5. Decreased level of consciousness (e.g., stupor, coma) 6. Erythematosus skin lesions or blisters

Withdrawal:
The typical signs and symptoms associated with acute withdrawal include:
Insomnia, anxiety, an increase in autonomic signs (e.g. sweating, tremor), hyperaesthesia, hyperacusis, headache, photophobia and withdrawal fits. The withdrawal symptoms are most severe in the first week after stopping benzodiazepine. Those sedatives that have shorter half live are associated with a more rapid onset of withdrawal as compared to drugs with a longer half life.

Sedative, hypnotic, or anxiolytic use disorder
The DSM-5 diagnostic criteria state that there must be problematic usage of sedative, hypnotics or anxiolytic use that has led to significant impairments in terms of functioning over a 12 month period.
The rest of the criteria are similar to that of alcohol use disorder.

Intoxication
The DSM-5 diagnostic criteria state that there must be recent usage that has led to at least one of the following signs and symptoms developing:
a. Slurring of speech
b. Incoordination with regards to movements
c. Unsteadiness of gait
d. Nystagmus
e. Impairments in cognition, attention and memory
f. Stupor or coma

Withdrawal
The DSM-5 diagnostic criteria states that there must be recent cessation of (or reduction in) the usage of sedative, hypnotic, or anxiolytic, accompanied by at least 2 of the following signs and symptoms:
a. Sweating or tachycardia (autonomic hyperactivity)
b. Hand tremors
c. Difficulties with falling asleep
d. Gastrointestinal disturbances - nausea or vomiting
e. Transient visual, tactile, or auditory hallucinations or illusions
f. Psychomotor changes - particularly agitation
g. Anxiety
h. Seizures

Treatment:
In treatment abusers of sedative, it is crucial to obtain a detailed drug history and also obtain urine and blood samples for drug and comorbid substance usage like alcohol. The drug history and the serum drug levels would help the clinician to determine the appropriate levels of benzodiazepine required for stabilization.
Detoxification involves the following processes:
1. Switching over to a longer acting benzodiazepine (e.g. diazepam) for patients to gradually undergo detoxification.
2. Once the patient is comfortable and stabilized with the longer acting drug, the dosage of the drug is then gradually reduced by 30% on the subsequent days, as tolerated.
3. Adjunctive medications might need required for stabilization, especially for individuals who have been consuming a supra-therapeutic dose previously.
4. Psycho-education and psychological intervention would help patients in the detoxification process.

Caffeine

Chemical properties: Caffeine is a methylxanthine (1,3,7-trimethylxanthine).

Neurochemistry of caffeine: The main action of caffeine is the competitive antagonism of adenosine A_1 and A_2 receptors which contribute to the neuropsychiatric effects such as psychosis in caffeine intoxication due to release of dopamine. Higher doses cause inhibition of phosphodiesterases, blockade of $GABA_A$ receptors and release of intracellular calcium. It reaches its peak blood levels after 1 to 2 hours and reduces cerebral blood flow although it is a stimulant.

Caffeine Intoxication
The DSM-5 diagnostic criteria states that there must be recent consumption of a typically high dose of caffeine, usually well in excess of 250 miligrams. There must be presence of at least 5 or more of the following signs and symptoms:
a. Feelings of restlessness
b. Feeling anxious
c. Feeling excited
d. Difficulties with falling asleep
e. Flushed face
f. Diuresis
g. Gastrointestinal discomfort and disturbances
h. Twitching of muscles
i. Rambling flow of thoughts and speech
j. Increased heart rate
k. Periods of inexhaustibility
l. Psychomotor agitation

Caffeine Withdrawal
The DSM-5 diagnostic criteria states that there must be abrupt cessation or reduction in the usage of caffeine after prolonged daily usage. This is usually followed within 24 hours, by at least 3 of the following symptoms:
a. Headache
b. Marked reduction in energy
c. Dysphoric mood, depressed mood or irritability
d. Difficulties with concentrating
e. Nausea, vomiting or muscle pain or stiffness

Adverse effects of caffeine:
1. **CNS**: migraine
2. **CVS**: Caffeine may precipitate sinus tachycardia but does not causing cardiac arrhythmias.
3. **GI tract**: Caffeine relaxes the lower oesophageal sphincter and can predispose to gastro-oesophageal reflux disease. It also causes hypersecretion of gastric acid and increases the risk of gastric ulcer.
4. Renal: Caffeine cause diuretic effect and people are advised to abstain from consuming caffeine in situations where dehydration may be significant.
5. **Pregnancy:** low birth weight and miscarriage. Caffeine enters amniotic fluid and breast milk. This will affect infants as caffeine is metabolized slowly.

Dependence and withdrawal
1. 10% of caffeine users experience withdrawal effects (e.g. more than 6 cups per day). Withdrawal starts at 1-2 hours post ingestion and becomes worst at 1-2 days and recede with a few days.
2. Common withdrawal effects include: headache, irritability, sleeplessness, anxiety, tremor and impairment of psychomotor performance.

Treatment: Largely symptomatic treatment with the usage of a short course of benzodiazepine for treatment of associated restlessness and anxiety.

Anabolic steroids

People misuse anabolic steroids to increase muscle growth and body bulk. It can be swallowed or injected. Adverse effects include gynaecomastia in men and clitoral enlargement in women, bone hypertension, cardiac disorders, liver (e.g. drug induced hepatitis) and renal impairment, shrinking of testicles and priapism. It will lead to aggression and irritability. Death may occur due to overdose and severe infection due to repeated intramuscular injection.

Gambling Disorder

The DSM-5 diagnostic criteria specify that there must be persistent and recurrent problematic gambling behaviour over the past 12 months that has led to significant impairment and distress.

This is accompanied by at least 4 of the following:

a. Needing to gamble with increasing amount of money in order to achieve the same level of excitement
b. Feelings of restlessness or irritability when attempting to cut down or stop gambling
c. Has had made repeated unsuccessful attempts at cutting down or stopping
d. Often preoccupied with gambling
e. Tendency to gamble when feeling distressed
f. Even after losing money, would often return the next day to chase one's loses
g. Tendency to lie to minimize the extent of involvement with gambling
h. Gambling has had affected significant relationships and has caused the individual to miss opportunities
i. Often having to rely on others to help bail out of a difficult financial situation.

Clinicians need to be mindful to exclude the possibility of an underlying bipolar disorder.

MOH guidelines for treating pathological gambling [Lee et al 2011]

Assessment should include (for OSCE station):
- initiation
- progression
- current frequency (days per week or hours per day)
- current severity (money spent on gambling proportionate to income)
- types of games played
- maintaining factors
- features of dependence
- Consequences: financial, interpersonal, vocational, social and legal
- Reasons for consultation, motivation to change and expectations of treatment
- Assessment of suicide risk
- Assessment of Axis I and II comorbidities, including alcohol and substance use disorders

Management
- A comprehensive treatment plan that incorporates a multi-disciplinary and multi-modal approach should be developed for the management of pathological gambling.
- An opioid antagonist like naltrexone may be considered for reduction of gambling urges and thoughts in pathological gamblers.
- Fluvoxamine and paroxetine may be considered for reduction of gambling behaviour, urges and thoughts in pathological gamblers.
- Motivational enhancement therapy (face-to-face or telephone counselling) and self-help workbooks are recommended for the treatment of gambling disorders, especially for individuals who are ambivalent about quitting gambling or entering treatment, or who are not keen on long-term therapy.
- Psychological interventions utilising the components of CBT are recommended for the treatment of pathological gambling.
- Financial counselling, limiting access to money and restricting admission into gambling venues are complementary and practical approaches that should be considered for those who have gambling related problems.

Sample MCQs

1) A nurse informs you that she has seen your patient intoxicated with alcohol during the outing. Your patient denies it. Which test would you perform to confirm that your patient is drinking again?

- A. Electrolytes
- B. Gamma-glutamyl transferase (GGT)
- C. Mean corpuscular volume (MCV)
- D. Serum alcohol level
- E. Urea.

The answer is B.
GGT becomes elevated after acute or chronic alcohol use, and remains elevated for two to five weeks afterwards. GGT is commonly used as an objective indicator of relapse. Another objective test is carbohydrate-deficient transferring (CDT) which is very expensive.

2) A 35-year-old man with history of alcohol misuse comes to the hospital and requests you to prescribe a drug to help him to maintain alcohol abstinence. Which of the following drugs does not help in maintaining alcohol abstinence?

- A. Acamprosate
- B. Clonidine
- C. Disulfiram
- D. Naltrexone
- E. Topiramate.

The answer is B.
Clonidine is an α_2 adrenergic agonist and it can reduce withdrawal but not helping with maintenance.

3) A 24 year old woman is referred by obstetrician. She has an unplanned pregnancy and smokes marijuana 5 times per day. Her partner is concerned of the effects of cannabis on the foetus. She avoids eye contact and appears anxious. She states she eats and sleeps well. She emphasizes that she has a friend who delivers a healthy baby and her friend smoked cannabis throughout her pregnancy. Her partner is very keen to persuade her to quit cannabis, which of the following best describes the stage of change?

A. Action
B. Contemplation
C. Decision
D. Pre-contemplation
E. Maintenance.

The answer is D.
The patient does not acknowledge or accept the potential problems associated with cannabis misuse and the need to change.

4. You are working in the Children Emergency Department. A mother brought her 14-year-old son which was found inhaling glue in school. Which of the following is NOT a sign of inhalant intoxication?

- A. Euphoria
- B. Diplopia
- C. Dysarthria
- D. Nystagmus
- E. Hyperreflexia.

The correct answer is E. Inhalants diminish reflex.
The signs of inhalant intoxication include dizziness, nystagmus, incoordination, slurred speech, unsteady gait, lethargy, depressed reflexes, psychomotor slowing, tremor, generalized muscle weakness, blurred vision, diplopia, stupor and euphoria.

5. A 50-year-old businessman has been injecting anabolic steroid to increase his muscle bulk. The following are all psychiatric complications of anabolic steroid misuse EXCEPT:

- A. Anxiety
- B. Apathy
- C. Depression
- D. Euphoria
- E. Irritability.

The answer is B.
Anabolic steroid misuse can cause irritability, increased aggression, mood swings, distractibility, forgetfulness, and confusion.

Sample MEQ

You are the resident in the medical department. You are asked to carry out a pre-admission assessment of a 40-year-old man who has extensive history of excessive alcohol consumption. He has been referred by his GP for management of alcohol withdrawal. According to his wife, he has been drinking a bottle of Chinese wine every day for the last 3 years. His wife also says that he has been drinking particularly heavily over the past 3 months and over this time has eaten only occasional meals.

When you see the patient, you are convinced that he is withdrawn from alcohol. He also appears malnourished. He does not have any psychotic symptom such as visual or auditory hallucination. You have completed the assessment, have conducted a physical examination and are charting initial medication.

1. Name two medications that you would order for this patient and its route of administration?

1) Diazepam (oral)
2) Parenteral thiamine (IM or IV).

2. Outline the reasons you need to prescribe the two medications for question 1. Give one reason per medication.

Diazepam (oral): to reduce withdrawal symptoms.

Parenteral thiamine: to offer prophylaxis or prevent Wernicke's encephalopathy.

3. The nurse informs you that he has hypoglycaemia and requests for dextrose saline infusion. What do you advise the nurse to check before giving the dextrose saline?

The nurse has to make sure that the thiamine was administered before giving the glucose infusion because the metabolism of glucose would further deplete the store of thiamine.

4. Name 4 signs or symptoms you would look for and monitored in alcohol withdrawal

Name 4 of the following:
- Agitation
- Anxiety
- Delirium or confusion
- Disorientation
- Labile affect
- Hypertension
- Restlessness
- Sweating
- Tachycardia
- Tremor
- Visual hallucinations.

5. The patient has recovered from alcohol withdrawal after one week of hospitalisation with diazepam treatment. Name three non-pharmacological interventions which can help this patient after discharge.

1) Motivational interviewing.
2) Alcohol anonymous
3) Cognitive behaviour therapy.

EMIs:
Withdrawal syndromes

A- Alcohol
B- Amphetamines
C- Benzodiazepines
D- Cannabis
E- Cocaine
F- Heroin
G- LSD
H- Nicotine
I- Tobacco

1. Cough, mouth ulcers and marked irritability – Nicotine
2. Yawning, sneezing and sweating – Heroin
3. Vivid dreams, depression and irritability - Benzodiazepines

References

1. Cook C (1994) Aetiology of alcohol misuse In Chick J and Cantwell R (1994) Seminars in Alcohol and Drug Misuse. Gaskell: London.

2. World Health Organisation (1992) *ICD-10: The ICD-10 Classification of Mental and Behavioural Disorders : Clinical Descriptions and Diagnostic Guidelines.* Geneva: World Health Organisation.

3. Babor TF, Higgins-Biddle JC, Saunders JB & Monteiro MG (2001) *The Alcohol Use Disorders Identification Test Guidelines for Use in Primary Care (2nd ed)*.Geneva: World Health Organization.

4. Cantwell R & Chick J (1994) Alcohol misuse: clinical features and treatment In Chick J and Cantwell R (1994) Seminars in Alcohol and Drug Misuse. Gaskell: London.

5. Gelder M, Mayou R and Cowen P (2001) Shorter Oxford Textbook of Psychiatry. Oxford: Oxford University Press.

6. American Psychiatric Association (1994) *DSM-IV-TR: Diagnostic and Statistical Manual of Mental Disorders (Diagnostic & Statistical Manual of Mental Disorders) (4th edition).* Washington: American Psychiatric Association Publishing Inc.

7. Sadock BJ, Sadock VA. (2003) *Kaplan and Sadock's Comprehensive Textbook of Psychiatry.* (9th ed.) Philadelphia (PA): Lippincott Williams and Wilkins.

8. Taylor D, Paton C, Kapur S (2009) *The Maudsley prescribing guideline*.10th edition London: Informa healthcare.

9. Johnstone EC, Cunningham ODG, Lawrie SM, Sharpe M, Freeman CPL. (2004) *Companion to Psychiatric studies.* (7th edition). London: Churchill Livingstone.

10. Hulse G (2004) Alcohol and drug problems. Melbourne: Oxford University Press.

11. Johns A and Ritson B (1994) Drug and alcohol-related problems In Paykel ES, Jenkins R (1994) *Prevention in Psychiatry.* Gaskell: London.

12. Jones S & Roberts K (2007) Key topics in psychiatry. Edinburgh: Churchill Livingstone Elsevier.

13. Vanessa Crawford (2005) Addiction psychiatry: dual diagnoses. MRCPsych Part II. Guildford MRCPsych Course.

14. Regier, D.A., Rae, D.S., Narrow, W.E. et al (1998) Prevalence of anxiety disorders and their comorbidity with mood and addictive disorders *British Journal of Psychiatry* **173**, 24-28.

15. Farrell, M., Howes, S, Taylor, C et al, (1998) Substance misuse and psychiatric comorbidity: an overview of the OPCS national psychiatric comorbidity survey. *Addictive Behaviours* **23**, 909-918.

16. Marsden, J., Gossop, M., Stewart, D et al (2000) Psychiatric symptoms among clients seeking treatment for drug dependence Intake data from the National Treatment Outcome Research Study. *British Journal of Psychiatry* **176**,285-289.

17. Caldwell, CR & Gottesman, II (1991) Sex differences in the risk for alcoholism: a twin study. *Behaviour Genetics*, **21**, 563.

18. Goodwin, D. W., Hermansen, L., Guze, S. B., et al (1973) Alcohol problems in adoptees raised apart from alcoholic biological parents. *Archives of General Psychiatry*, **28**, 238-243.

19. Li H, Borinskaya S, Yoshimura K,Kal'ina N, et al (2009) Refined Geographic Distribution of the Oriental *ALDH2*504Lys* (nee *487Lys*) Variant. *Annals of Human Genetics* **73**,335–345

20. Mythily S., Abdin e, Vaingankar K, Phua AM, Tee J, Chong SA. Prevalence and correlates of alcohol use disorders in The Singapore Mental Health Survery. *Adddiction 2012 [Epub]*

21. Reported drug abusers by sex by common type of drug abused. Central Registry of Drug Abuse, Hong Kong (2011). Extracted from: http://www.nd.gov.hk/statistics_list/doc/en/t15.pdf

22. American Psychiatric Association (2013) *DSM-5: Diagnostic and Statistical Manual of Mental Disorders (Diagnostic & Statistical Manual of Mental Disorders* (5th edition). Washington: American Psychiatric Association Publishing Inc.

Chapter 8: Eating disorders and metabolic syndrome

Anorexia nervosa (AN)	Bulimia Nervosa (BN)
Epidemiology	Epidemiology
Ateiology	Ateiology
ICD-10 / DSM-5 diagnostic criteria	ICD-10 / DSM-5 diagnostic criteria
Differential diagnosis	Differential diagnosis
Physical examination and laboratory findings	Physical examination and laboratory findings
Management	Management
Prognosis	Prognosis
	Eating disorders (NOS – Not otherwise specified)
	Obesity
	Clinical OCSE
	Revision MCQs, MEQ & EMIs

Anorexia nervosa (AN)

Epidemiology
- AN is ranked as the third most common type of chronic illness amongst adolescent females.
- The prevalence rate of AN have been estimated to be 0.5-3.7%. Incidence rates are higher amongst adolescent and early young adult women.
- Studies have shown that the incidence rates are higher among certain groups of individuals (e.g. ballet dancers and gymnasts).
- In Singapore, the prevalence rates of eating disorders have been demonstrated to be around 7.4% based on a study done in 2006.

Aetiology

1. Biological causes:
   - **Genetic causes:** Relatives of AN patients have an increase in risk in developing AN by 10 fold. MZ:DZ = 65%:32%
   - **Birth trauma**: Cephalo-haematoma, premature birth and small for gestation age are predisposing factors for AN.
   - **Hypothalamic dysfunction.**

2. Psychological causes:
   - **Development:** Failure of identity formation and psychosexual development in adolescence.
   - **Personal events**: Childhood obesity
   - **Family factors:** Young AN patients may use the illness itself to overcome rigidity, enmeshment, conflict and overprotection in the family.
   - **Underlying personality traits**: Perfectionistic and neurotic traits are predisposing factors.

3. Socio-cultural causes:
   - **Changes in nutritional knowledge and dietary fashion in the society**
   - **Cult of thinness**
   - **Changed roles and images in women** to pursue thinness

ICD-10 & DSM-5 diagnostic criteria

- **Considerable weight loss** (at least 15% below expected weight) which is **self-induced** by avoidance of 'fattening foods'. Self-induced vomiting, purging (laxatives or enemas) and excessive exercise are supportive features but not essential elements based on the ICD-10.
- **Self-perception** involves an overvalued idea e.g. a dread of fatness and a self-imposed low weight threshold.
- **Evidence of disorder in the HPA axis:** ♀: amenorrhoea, ♂: loss of sexual interest and potency.
- **In ICD-10 F50.1 atypical AN**, the key features of AN are absent or present only to mild degree.

Anorexia Nervosa

The DSM-5 diagnostic criteria states the following:

1. Restriction of input relative to exact requirements, thus leading to significantly low body weight. Low body weight is now defined as a weight that is less than what is considered to be minimally normal for adults, and for children and adolescents, less than what is considered to be minimally expected.
2. Marked and excessive fear of putting on weight, or of becoming fat. Repetitive behaviours are carried out to prevent weight gain, despite the already low weight.
3. Distortions in an individual's self perception of body weight or shape, associated with the lack of recognition of the serious consequences of the current low body weight.

The DSM-5 has 2 main subtypes:

a. Restricting type - Where over the past 3 months, weight loss is achieved mainly by means of dieting, fasting or excessive exercises

b. Binge-eating or purging type - Where over the past 3 months, there has been recurrent episodes of binge eating or purging behavior (e.g. vomiting, use of laxative, diuretics and enema).

Severity markers: BMI Range
Mild- 17, Moderate 16-16.99, Severe 15-15.99, Extreme <15

Differential diagnosis

1. Other psychiatric disorders (e.g. depression, schizophrenia, obsessive-compulsive disorder and psychotic disorder).
2. Medical disorders (e.g. hypopituitarism, thyrotoxicosis, diabetes mellitus, neoplasia, reticulosis, malabsorption).

Physical examination findings

- **CNS**: impaired cognition, poor concentration, seizures, syncope, depression, obsessive and compulsive behaviours.

- **CVS**: bradycardia (30-40 beats/minute), hypotension (systolic <70mmHg), prolonged QTc, arrhythmia, mitral valve prolapse, pericardial effusion, cardiomyopathy (Echocardiogram may be indicated).

- **GIT**: delayed gastric emptying and severe constipation, painful and distended abdomen and nutritional hepatitis.

- **Renal:** nocturia and renal stones.

- **Reproductive system:** prepubertal state: amenorrhoea, small ovaries and uterus, infertility and breast atrophy.

- **Musculoskeletal system:** Cramps, tetany, muscle weakness, osteopenia and stress fractures.

- **Peripheral nervous system**: peripheral neuropathy and impaired autonomic function.

- **Dermatological**: dry skin, brittle nail, loss of head hair, increase in body hair (lanugo hair), pallor (anaemia), Raynaud's phenomenon: discolouration of fingers and toes, peripheral cyanosis.

- **Hypothermia.**

Core laboratory findings

- **Full blood count**: Anaemia (usually normochromic but Fe/B12 deficiency is possible), leukopenia, ↓ESR, thrombocytopenia, and ↓ complements.

- **Electrolyte disturbances:** ↓ in K, Ca^{2+}, Na, PO_4, Mg^{2+}; Liver and renal failure: ↑ amylase isoenzyme, ↓albumin, ↓glucose, ↓insulin, ↑lipid (due to ↓ oestrogen), metabolic acidosis (diarrhoea) and metabolic alkalosis (vomiting).

- **HPA Axis**: ↑CRH, normal ACTH, ↑cortisol.

- **Other hormones:** ↓ FSH and LH, ↓ oestrogen, ↓ testosterone, ↓T4/T3 → ↑ cortisol and ↑ growth hormone.

- Brain pseudoatrophy, ↓ in bone mineral density and abnormal EEGs.

Management

Consideration for hospitalization: There are several factors that should be considered prior to acute admission of any patient. Typically, if the patient has a BMI < 13, heart rate < 40 beats per minute, failure of outpatient treatment and high suicide risk, then they are typically recommended for admission. The average duration of hospitalization is between 1 to 3 months.

When they are admitted as an inpatient, the treatment program adopted utilizes both pharmacological and non-pharmacological methods.

1. Pharmacological methods – No drugs have been proven to be effective in the treatment of AN. Studies have been looking into the usage of fluoxetine, which has demonstrated improvement in mood and reduce obsessions in checking their body weight in some patients. There have been previously published accounts of how atypical antipsychotic agents like olanzapine might provide benefits. Olanzapine may reduce the rigidity of thinking and increase the appetite of the patient.

2. Psychological methods – A variety of psychotherapeutic methods have been adopted in the treatment of these patients.
These include cognitive-behavioural therapy, interpersonal psychotherapy and family therapy.

- **Cognitive-behavioural therapy (CBT)** would be useful for patients, as they are being taught how to monitor their eating habits, develop skills needed to deal with interpersonal relationships and more importantly, identify their underlying negative automatic thoughts (e.g. I am too fat), cognitive errors (selective abstraction, just focusing on body weight and forget about other aspects of life) and the therapeutic process would help them challenge those thoughts accordingly.
- **Interpersonal psychotherapy (IPT)** might be an alternative therapy, however, it is more suitable for patients who encounter interpersonal problems which is the main predisposing factor for eating disorder. IPT involves the application of interpersonal inventory to allow the therapist to assess patient's interpersonal relationship. The focus of IPT includes loss of roles, change in roles, role transition and interpersonal problems. This may be relevant for patients changing from one environment to another (e.g. changing from secondary school to university or from university to work).
- **Family therapy** would be useful for patients who stay with a dysfunctional family. It is highly effective when AN serves a role to draw attention from family members and the patient is in a triangular relationship with other family members and AN is the focus of the dysfunctional family. Family therapy is useful for family which does not have a clear hierarchy and boundary among family members.

Psycho-education is an important aspect of treatment. It is of importance that the patient gets informed about the benefits of achieving an adequate weight is important in order to reverse the effects of prolonged starvation.

The reasonable target weight gain would be to achieve an increment of around 0.5 to 1 kg per week.

Prognosis

- One-third of AN patients may attempt suicide or self harm.
- Mortality: 10%

Poor prognostic factors:	Good prognostic factors:
1. Relatively late age of onset. 2. Longer duration of illness (especially long untreated illness). 3. Dysfunctional family. 4. Personality disorder. 5. Presence of very frequent vomiting and very severe weight loss. 6. Extreme treatment avoidance. 7. Male gender.	1. Absence of severe weight loss and serious medical complications. 2. Good motivation to change. 3. Supportive family.

Bulimia nervosa (BN)

Epidemiology
- The prevalence rates of BN in the United States have been estimated to be around 1-3% of the general population. The mean age of onset is around 16 to 18 years old. There is a predominance of females being affected, with the female to male ratio being 10:1.
- In Singapore, the prevalence rates of eating disorders has been estimated to be around 7.4% based on a study (Ho et al, 2006).

Aetiology

1. Biological causes:
- **Genetic causes:** The risk of relatives of BN patients developing BN is 4 times higher than non-relatives. History of weight loss and strict diet are common.
- **Neurochemistry:** ↓5HT

2. Psychological causes:
- Preoccupation with weight and body shape as a result of personal history of obesity and transgression of self-imposed dietary rules.
- Poor impulse control.
- Binge eating as a maladaptive way for coping with stress.

3. Socio-cultural causes:
- Peer influence and easy access to junk food.

ICD-10 & DSM-5 diagnostic criteria

- Recurrent episodes of overeating (at least twice a week over a period of 3 months)

- Preoccupation with food and strong sense of compulsion to eat.

- Attempts to counter the "fattening" effects of food by induction of vomiting, abusing purgatives, alternating starvation and use of drugs such as appetite depressants and diuretics

Bulimia Nervosa
The DSM-5 diagnostic criteria states the following:
1. Recurrent episodes of binge eating. Binge eating refers to eating in a fixed duration of time, an amount definitely larger than what most individuals would eat in a similar situation. There is a lack of control with regards to eating during these episodes.
2. Repetitive compensatory behaviours such as self-induced vomiting, usage of laxatives, or other medications or prevent weight gain
3. These episodes occur at least once a week for 3 months.
4. Self-esteem is affected by self-evaluation of body weight and shape.
The DSM-5 states that this disturbance does not occur exclusively during episodes of anorexia nervosa.

Differential diagnosis
Frontal lobe syndrome, Prader-Willi syndrome and Kleine-Levine syndrome, gastrointestinal or brain tumours, iatrogenic increase in appetite.

Physical examination findings
- **CNS:** epilepsy.

- **Oral and oesophagus:** parotid gland swelling, dental erosions, oesophageal erosions and Mallory-Weiss tear.

- **CVS:** arrhythmias and cardiac failure leading to sudden death.

- **GIT:** gastric perforation, gastric/duodenal ulcers, constipation and pancreatitis
- **Tetany and muscle weakness**
- **Russell's sign:** abrasions over dorsal part of the hand because fingers are used to induced vomiting.

Core laboratory findings
- **FBC:** leukopenia and lymphocytosis.
- **U&Es:** ↓ in K^+, Na^+, Cl^-, ↑bicarbonate
- ↑ in serum amylase
- **Metabolic acidosis** due to laxative use
- **Metabolic alkalosis** due to repeated vomiting.

Management
Most patients who are diagnosed with BN usually do not require acute inpatient hospitalization. Outpatient treatment programs could be applied instead. The consideration for inpatient treatment program would be warranted should there be additional associated psychiatric comorbidity, which would require immediate hospitalization for stabilization of symptoms.

Both pharmacotherapy and psychotherapy could be considered.
Pharmacological treatment: antidepressants (SSRIs such as fluoxetine or fluvoxamine) have been shown in several studies to be effective in treatment of BN. It would be able to help in reduction of binge eating and also the associated impulsive behaviour. Dosages that are used in the treatment of BN may be higher than the dosages that are used for the treatment of depressive disorders. Evidence for the usage of mood stabilizers in the treatment of BN is lacking; but they are occasionally considered for use in the context of BN associated with borderline personality disorder.

Psychological treatment – Both cognitive behavioural therapy (CBT) and interpersonal psychotherapy (IPT) have been used. CBT has been shown to be highly effective for BN. It helps by enabling individuals to recognize their underlying pathological behaviour pattern (e.g. bingeing as a way of coping) and also helps by enabling individual to recognize their distorted beliefs regarding self and body image. The behaviour diary can help to monitor the frequency of binging and helps the patient to reduce its frequency and replace by more adaptive behaviour. IPT can be applied as described previously in AN.

Prognosis and comorbidity

Poor prognostic factors:	Psychiatric comorbidity:
1. Low self-esteem 2. Severe personality disorder	Depression, anxiety, borderline personality, and poor impulse control: self mutilation (10%), promiscuity (10%), shoplifting (20%), suicide attempts (30%) and alcohol misuse (10-15%).

Avoidant / Restrictive Food Intake Disorder
The DSM-5 diagnostic criteria states that the eating abnormalities must be such that there must be persistent failure to meet appropriate nutritional and energy needs, in association with the following:
a. Marked weight loss
b. Significant deficiency in nutrition
c. Needing to depend on enteral feeding or oral supplements
d. Marked impairments in functioning.

Binge Eating Disorder
The DSM-5 diagnostic criteria states the following:
1. Recurrent episodes of binge eating. Binge eating refers to eating in a fixed duration of time, an amount definitely larger than what most individuals would eat in a similar situation. There is a lack of control with regards to eating during these episodes.
2. The binge eating episodes are associated with the following:

a. Eating till feeling uncomfortably full
b. Eating more rapidly than normal
c. Eating large amounts even when not physically hungry
d. Eating alone due to feelings of embarrassment by how much oneself is eating
e. Feeling disgusted, and guilty after the episodes

These binge eating occurs at least once a week for the past 3 months.

Other specified Feeding or Eating Disorder include:
1. Atypical Anorexia Nervosa
2. Bulimia Nervosa (of low frequency and limited duration)
3. Binge-eating disorder (of low frequency and / or limited duration)
4. Purging disorder

OSCE (Assess AN)

Name: Sally **Age**: 15-year-old

Sally, a secondary school student with a two-year history of anorexia nervosa, is admitted to the hospital following a seizure after prolonged fasting. On admission, her BMI is 10 and her heart rate is 35 beats per minutes. You are approached by her parents who beg you to save Sally.

Task: To take a history from Sally to establish the aetiology and course of anorexia nervosa

A) Severity of AN symptoms	A1) Dietary history	A2) Longitudinal weight history	A3) Methods to lose weight and binge eating	A4) Body image distortion	A5) Serious medical complications in the past:
	"Hello Sally, I am Dr. Tan. Can you take me through your diet habit on a typical day?" Look for the number of meal times, the content of food. "How long have you been eating in this way?" "Where do you learn this diet habit from?"	Take a history on Sally's weight. E.g. the lowest, highest and average weight in the past 2 years. "What is your ideal weight?"	Explore the methods used by Sally (e.g. avoidance of 'fattening foods', self-induced vomiting, purging and excessive exercise) Although Sally presents with AN, it is important to ask about binge eating	Assess how fixated Sally is on her overvalued idea (e.g. dread of fatness) and find out her self-imposed weight threshold. "How do you feel when you look into the mirror?" "Your BMI is only 10. How do you feel about it?" If patient still thinks she is too fat, gently challenge her belief and check her rationales.	Explore common neuropsychiatric complications (e.g. slowing of mental speed, fit), gastrointestinal (GI bleeding), and endocrine systems (no menstruation). Severe weight loss, very low heart rate and metabolic complications such as very low potassium or anaemia. Explore relevant past medical history, e.g. childhood obesity.
B) Aetiology of Sally's illness	B1) Identify predisposing factors Family dysfunction including marital disharmony and	B2) Identify precipitating factors e.g. Sally may use her illness to get more attention from her parents	B3) Identify maintaining factors Identify the role of family in reinforcing and maintaining her abnormal eating	B4) Development in adolescence Explore her cognitive and psychosexual development. Focus on common	B5) School and peers Explore her interests and hobbies (e.g. ballet dancing, athletes) and academic performance. Explore her peer and

	sibling rivalry. Enmeshment, child abuse and rigidity in parenting may be present. Explore the family's views on food and weight.	and prevent them from arguing. This will positively reinforce her illness.	behaviour.	issues such as individuation.	romantic relationships (previous bullying or rejection due to body image)
C) Course of Sally's illness, comorbidity and risk	C1) Previous treatment Explore both outpatient and inpatient treatment being offered to Sally. Explore previous use of medication (e.g. antidepressant, antipsychotics) and adherence to psychotherapy sessions.	C2) Outcomes of previous treatments Focus on the weight restoration and identify reasons resulting in failure (e.g. engagement difficulties with Sally)	C3) Sally's insight and feeling towards her illness Sally may have impaired insight and denies any illness. She would be aggrieved by repeated attempts by her family to get her to seek help.	C4) Explore comorbidity e.g. depression, anxiety, OCD, substance abuse and perfectionistic personality Explore how the comorbidity influences the response to treatment.	C5) Risk assessment: History of suicide and deliberate self harm.

OSCE (Management of AN)

> After assessing Sally, she is keen for hospitalisation.
>
> Task: Discuss the immediate and short-term management with her parents.

Approach to this OSCE station (including the recommendations from the NICE guidelines - UK)

Overview of immediate, short and long term management.

Immediate management:
1. Acknowledge the severity of Sally's situation (i.e. the significance of her low BMI and possible mortality), the doctor's duty of care and the need for her to remain in treatment as this will be in her best interest.
2. A detailed assessment for Sally is required with necessary investigations
   (e.g. FBC, LFT, RFT, electrolytes, hormone profiles, fasting venous glucose to rule out DM, ECG and EEG). A detailed mental state examination including cognitive assessment and further risk assessment should be conducted
   Physical examination will focus on signs of AN.
3. Inform Consultant on call and make referrals for co-management with other disciplines. Discuss with the consultant with regards to the immediate management steps to adopt and whether oral, naso-gastric feeding or intravenous fluid replacement would be required. The decision will also be based on the input from the paediatrician, the potential re-feeding syndrome and the relative difficulties for administration as Sally may not cooperate.
4. If feeding is commenced, it is important to watch out for refeeding syndrome. Re-feeding syndrome refers to both the symptoms which occur when the patient is renourished (e.g. dependent oedema, aches and pains), the electrolyte deficiencies (e.g. K, Mg^{2+}, PO^{4+}) and cardiac decompensation (congestive heart failure) as a consequence of re-feeding. The preferred re-feeding method should be designed by a dietician to increase daily caloric intake slowly by 200-300 kcal every 5 days until sustained weight gain of 1kg per week is achieved. U&Es will be checked every 3 days in the first 1 week and then weekly during the re-feeding period. Diazepam will be administered per rectal or IV route when seizure occurs.

Short-term management:
1. A weight target should be set (0.5 to 1kg per week). Complete bed rest and regular nursing monitoring is advised.

2. Safety issues such as risking her life through further starvation or suicide must be considered

3. Inpatient treatment programme will involve nutritional rehabilitation and a structured protocol based on operant conditioning model with a balance of positive and negative reinforcers. There will be supervision and control of Sally's eating behaviour and regular monitoring of her weight. The NICE guidelines also recommend the gradual reduction of laxative use. Psychological treatment should be provided which has a focus both on eating behaviour and attitudes to weight and shape, and on wider psychosocial issues with the expectation of weight gain. Pharmacotherapy (e.g. antidepressants) is used to treat comorbidity such as depression or obsessive compulsive features. The use of fluoxetine to improve outcome and olanzapine to reduce rigidity of abnormal thoughts. The use of multivitamins, calcium and vitamin D are recommended but not hormonal treatment for osteoporosis.

Long-term management:
1. For post-hospitalisation management, the NICE guidelines recommend the monitor of growth and development as Sally is an adolescent. The NICE guidelines also recommend psychological treatment such as CBT, IPT(Inter-personal therapy) or family interventions. The choice of psychotherapy is based on Sally and her parents' preferences. The aims of psychological treatment are to reduce risk, to encourage weight gain and healthy eating, to reduce other symptoms related to an eating disorder and to facilitate recovery. The duration of psychological treatment is at least 12 months.

Metabolic syndrome

Metabolic syndrome is an important topic in psychiatry as second generation antipsychotics like olanzapine cause metabolic syndrome.

The US National Cholesterol Education Program Adult Treatment Panel III (2001) requires at least three of the following [11]:

- Dyslipidemia: TG $\geq$ 150 mg/dL (1.7 mmol/L)
- Blood pressure $\geq$ 130/85 mmHg
- Fasting plasma glucose $\geq$ 110 mg/dL (6.1 mmol/L)
- Central obesity: waist circumference $\geq$ 102 cm (male), $\geq$ 88 cm (female)
- Dyslipidemia: HDL-C < 40 mg/dL (male), < 50 mg/dL (female)

Management
- Dietary control and regular exercise
- Prescription of lipid lower agent, antihypertensive or oral hypoglycemics.
- Change antipsychotics. E.g. aripiprazole has the lowest chance to cause metabolic syndrome but it is expensive. First generation antipsychotics such as haloperidol have lower chance to cause metabolic syndrome as compared to second generation antipsychotics.

Revision MCQs

1. A 20-year-old woman with eating disorder is referred for poor impulse control. Physical examination reveals calluses on the knuckles. What is this sign called?

A. Crichton–Browne sign
B. Hoover sign
C. Lombardign
D. Russell sign
E. Waddell sign

Answer: D
Russell sign refers to calluses on the dorsum of the hand that occur due to induced vomiting over an extended period of time. Russell sign is seen in patients with bulimia nervosa and anorexia nervosa. Additionally, there is a close relationship between bulimia nervosa and borderline personality disorder.

2. A 20-year-old woman suffers from anorexia nervosa. Her parents are concerned about her outcome. Which of the following factors indicates a poor prognosis based on the medical literature?

A. Early age of onset
B. Family history of anorexia nervosa
C. Female gender
D. Family history of bulimia nervosa
E. Later age of onset

Answer: E
Later age of onset is a poor prognostic factor. Other poor prognostic factors include long duration of illness, severe weight loss, substance misuse and obsessive-compulsive personality.

3. A 28-year-old school teacher was referred by her GP for assessment of depression. She appears to be very thin but does not know her BMI. She insists that she was too fat in the past, which resulted in interpersonal problems. She eats three meals a day but is not able to describe her diet in detail. She denies excessive exercise but induces vomiting if she eats too much. She complains of amenorrhoea and alopecia. She has been irritable throughout the interview and emphasizes that she suffers from depression but nothing else. She is only keen to continue fluoxetine given by her GP but no other treatment. She emphasizes that she is in a good physical condition and is able to teach. After the interview, the nurse measures her BMI and the result is 13 kg/m^2. What is the most likely diagnosis?

A. Anorexia nervosa
B. Borderline personality disorder
C. Bulimia nervosa
D. Depressive disorder
E. Hypomania

Answer: A
In clinical practice, it is not uncommon to encounter patients with anorexia nervosa minimizing symptoms of eating disorder and attributing their low body weight to something else. It is often more useful to pay attention to objective signs such as low BMI and amenorrhoea to establish the diagnosis of anorexia nervosa.

4. A 20-year-old woman is found to have an enlarged parotid gland on physical examination. Which of the following diagnosis is likely?

A. Anorexia nervosa
B. Bulimia nervosa
C. Erotomania
D. Kleptomania
E. Trichotillomania

Answer: B
The frequent self-induced vomiting in bulimia nervosa leads to reflux of gastric acid and causes inflammation in the parotid gland. Erotomania refers to delusion of love. Kleptomania is compulsive stealing. Trichotillomania refers to the tendency of hair pulling.*

5. A 20-year-old woman suffering from anorexia nervosa presents with hypokalemic alkalosis. Which of the following behaviours is most likely to contribute to this finding?

A. Binging
B. Exercising
C. Fasting
D. Inducing diarrhoea
E. Vomiting

Answer is E.
Self-induced vomiting contributes to hypokalemic alkalosis.

Revision MEQ

You are the resident in the Children Emergency Department. A GP has referred to you a 14-year-old girl who has been losing weight in the past 4 months. In the last 4 months, she has become increasingly 'fussy' about her food, measures the calories she consumes and is 'obsessed with monitoring her weight'. Her parents cannot manage her at home as she refuses to eat. Her BMI is now 13. She has lost 18Kg due to diet and uncontrolled exercising in the past 3 months. Her mood is stable and she does not have suicidal thought. Her ECG now shows a regular heart rate of 30/min, with some flat and inverted T waves.

Question 1

What is the most likely diagnosis?

Anorexia nervosa

Question 2

What information you would ask in this history to confirm the most likely diagnosis which you have stated in question 1?

History of amenorrhoea

Question 3

List 3 COMMON FINDINGS on physical examination (other than low weight) which can occur with her condition

1. Cyanosis
2. Dehydration
3. Lanugo hair
4. Muscle atrophy
5. Hyporeflexia
6. Hypotension
7. Peripheral neuropathy
8. Reduced secondary sexual characteristics

Question 4

Her parents ask to admit the patient for inpatient treatment. Give THREE REASONS to support inpatient treatment in this case.

1. Low BMI
2. ECG abnormalities
3. Rapid weight loss in short period of time
4. Refusal to eat at home

Question 5

She also induces vomiting. State ONE BIOCHEMICAL ABNORMALITY which is likely to be found?

Hypokalaemia

EMIS:
Eating Disorders

A – Lanugo hair
B – Dilated Pupils
C – Constricted Pupils
D – Xanthelasma
E – Goiter
F – Lemon Stick appearance
G - Parotid swelling
H – Russell Sign

Question 1: A girl tends to hide food in the cupboard, and also refused to sit with others to have a meal. She has lost a lot of weight over the past 6 months. She used to be doing well in school, but recently her performance has been declining. She does not use anything to induce vomiting, but she exercises up to 3 times per day – Lanugo hair

Question 2: A women has a known history of eating disorder not otherwise specified and now comes into clinic vocalizing a history of uncontrollable episodes of overeating, which usually result in her purging and vomiting. She claimed that she has been maintaining her weight, but has lost control over eating again. – Parotid swelling, Russell sign

References

Books

[1] World Health Organisation (1992) *ICD-10 : The ICD-10 Classification of Mental and Behavioural Disorders : Clinical Descriptions and Diagnostic Guidelines*. Geneva: World Health Organisation.

[2] Taylor D, Paton C, Kapur S (2009) *The Maudsley prescribing guidelines*.10th edition London: Informa healthcare.

[3] Treasure J (2007) Anorexia Nervosa and bulimia nervosa in Stein G & Wilkinson G (2007) *Seminars in general adult psychiatry*. London: Gaskell.

[4] Birmingham CL and Beumont P (2004) *Medical Management of Eating Disorders*. Cambridge: Cambridge University Press.

[5] American Psychiatric Association (1994) *DSM-IV-TR: Diagnostic and Statistical Manual of Mental Disorders (Diagnostic & Statistical Manual of Mental Disorders)* (4th edition). Washington: American Psychiatric Association Publishing Inc.

[6] Semple D, Smyth R, Burns J, Darjee R and McIntosh (2005)

Oxford handbook of Psychiatry. Oxford: Oxford University Press.

[7] Firth JD, Collier JD (2001) *Medical Masterclass: Gastroenterology and hepatology*. London: Royal College of Physicians

[8] Olumoroti OJ & Kassim AA (2005*) Patient Management Problems in Psychiatry*. London: Elsevier Churchill Livingstone.

[9] Johnstone EC, Cunningham ODG, Lawrie SM, Sharpe M, Freeman CPL. (2004) *Companion to Psychiatric studies*. (7th edition). London: Churchill Livingstone.

[10] Ho TF, Tai BC, Lee EL, Cheng S, Liow PH. Prevalence and profile of females at risk of eating disorder in Singapore. *Singapore Med J 2006*; 47(6):499

[11] Expert Panel On Detection, Evaluation, And Treatment Of High Blood Cholesterol In Adults (May 2001). "Executive Summary of the Third Report of the National Cholesterol Education Program (NCEP) Expert Panel on Detection, Evaluation, and Treatment of High Blood Cholesterol in Adults (Adult Treatment Panel III)". *JAMA: the Journal of the American Medical Association* **285** (19): 2486–97

American Psychiatric Association (2013) DSM-5: Diagnostic and Statistical Manual of Mental Disorders (Diagnostic & Statistical Manual of Mental Disorders) (5th edition). American Psychiatric Association Publishing Inc: Washington.

Websites:

[10] NICE guidelines for eating disorders
http://guidance.nice.org.uk/CG9

[11] NICE guidelines for antenatal and postnatal mental health
http://guidance.nice.org.uk/CG45

Chapter 9 Psychiatric emergencies

Neuroleptic malignant syndrome	Suicide / deliberate self-harm and assessment
Serotonin syndrome	Drug overdose – paracetamol
	Drug overdose – benzodiazepines

Neuroleptic malignant syndrome (NMS)

Prevalence:
The incidence of NMS is 0.1-0.2%. Its incidence is lower among second-generation antipsychotics. There is mo geographical variation.

Aetiology:

Patients
Younger patients, agitation, and individuals who are physically exhausted or dehydrated are predisposing factors. Family history of NMS would predispose an individual to NMS as well.

Medication:
Medication factors that predispose individuals to NMS include the usage of high dose of antipsychotics; potent antipsychotics (e.g. haloperidol) and antipsychotics that are given via the intravenous or intramuscular routes.

Individuals who are either antipsychotic naïve or who are treated by multiple antipsychotics are at increased risk for NMS.

Underlying disorders/conditions:

Several underlying conditions predispose an individual to be at an increased risk for NMS. These conditions include:
- Underlying medical illnesses
- Catatonia
- Lewy body dementia
- Basal ganglia dysfunction
- Head injury
- Epilepsy
- Learning disability
- Low serum iron (Iron plays a key role in dopaminergic function)
- High CK level.

Psychiatric comorbidity
- Substance misuse (LSD is associated with hyperthermia syndrome).

Environmental factors:
- High ambient temperature and humidity

Neurochemistry underlying NMS:

NMS is an idiosyncratic reaction towards antipsychotics and caused by a sudden hypodopaminergic state that affects the hypothalamus. It will results in hyperthermia, catatonia, autonomic dysfunction, rigidity and clouding of consciousness.

Signs and symptoms of NMS:

Criteria A: 1) Muscle rigidity 2) Fever Criteria B: 1) Altered consciousness 2) Mutism 3) Dysphagia 4) Diaphoresis 5) Tachycardia 6) Labile blood pressure 7) Tremor 8) Incontinence 9) Leucocytosis 10) Laboratory evidence of muscle injury: increase in creatinine kinase levels.	Criteria C: 1) Not due to other medical causes such as viral encephalitis Criteria D: 1) Not due to other psychiatric disorders

Differential diagnosis:

1) CNS infection e.g. meningitis
2) Septicaemia
3) Serotonin syndrome
4) Intoxication of other drugs: lithium, cholinergic drugs, MAOIs, amphetamine and anticholinergic drugs
5) Lethal catatonia (prodrome of psychotic symptoms for 2-8 weeks, intense motor excitement for several days, lead to autonomic dysfunction, catatonia, stupor, coma and death).
6) Catatonia (both NMS and catatonia can lead to an increase in creatinine kinase. Catatonia presents with echolalia, echopraxia, ambitendency and abnormal posturing but not NMS)
7) Heavy metal poisoning (thallium or arsenic)
8) Myocardial infarction
9) Heat stroke (during the heat wave and mainly affect elderly in residential care)
10) Malignant hyperthermia as hypersensitive reaction to certain anaesthetics in genetically predisposed individuals)
11) Tetanus infection
12) Thyroid storm
13) Acute intermittent porphyria.

Relevant Investigations:

1) FBC: leucocytosis (>95%)
2) Disseminated intravascular coagulation (DIC) screen: PT, APTT, INR and fibrinogen and check peripheral blood smear.
3) RFTs, LFTs, TFTs, ammonia level, ABG, iron levels (low iron levels predict poor prognosis)
4) Creatinine kinase (↑ both NMS and serotonin syndrome)
5) Blood cultures
6) Urinalysis (to look for myoglobin)
7) Urine cultures
8) CXR
9) CT (brain)
10) EEG to if seizure is suspected
11) Lumbar puncture if CNS infection is suspected
12) Abdominal / pelvic CT if abscess is suspected.

Treatment of NMS:

NMS is a clinical emergency. It is important to stop the antipsychotics immediately. Supportive measures such as bed rest and controlling the hyperthermia by rapid cooling with the help of tepid water spray and via direct fluid replacement should be started immediately. Consideration of ventilator support or intubation would be necessary if the patient has severe breathing difficulties.

Pharmacological treatment:

Dopamine agonists
Bromocriptine 2.5 mg TDS could be given. It may worsen the underlying psychosis.

Dantrolene:
Dantrolene is administered parenterally (50-75mg immediately then every 6 hours until the maximum dose of 10mg/kg/day) It inhibits the ionised calcium release and used for treatment of malignant hyperreflexia. It should be discontinued when symptoms resolve.

Benzodiazepine:
Lorazepam (Oral or IV) can be administered up to 8-24mg/day. GABA-mimetic activity may indirectly increase dopaminergic function in the basal ganglia.

ECT:
ECT can treat malignant catatonia and NMS and patients usually respond after 4 sessions of ECTs.

**Reintroduction of antipsychotics from another class after 2 weeks or those with low dopamine receptor affinity e.g. quetiapine or aripiprazole. Begin with very small dose and increase very slowly with close monitoring of temperature, pulse and blood pressure.

Prognosis:

There might be complication such as renal failure or respiratory failure that might lead to mortality.
Current mortality rate: < 10%
NMS can last from a few days to a month.

Serotonin syndrome

Prevalence: 15% of people who took an overdose of a combination of SSRIs.

Aetiology

Combination of medications
1) MAOIs + SSRIs
2) Switching from one antidepressant to another without adequate washout period
3) Overdose of SSRIs.

Neurochemistry

Serotonin syndrome is a clinical emergency due to the increased in 5-HT and the stimulation of $5HT_{1A}$ and $5HT_{2A}$ receptors.

Clinical signs and symptoms

Clinical triads
1) Mental status change: anxiety/ agitation to extreme confusion.

2) Autonomic hyperactivity: tachycardia, tremor, flushing, hyperthermia and excessive sweating.
3) Neuromuscular abnormalities: generalised hyperreflexia, clonus (ankle or ocular), myoclonus and rigidity.

Initial symptoms involve akathisia, agitation, tremor, tachycardia, autonomic instability, increased bowel sounds, diarrhoea and mydriasis.

Serotonin syndrome has more rapid onset and development compared to NMS. Serotonin syndrome causes less rigidity compared to NMS. Serotonin syndrome may lead to hyperkinesia.

Differentials and Investigations
The differentials and investigations are largely similar to NMS. Please refer to the previous section.

Treatment of serotonin syndrome
Given that this is a clinical emergency, it is important to stop the antidepressants immediately. Supportive measures should be adopted.

Benzodiazepines:
Lorazepam (Oral or IV) as stated under NMS.

Other alternative treatment:
ECT is indicated for patients who are severely depressed.

Re-starting antidepressant
It may not be necessary to avoid the offending agents which cause serotonin syndrome.

Suicide / Deliberate self-harm

Epidemiology of suicide in Singapore

Suicide trends from 1955 to 2004 (Chia et al 2010)

Overall, suicide rates in Singapore remained stable between 9.8-13.0/100,000 over the last 5 decades. Rates remain highest in elderly males, despite declines among the elderly and middle-aged males in recent years. Rates in ethnic Chinese and Indians were consistently higher than in Malays. While the rates among female Indians and Chinese have declined significantly between 1995 and 2004, some increase was noted in female Malays. Although there was no increase in overall suicide rates, risk within certain population segments has changed over time.

Suicide methods (Chia et al 2011)

Common methods used in Singapore between 2000 and 2004 were jumping (72.4%), hanging (16.6%), and poisoning (5.9%). Those who jumped were more likely to be young, single, female, and to have had a major mental illness. By comparison, those who hung themselves were more likely to be older, Indian, and to leave a suicide note. Those who used poison were more likely to be married, to be on antidepressants, to have previously attempted suicide, and to leave a letter.

Young people (10-24 years) (Loh et al, 2012):
The suicide rate is around 6 per 100, 000, with gender ratio of 1:1 and higher rates among ethnic Indians. Psychosocial stressors and suicide by jumping from height were common. Mental health service use was associated with unemployment, previous suicide attempts, family history of suicide, more use of lethal methods, lack of identifiable stressor, and less suicide notes.

Old people (> 60 years) (Ho et al, 2012)
Elderly victims with past history of suicide attempts were more likely to suffer from major mental disorders, encounter social problems in life, have alcohol detected in the blood toxicology report at autopsy, receive psychiatric treatment in the past, presence of antidepressant in the blood toxicology report at autopsy and be admitted to a mental hospital with gazetted

wards. Conversely, those without past history of suicide attempts were more likely to have pre-suicidal plan for the final suicide act and receive medical or surgical treatment in the past.

Predisposing factors:

Biological causes:

1) **Serotonin (5HT):** Through estimates of 5-HT metabolites, 5HIAA (5-hydroxyindoleacetic acid), lower 5-HT levels were found in suicide attempters and in post-mortem brain tissue of persons who had committed suicide and in genetic association of family studies.

2) **Genetics**: genetic factors accounted for 45% of variance in suicidal thoughts and behaviour.

3) **Physical illness:** HIV/AIDS, malignancy (head and neck cancers), Huntington's disease, multiple sclerosis, peptic ulcer, renal disease, spinal cord injury and SLE. Pregnancy and puerperium have protective value and decrease the suicide risk of women.

Social factors:

1) **Durkheim:** In 1897, Durkheim suggested suicide results primarily from social factors. There are four types of suicide: egoistic (poor integration into society), altruistic (over integration into society e.g. political hunger strike), anomic (loosening bonds between people, e.g. in inner city), fatalistic (excessive regulation by society and no personal freedom, e.g. suicide of slaves).

2) **Cultural:** Gothic culture in the UK has higher risk of deliberate self-harm.

Other aspects:

1) **SSRIs induced suicide:** There is no evidence of increased risk of suicide with SSRI usage in adults but a modest increase in risk in children and adolescents. More data is emerging that SSRIs may be safe to use in depressed children and adolescents become completed suicide is rare. Doctors should monitor patients closely in the beginning stage as improvement in energy level may allow patient to carry out suicide.

2) **Increase in adolescent male suicide rates:** due to increase in incidence of alcohol and drug misuse.

3) **Suicide pacts and internet:** Suicide pacts refer to two or more persons agree to commit suicide together. With the development of internet, there are suicide websites which teach people methods to commit suicide and blogs to allow unknown people to exchange suicidal ideas and commit suicide together.

4) **Suicide terrorism** is a result of religious fundamentalism but they are not psychiatrically ill.

5) **Homicide following suicide:** committed by young men with intense sexual jealousy, depressed mothers, despairing elderly men with ailing spouses.

6) **Media coverage and copycat suicides**: charcoal burning has become a popular method of committing suicide in South East Asia with irresponsible media coverage who did not follow WHO guidelines.

7) **Access to firearms:** In countries with easy access to firearms, this is a common method to commit suicide. In countries where access is difficult, hanging and jumping from height are the common methods.

Factors associated with repeated suicidal attempts include:

Demographic characteristics	Psychiatric history	Personal characteristics
1) separation / divorce 2) low social class 3) living alone 4) poor social support 5) unemployment.	1) previous psychiatric treatment (inpatient or outpatient) 2) alcohol and drug abuse 3) past history of self-harm (especially if associated with hospital admission and following multiple episodes) 4) family history of self-harm or suicide 5) Personality disorder, especially antisocial personality.	1) impulsivity 2) criminality 3) high levels of hostility 4) refusal of help: lack of co-operation with helping agencies 5) poor coping skills or problem-solving abilities 6) hopelessness 7) High levels of intercurrent social stress, particularly relationship problem.

Deliberate self harm

Predisposing and precipitating factors

1) More common in young females (age between 15 to 24 years).
2) Relationship or interpersonal problem (e.g. argument with someone) is a common precipitating factor.
3) Association with low socioeconomic class.
4) Association with poor impulse control and borderline personality.

Definitions of deliberate self harm

- **Non-fatal Deliberate Self Harm (DSH)** (Morgan 1979) : "Deliberate non-fatal act known to be potentially harmful, or if an overdose, that the amount taken is excessive"
- **Parasuicide** (Kreitman 1977) : "Behavioural analogue of suicide without considering psychological orientation towards death"
- **Attempted Suicide** (Stengel & Cook) : "Every act of self injury consciously aimed at attempts to kill themselves. But it acknowledges the gravity of the situation"
- **Deliberate self-poisoning** (Kessel and Grossman 1965) : 'deliberate self injury' substituted for 'attempted suicide' because many patients 'performed their acts in the belief that they were comparatively safe'

Clinical OSCE on suicide assessment

Name: Miss C   **Age:** 22 – year –old   University student

Miss C is referred to you by her general practitioner after she took 30 tablets of paracetamol. She states that she lacks motivation in life. Life appears to be meaningless. Her existence is only postponing the inevitability of death. She has a history of repeated self-injury and she had two previous psychiatric admissions with her discharging herself. She claims that she has feels this way throughout her life

Task: 1) Perform a suicide risk assessment. 2) Explore the underlying cause for her suicidal ideation.

A) Risk assessment	A1) Empathy statement: "I can imagine that you have gone through a difficult period. I am here to help you and listen to you."	A2) Current suicidal intention 1) Do you wish that you were dead? 2) Do you still have thoughts of ending your life? (If so, are they intermittent or more persistent?) 3) How often do you act on these ideas? 4) How strongly are you able to resist those thoughts?	A3) Detailed assessment of suicide plan 1) Intent: 'Did you intend to end your life by taking an overdose?' 2) Detailed plans made: 'Did you plan for this suicide attempt? If yes, how long did you plan for it?' 3) Method considered and available: 'Besides overdose, did you harm yourself in other ways?' 4) 'Did you act alone or in front of the others?' 5) Did you inform anyone prior to suicide attempt? 6) Post-suicide attempt: 'Did you try to avoid discovery? Did you seek help?'	A4) Negative aspects of life 1) 'Have you ever felt despaired about things?' 2) 'Have you ever felt life is a burden?' 3) 'Have you ever felt entrapped, defeated or hopeless?'	A5) Positive aspects of life 1) 'Do you hope that things will turn out well?' 2) 'Do you get pleasure out of life?' 3) 'Can you tell me more about your support system?' 4) 'Do you have any spiritual support? E.g. religion?'
B) Underlying causes	B1) Current life stressors e.g. adjustment to university life, study load and relationship problems.	B2) Assessing problems from a developmental perspectives e.g. childhood physical abuse, separation from parents, marital discord of parents, witnessing domestic violence or witnessing someone committing suicide in the	B3) Existential crisis 'Do you feel isolated?' 'How do you see the world? Do you feel the world is hostile and meaningless? If so, is suicide your final destiny?'	B4) Past suicide attempts and self harm History of self harm and suicide attempts. Explore common precipitating factors of previous suicide attempts.	B5) Psychosocial problems e.g. unplanned pregnancy, financial problems and poor coping.

C) Psychiatric comorbidity	C1) Depression and previous therapeutic relationships Explore depressive symptoms in details (e.g. low mood, guilt, insomania, loss of interest).	C2) Substance abuse 1) Do you take recreational drugs to cope with life? 2) How about alcohol or smoking?	C3) Eating disorder 1) How do you see your body image? 2) Have you put yourself on diet? 3) How about binge eating?	C4) Explore personality disorder 1) Borderline personality traits: chronic feeling of emptiness, unstable emotion, impulsiveness.	C5) Early psychosis or schizophrenia 1) Command hallucination (e.g. have you ever heard voices to ask you to harm yourself?)
	family and risk of developing post traumatic stress disorder				

NICE guidelines recommendations on the management of self-harm

Adults
1) For adults with insufficient capacity, offer interventions under the common law if the benefits outweigh consequence of not intervening.
2) Discuss treatment options and consider the patient's preference if he or she has capacity.
3) For adults who repeatedly self-poison, consider discussing the risks of self-poisoning with the patients and carers where appropriate. Do not offer harm minimisation advice as there is no safe limit for overdose.
4) For adults who repeatedly self-injure, consider giving advice on self-management of superficial injuries (e.g. providing tissue adhesive), appropriate alternative coping strategies, harm minimisation and dealing with scar tissue.
5) Consider offering an intensive therapeutic intervention (greater access to a therapist, home treatment) combined with outreach to people who have self-harmed with high risk of repetition. The duration of intensive intervention is 3 months.
6) Refer people with borderline personality disorder for dialectical behaviour therapy.

Children and adolescents
1) All children and young people should normally be admitted into a paediatric ward under the overall care of a paediatrician and assessed fully the following day after obtaining consent from the young person or parents.
2) If the young person is 14 years or older, consider an adolescent paediatric ward.
3) During admission, the Child and Adolescent Mental Health services team should provide consultation for the young person, their family, the paediatric team, social services, and education staff.
4) For young people who have self-harmed repeatedly, consider offering developmental group psychotherapy with other young people. This should include at least six sessions but can be extended by mutual agreement.

Old people
1) All acts of self-harm in people over the age of 65 years should be taken as evidence of suicidal intent until proven otherwise.
2) Always consider admitting the patient for mental health assessment, risk and needs assessment. Admission will allow monitoring changes in mental state and assess the levels of risk.

Drug Overdose: paracetamol

Paracetamol is the most commonest reported poisoning agent since it is available without prescriptions. It is also a common co-ingestant which causes a significant number of morbidities, namely hepatic injury.

For acute overdose, a single ingestion of more than 150mg/kg is considered to be potentially toxic, for repeated overdose, which is a history of more than 4g of paracetamol over 24 hours in adult or more than 90mg/kg over 24 hours in children may suggest potentially toxicity.

Since it takes around 4 hours for the paracetamol levels to reach the peak serum concentration in overdose situations, time for checking the first drug level should be at least 4 hours (but not later than 15 hours) post ingestion. Liver enzyme levels (aspartate aminotransferase), clotting profile, blood gas and lactate should be taken as baseline.

For the initial management, gastric lavage is not a must for all patients; it is only indicated in those who present with massive overdose (more than 1g/kg) or those with severe toxic co-ingestant. Activated charcoal may help in decreasing the ongoing absorption when given within 1 hour post ingestion.

N-acetylcysteine (NAC), which is a well known effective antidote for paracetamol overdose, prevents the covalent binding of NAPQI (toxic metabolite of paracetamol) to hepatocytes. NAC is 100% protective when given within 8 hours of ingestion, and is known to be effective when given within 24 hours. In adults, it is given intravenously by an initial loading dose of 150mg/kg in 200ml D5 in 1 hour, followed by 50mg/kg in 500ml D5 over 4 hours and then a further 100mg/kg in 1000ml D5 over 16 hours.

In patients who receive NAC within 8 hours post ingestion, a 21-hour course provides a full protection, but for those who were put on NAC more than 8 hours post-ingestion, a continuous NAC infusion is recommended until the liver injury resolve and the paracetamol level is not detectable. Please contact Emergency Medicine toxicology team at your hospital for further advice in this case.

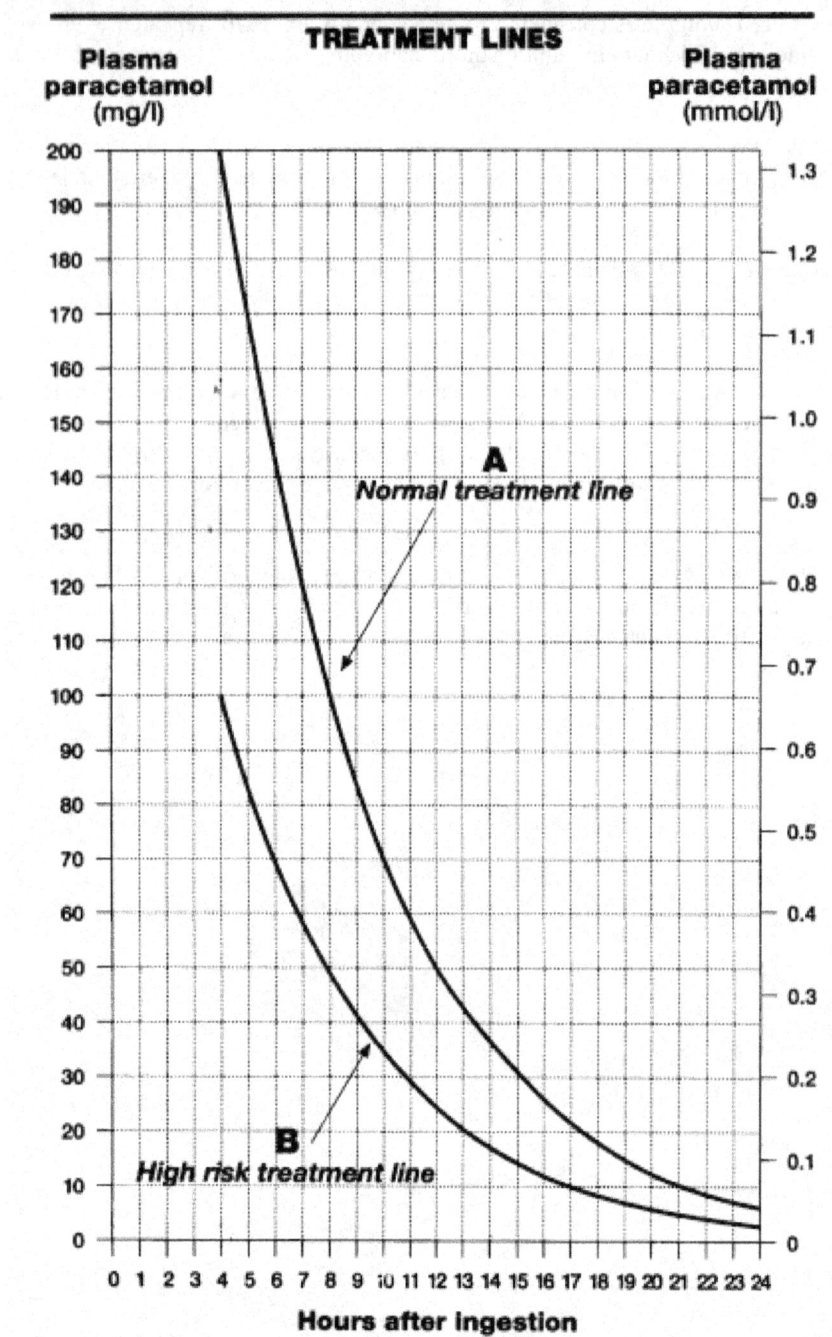

The Prescott nomogram was developed in adults and guide the plasma paracetamol level concentration in which NAC should be given to poisoned patients. High risk patient included those with induced liver enzymes arising either from chronic alcohol consumption or those who on anticonvulsant drugs or those with depleted glutathione store. Beware that this nomogram is not valid in case of staggered overdose or those who presented over 24 hours post-ingestion.

Drug Overdose: Benzodiazapines

Benzodiazepine, zopiclone and zolpidem are commonly abused hypnotics world-wide, death from pure overdose is relatively rare due to their good safety margin. These two groups of hypnotics are similar in terms of presentations and management approach. Both of them will act on the GABA receptor in neuron and lead to depression of the CNS. Patients who take an overdose on medication will present with excess sedation, lethargy, in coordination, slurred speech and impaired cognitive function, but most patients are arousable and can maintain a patent airway.

In Hong Kong, there are some local case series of methaemoglobinemia after massive zopiclone (more than 100 tablets) ingestions. The methaemoglobinemia level peaked at 13-18 hours and may rise up to 90 hours post ingestion. Methylene blue injection was needed in some cases.

Drug levels are not useful in the management. Investigations for any undeclared co-ingestions, e.g. blood paracetamol level, ECG to look for QRS for any possible co-ingestions of sodium channel blocking agents such as tricyclic antidepressants should be considered. Alternative medical causes for CNS depression should be ruled out in patients with deep coma.

Management is mainly supportive, and make sure that the patient is able to maintain a patent airway with adequate ventilation and oxygenation. Gastric lavage is rarely necessary, activated charcoal should only be considered if the patient is presented within 1 hour of ingestion and those with a patent airway, otherwise aspiration pneumonia will be a serious iatrogenic complication.

Flumazenil, a specific benzodiazepine antagonist, is not recommended to use routinely in every coma patients presented with benzodiazepine overdose. It will precipitate convulsion and withdrawal symptoms in chronic benzodiazepine users and those with known history of epilepsy, it may unmask the toxicity of other pro-convulsant co-ingestions (such as TCA and anti-psychotics). Flumazenil should only be considered after a comprehensive assessment of patient's respiratory status and it is used in patients with evidence of respiratory depression likely to lead to ICU admission with endotracheal intubation.

Sample MCQ

1. A 20-year-old woman was given an intramuscular injection of clopixol acuphase. She seems to develop neuroleptic malignant syndrome (NMS). Which of the following clinical features is the least important to establish the diagnosis of NMS?

A. Difficulty in swallowing
B. Onset of symptoms occur after 1 day
C. Hyperthermia
D. Incontinence
E. Labile blood pressure

Answer: B
The symptoms of NMS may occur after a few days or weeks after intramuscular administration of antipsychotic agent. Hence, the onset is the least important criteria.

2. You are teaching depressive disorder to a group of medical students. They want to know what percentage of patients admitted to the university hospital will have recurrence and require further admission in long run without committing suicide. Your answer is:

A. 20%
B. 30%
C. 40%
D. 60%
E. 80%

Answer: D
An old British study showed that approximately 60% of patients had been re-admitted at least once. Only 20% had recovered fully with no further episodes and 20% were incapacitated throughout or died of suicide.**

Reference: Lee AS, Murray RM. The long-term outcome of Maudsley depressives. *British Journal of Psychiatry* 1988; **153**: 741–51.

Sample MEQ
You are the resident working in the emergency department. A 25-year-old woman took 40 tablets of paracetamol to end her life.

1. You would like to assess her suicide risk. State Ten (10) questions you would like to ask to assess her current suicide risk.

1. Did you take the overdose on impulse or plan for a long time?
2. Did you try to avoid discovery before you took the overdose?
3. Did you write a suicide note or send a SMS / email before you attempt suicide?
4. Did you write a will or have arrangement after your death?
5. Did you expect that you would die by taking 30 tablets of paracetamol?
6. Did you mix with alcohol?
7. Did you cut yourself or use other means to harm youself?
8. Did you seek help or avoid seeking help after the overdose?
9. Do you feel remorseful of the suicide attempt
10. Do you still want to die?
11. History of past suicide attempts.

2. You have decided to admit this patient to the medical ward for management. What advice would you give to the nursing staff.

Suicide caution or suicide precaution or close monitoring

3. You would like to assess her mood. Name Ten (10) diagnostic criteria for depressive disorder (based on DSM-IV-TR or ICD-10 criteria).

1. Delusion – mood congruent
2. Hallucinations – mood congruent, usually auditory
3. Low mood
4. Low energy or tiredness
5. Low sexual drive
6. Pessimism or negative thinking or cognitive distortion
7. Poor concentration or poor attention
8. Poor appetite
9. Guilt
10. Suicidal thoughts or plan
11. Weight loss

4. After treatment in the medical ward, she is medically stable. She is interested to take antidepressants to treat her depressive disorder. Name ONE (1) class of antidepressants which you would recommend as first-line and give ONE (1) example of antidepressant under the class you have chosen.

Selective serotonin reuptake inhibitors (accept SSRIs)

Fluoxetine (Prozac), or fluvoxamine (faverin) or

escitalopram (lexapro), sertraline (Zoloft), or paroxetine (seroxat)

5. Patient is interested to try fluoxetine but she wants to find out the side effects. Name Six (6) common side effects.

1. Anxiety
2. Nausea
3. Diarrhoea
4. Headache
5. Insomnia
6. Sexual dysfunction

EMIS:

A-Agitation
B - Clouded consciousness
C - Diarrhoea
D-Fluctant blood pressure and pulse
E-Hallucination

Characteristic features of delirum tremens include all except – Diarrhoea

References

Books

Taylor D, Paton C, Kerwin R (2007) *The Maudsley prescribing guideline*. London: Informa healthcare.

American Psychiatric Association (1994) *DSM-IV-TR: Diagnostic and Statistical Manual of Mental Disorders (Diagnostic & Statistical Manual of Mental Disorders* (4th edition). Washington: American Psychiatric Association Publishing Inc.

Gelder M, Mayou R, Cowen P (2001) *Shorter Oxford Textbook of Psychiatry*. Fourth Edition. Oxford: Oxford University Press.

Vassilas CA, Morgan G, Owen J, Tadros G (2007) Suicide and non-fatal deliberate self harm in Stein G & Wilkinson G (2007) *Seminars in general adult psychiatry*. London: Gaskell.

Website

NICE guidelines on management of self harm:
http://guidance.nice.org.uk/CG16/QuickRefGuide/doc/English

www..trickcyclists.co.uk

Journals

Ahuja N & Cole AJ (2009) Hyperthermia syndromes in psychiatry. *Advances in psychiatric treatment*; **15**: 181-191.

Caroff SN & Mann SC (1993) Neuroleptic malignant syndrome. *Medical Clinics of North America*; **77**: 185-202.

Chia BH, Chia A, Ng WY, Tai BC. (2011) Suicide methods in singapore (2000-2004): types and associations. Suicide Life Threat Behav. 2011 Oct;41(5):574-83.

Chia BH, Chia A, Yee NW, Choo TB. Suicide trends in Singapore: 1955-2004. Arch Suicide Res. 2010 Jul;14(3):276-83.

Hall RCW, Hall RCW & Chapman M (2006) Neuroleptic Malignant Syndrome in the elderly: diagnostic criteria, incidence, risk factors, pathophysiology and treatment. *Clinical Geriatrics*; **14**: 39-46.

Ho RC, Ho EC, Tai BC, Ng WY, Chia BH (2012) Compare and contrast elderly suicide with and without history of past suicide attempts: an implication for suicide prevention and management. Submitted for publication.

Isbister GK, Buckey NA, Whyte IM (2007) Serotonin: a practical approach to diagnosis and treatment. *Medical Journal of Australia*: **187**: 361-365.

Simon GE, Savarino J (2007) Suicide attempts among patients starting depression treatment with medications or psychotherapy. *American Journal of Psychiatry*, **164**(7):1029-34.

Gibbons RD, Brown CH, Hur K, Marcus SM, Bhaumik DK, Mann JJ (2007) Relationship between antidepressants and suicide attempts: an analysis of the Veterans Health Administration data sets. *American Journal of Psychiatry*, **164**(7):1044-9.

Statham DJ, Heath AC, Madden PA, et al (1998) Suicidal behaviour: an epidemiological and genetic study. *Psychological Medicine*, **28**:839-855.

Lee E, Leung CM. (2008) Clinical predictors of psychiatric and medical morbidities of charcoal-burning suicide attempt in Hong Kong. *Gen Hosp Psychiatry*. **30(6)**:561-3.

Fung HT, Lai CH, Wong OF, Lam KK, Kam CW. (2008) Two cases of methemoglobinemia following zopiclone ingestion. *Clin Toxicol*; **46(2):**167-70.

Kung SW, Tse ML, Chan YC, Lau FL, Tsui SH, Tam S, et al. (2008) Zopiclone-associated methemoglobinemia and renal impairment. *Clin Toxicol* ; **46(10):**1099-100.

Loh C, Tai BC, Ng WY, Chia A, Chia BH. 2012 Suicide in young singaporeans aged 10-24 years between 2000 to 2004. Arch Suicide Res. Apr;16(2):174-82.

Chapter 10   Psychotherapies

Interactions between psychotherapist and client	Interpersonal therapy
Supportive psychotherapy	Family therapy
Defence mechanisms	Group therapy
Psychodynamic psychotherapy	Couple therapy
Cognitive behavioural therapy	Eye movement and desensitisation reprocessing
Dialectical behavioural therapy	Grief counselling
Cognitive analytic therapy	Motivational interviewing

Interactions between psychotherapist and client in psychotherapy

Interactions in supportive psychotherapy and counselling

Therapists can perform the following:

General techniques: Listen and show concerns, restore morale, an element of suggestion may be helpful towards improvement.

Giving advice and information on the condition and how to behave in a specific situation.

Affirmation: confirmation of the validity of a prior judgment and/or behaviour.

Praise to reinforce certain behaviours and thoughts.

Explanation of medication compliance, family intervention, aspects of illness behaviour or alternative forms of psychotherapy

Observation: attention to non-verbal behaviour, emotions and other patterns of communication displayed by patient.

Therapist supports defences, minimises anxiety and regression. Dreams and transferences <u>are not</u> explored.

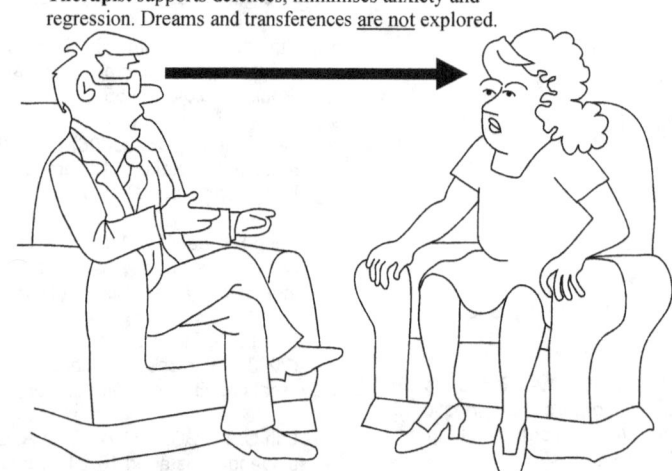

Examples of counselling:

Counselling: giving advice, allowing patient to release emotion and providing knowledge of the disease.

Problem – solving counselling: List problems, target at one problem, consider solutions, try out solutions and review outcome.

Interpersonal counselling: Identify relationship problems and consider ways to cope with difficulties.

Crisis intervention: focus on current problem, reduce arousal, allow expression of emotion, reassurance and consider solutions.

Supportive psychotherapy

Aim: Supportive psychotherapy is a unique psychotherapy solely focuses on the needs of the patient.

Techniques
1. Therapist listens, allows emotional release, provides information and encourages hope.
2. Therapist helps client to develop insight into problems as needed to improve adaptive responses.
3. Therapist encourages patients to develop positive feelings to help maintain therapeutic alliance.
4. Transference is not discussed in supportive psychotherapy.

Indications: most situations (e.g. adjustment issues, acute stress reaction, relationship problems etc).

Contraindications:
1. When psychotherapy itself is contraindicated (e.g. advanced dementia delirium and intoxication)
2. Poor motivation
3. When other psychotherapies are more appropriate (e.g. cognitive behaviour therapy for obsessive compulsive disorder).

Interactions in exploratory or psychodynamic psychotherapy

Therapist can perform the following:

Clarification: therapist wants to check whether his or her understanding is correct and help patients to recognise repetitive patterns in patient's life.

Interpretation: statements made by the therapist to help explain the patient's thoughts, feelings, behaviours or symptoms.

Empathetic validation: therapist puts himself or herself into patient's shoes and tries to understand patient's inner state.

Countertransference: involves unconscious emotional needs, wishes or conflicts arising from the therapist's prior relationship experiences being evoked by the patient during psychotherapy.

Observation of non – verbal behaviour

Therapist confronts defences, analyses anxiety, transference and allows regression. Dreams and transferences are explored.

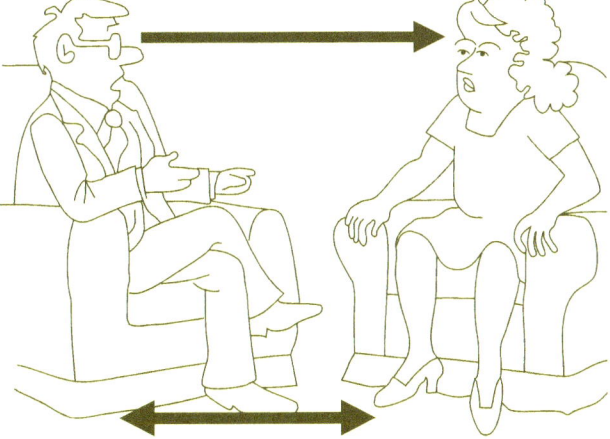

Therapeutic alliance: an agreement between patient and therapist to work together on psychological problems with patient's active commitment

Phenomenon associated with patient:

Suitability: 1) curiosity about self and wish to understand; 2) capacity to maintain an area of objectivity about oneself; 3) being psychologically minded; 4) capacity to enter into a relationship and tolerate frustration; 5) not suffering from active addictions or psychotic symptoms

Transference: patient unconsciously relates the therapist to someone from her past.

Resistance: patient is ambivalent about getting help and may oppose attempts from the therapist to help. This may manifest in the form of silence, avoidance or absences. These can reveal a great deal about significant relationships in the past.

Issues related to countertransference

Transient and mild countertransference reactions that are easily recognised may have little impact on the therapy. Persistent and generalised countertransference may promote 'selection' or 'rejection' of certain types of clients to be taken on into therapy. More persistent or recurrent reactions may need to be closely examined during psychotherapy supervision. The clients' evoking countertransference reactions may lead the psychotherapists to reflect on and improve their awareness of their own countertransference vulnerabilities, causing them to deal with the countertransference by adjusting their practices accordingly, seeking further supervision from colleagues or psychotherapeutic assistance for themselves.

Failure to detect persistent and recurrent countertransference may lead to the following negative therapeutic consequences:
1. Failure to institute appropriate treatment
2. Sexual contact with the client
3. Hostility or abusiveness towards the client
4. Tolerating abusive behaviour by the client
5. Fostering undue dependency and interminable treatment
6. Over-interpretation as a result of projection of the therapist's own conflicts on to material presented by the client.

Defence mechanisms

Table 1. Definition of defence mechanisms

Repression	Unconscious forgetting of painful memories and impulses.
Regression	Revert to functioning of a previous maturational point.
Denial	Refusal to consciously acknowledge events or truths that are obvious.
Projection	Attributing one's own unacceptable ideas or impulses to another person.
Projective Identification	One person projects a thought, belief or emotion to a second person. Then, there is another action in which the second person is changed by the projection and begins to behave as

	though he or she is in fact actually characterized by those thoughts or beliefs that have been projected.
Reaction formation	The expression externally of attitudes and behaviours which are the opposite of the unacceptable internal impulses.
Displacement	Transferring the emotional response to a particular person, event, or situation to another where it doesn't belong but carries less emotional risk.
Rationalisation	Justifying behaviour or feelings with a plausible explanation after the event, rather than examining unacceptable explanation.
Sublimation	Regarded as healthy defence mechanism, The external expression of unacceptable internal impulse in socially acceptable way.

Table 2. The roles of defence mechanisms in psychiatric disorders

Disorders	Defence mechanisms
Schizophrenia	1. **Splitting** between internal world and reality due to breakdown of ego boundaries. 2. **Projection** of persecutory experience and fear of annihilation to objects in the external world.
Depressive disorder (often involves loss)	1. **Denial** of grief reaction toward the lost object, status or person. 2. **Introjection** of anger into oneself.
Bipolar disorder	1. Mania is a defence against depression and **denial** of depression. 2. **Manic defence** is used in narcissistic personality.
Anxiety disorders	1. **Denial** of one's anger and destructiveness 2. **Projection** of one's fear to the dangerousness of the external world.
Phobia	1. **Isolation** of unconscious fears. 2. **Displacement** of fear onto an external object.
OCD	1. **Projection** of unconscious wish to damage onto an external object such as dirt. 2. **Undoing** of obsession thoughts by compulsive behaviours. 3. Obsession for order is a **reaction formation** to unconscious wish of disorganisation.
Eating disorder	1. Anorexia involves **denial** of bodily needs to eat, having menstruation and sexual development 2. Bulimia involves **displacement** of fear and craving of intimacy onto food.

Psychodynamic psychotherapy

Aims:

The main aim of psychodynamic psychotherapy is to enable clients to improve their self-understanding and enabling them to engage in a reflection of themselves. The therapeutic process enables clients to develop better tolerance for frustration, and also helps to increase their awareness of their underlying maladaptive defence mechanisms that would require modification. Therapy aims to help them improve their interpersonal relationships and also helps them cope with their personal symptoms.

Indications:

Commonly used for individuals with depressive and anxiety related disorders, or for individuals who have had experienced childhood abuse and trauma or who are dealing with relationship and personality problems. Psychodynamic psychotherapy requires clients to have adequate ego strength and possess abilities to form and sustain relationships. More importantly, the clients must be motivated to change, and receptive to psychological therapy.

Contraindications:

Individuals with schizophrenia, tendencies for serious self harm, or with addiction related problems are not suitable for this modality of therapy. Those with extremely poor insight into their own conditions are also excluded.

Techniques:

1) Establish working alliance where the client and therapist agree to work on an emotional or psychological problem
2) Free association
3) Focus on identified conflict
4) Development of therapeutic alliance
5) Transference interpretation
6) Working through involves drawing previous maladaptive patterns or defences to conscious awareness
7) Enactment refers to playing out the psychological phenomenon such as regression in a safe setting to facilitate understanding
8) Containment of anxiety
9) Resolution of conflicts and avoid maladaptive defences
10) Addressing termination issues
11) Techniques used in supportive psychotherapy may be appropriate in times.

Phases:

The therapist engages in a discussion and clarification of the current problem when therapy commences. An exploration into the nature and the origins of the problem is undertaken. Therapeutic alliance is established between the therapist and the client, and the therapeutic relationship thus formed and enables the identification and confrontation of defences. Underlying unconscious motives are identified during the therapy itself and subjected to interpretation. The therapist helps the client in working through and resolves existing conflicts. Once that it achieved over several sessions, the therapy is then terminated.

Negative reactions:

1) **Acting out** refers to the poor containment of strong feelings triggered by the therapy (e.g. anger towards the therapist).
2) **Acting in** refers to the exploration of therapist's personal and private information.
3) **Negative therapeutic reaction** refers to the unexpected deterioration in the face of apparent progression (e.g. premature termination of therapy without explanation after apparent engagement).

Termination of psychodynamic therapy

Termination is a point where therapeutic endeavour comes to a defined end date set by the contract in the beginning but it is often a complex and powerful event. It can be examined at the intrapsychic, behavioural, cognitive, interpersonal and social levels (e.g. legal implications).

Cognitive behavioural therapy (CBT)

Cognitive therapy: **Beck** proposed that negative thinking in depression originates in earlier assumptions and plays central roles in the maintenance of depressive symptoms. The **Beck's cognitive model of depression** includes the effect of early experiences, core beliefs, assumptions, cognitive distortions, automatic thoughts and the negative cognitive triad. Depression can be treated by modifying one of the components of Beck's cognitive model.

Behaviour therapy: **Mowrer's two factor model** states that fear of specific stimuli is acquired through classical conditioning and clients try to reduce fear by avoiding the conditioned stimuli through operant conditioning. **Ayllon and Azrin's token economy model** is a ward environment where reinforcers were applied to systematically change clients' behaviour.

Aim: The aim of CBT is to improve emotion by identifying and changing dysfunctional thoughts and maladaptive behaviours by a present-oriented, time limited and highly structured therapy.

Indications: CBT suitable for the following conditions: mood disorders, anxiety disorders, eating disorders, phobias, schizophrenia (target at delusions and hallucinations, coping enhancement, enhance adherence to treatment and relapse prevention), personality disorders (refer to dialectical behavioural therapy), substance abuse and consultation liaison setting (chronic pain, chronic fatigue, physical illnesses).

General principles: CBT deals largely with the here and now. It is a time-limited therapy which involves psychoeducation, engagement and collaboration between the therapist and the client.

Techniques:

1) Identify negative automatic thoughts
2) Identify maladaptive belief and rate its strength
3) Restructure maladaptive belief
4) Formulate alternative positive belief
5) Rate impact of maladaptive belief on emotion
6) Rate the impact of new belief on emotion.

Phases:

Session 1-2: Assessment by **socratic questioning** (The use of questions to reveal the self-defeating nature of the client's negative automatic thoughts) and identify **cognitive triads** (automatic thoughts, cognitive distortion, faulty assumptions);
Session 3-4: case formulation and explain CBT model;
Session 5-7: keeping diaries and homework monitoring to look for cognitive errors, reattribution by reviewing evidence, challenging cognitive errors, cost benefit analysis, forming action plans and overcoming resistance;
Session 8 – 10: BT involves identification of safety behaviours, entering feared situation without safety manoeuvres, applying relaxation techniques, activity scheduling, assertiveness training and reviewing results;
Session 11 – 12: relapse prevention and termination.

The methods of measurement used in CBT including: direct observation, physiological measures, standardised instruments and self report measures such as the Beck Depression Inventory, the Beck Anxiety Inventory and Fear Questionnaire.

Dialectical behaviour therapy (DBT)

Indications: borderline personality disorder (BPD) and client's attempt repetitive self-harm

Techniques:
1) **CBT**
2) **Dialectical thinking:** advise the client not to think linearly. Truth is an evolving process of opposing views rather than extremes
3) **Zen Buddhism**: emphasize wholeness, see alternatives and engulf alternatives.
4) Use of **metaphors**: enhancing effectiveness of communication, discovering one's own wisdom and developing therapeutic alliance.

Components:
1) **Individual sessions**: 45-60 minutes on a weekly basis, to review diary cards in the past one week and discuss life threatening behaviours;
2) **Skills training group by a trainer**: weekly group for 2 hours, didactic in nature and manual-based;
3) **Brief out-of-hours telephone contact as part of treatment contract**

4) **Weekly consultation group** between the individual therapist and the skills trainer.

Interpersonal therapy (IPT)

Indications: Depressive disorder (equally effective as CBT), eating disorder (bulimia nervosa), interpersonal disputes, role transition, grief and loss

Techniques:

1) To create a **therapeutic environment** with meaningful therapeutic relationship and recognize the client's underlying attachment needs.
2) To develop an understanding of the **client's communication difficulties and attachment style** both inside and outside the therapy.
3) To identify the client's **maladaptive patterns** of communication and establishment of **insight**. The therapist can adopt three stances: **neutral stance, passive stance and client advocate** stance
4) To assist the client in building a better social support network and mobilise resources.

Family therapy

Commonly encountered family problems and strategies to resolve:

1) **Task accomplishment problems** (e.g. developmental tasks): identifying the task, exploring alternative approaches, taking action, evaluating and adjusting.
2) **Communication problems and triangulation**: introduction of humour, demonstration of warmth and empathy, role play and modifying both verbal and non-verbal communication.
3) **Role problems** (e.g. family scapegoat, parental child): identify problems and redefine roles. The **same-sex parent** functions as primary programmer and disciplinarian. This will promote maximum ego development by setting limits and higher level goals. The **opposite-sex parent** functions as the facilitator or mediator within the triangular relationship to correct inappropriate parenting from the same sex parent.
4) **Behaviour control problems** (e.g. conduct disorder in a child): engage family to deliver BT in home environment.
5) **Boundary issues between family members** (e.g. enmeshment): to promote communication and emotional interchange in disengaged relationship; to create necessary separation and independence in case of enmeshment.
6) **Suprasystem problems** (high crime rate in neighbourhood and affecting the family): Strengthening the boundary between the family and the suprasystem; to work with extended family and social agencies.

Types of family therapy:

Structural family therapy (Minuchin):

It identifies the set of unspoken rules governing the hierarchy, sharing of responsibilities and boundaries. The therapist will present these rules to the family in a paradoxical way to bring about changes. The therapist is actively in control of proceedings.

Systemic family therapy:

1. **Milan associates** often involve more than two therapists working in a team to maintain the systemic perspective. One therapist is always with the family while the team observes through the one-way screen or video camera. The team offers input to the therapist via telephone or during intersession breaks to discuss the family system. There are pre- and post-session discussions.

2. **Reframing** an individual's problem as family problem (e.g. daughter's borderline personality may be reframed as the parental; problem in providing care to the child). An internal problem can be reframed as external problem if there is unproductive conflicts in the family which exhausted everyone. (e.g. The family is under influence of anger rather than an individual member is an angry person).

3. Exploring the **coherence** and understand the family as an organised coherent system.

4. **Circular questioning** is used to examine perspective of each family member on inter-family member relationship. It aims at discovering and clarifying conflicting views. **Hypothesis** can be formed from the conflicting views and change can be proposed.

Strategic family therapy uses a complex plan rather than a simple directive to produce change in the family. It employs the following techniques:

1. **Positive connotation**: It is a form of reframing by ascribing positive motives to the symptomatic behaviour.

2. **Metaphors** allow indirect communication of ideas. (e.g. **Relationship metaphors** use one relationship between therapist and one family member as a metaphor for another relationship. **Metaphorical object** refers to the use of a concrete object to represent abstract ideas (e.g. a blank sheet of paper in an envelope to represent family secrets).

3. **Paradoxical interventions:** This method is used when direct methods fail or encounter strong resistance in some family members. The therapist will reverse the vector (i.e. rather than a top-down approach from the parents, a bottom-up approach from the child to the parents is allowed). This therapist will also disqualify anyone who is an authority on the problems including the parents. (e.g. children are allowed to challenge the parents and undesired behaviour is encouraged. Paradoxically, change and improvement will take place as family members cannot tolerate the paradoxical pattern).

4. **Prescribing family rituals**: **Rituals** refer to membership, brief expression and celebration. (e.g. the therapist passes a metaphor object to the family and any family member can use this object to call for a meeting at home). **Ritual prescription** refers to setting up a time table which assigns one parent to take charge on an odd day and a child to take charge on an even day. Ritual prescription is useful for family with parental child.

5. **Other strategies** include humour, getting help from a consultation group which observes through the one way mirror to offer advice and debate among family members.

Eclectic family therapy

It concentrates on present situation of the family and to examine how family members communicate with one another. It is flexible and allows time for the family to work together on problems raised in the treatment. It is commonly used in adolescents and their family.

Group therapy

Types of group therapy:

1. **Milieu group therapy**: **Main** developed therapeutic community where the whole community (e.g. the ward) is viewed as a large group. **Rapaport** described four characteristics in milieu GT: **democratisation** (equal sharing of power), **permissiveness** (tolerance of others' behaviour outside the setting), **reality confrontation** (confront with the views from others) **and communalism** (sharing of amenities).

2. **Supportive group therapy**: for clients with chronic psychiatric disorders such as schizophrenia to attend a day hospital with GT. It involves empathy, encouragement and explanation. (e.g. Schizophrenia Fellowship).

3. **Outpatient group therapy**: It may involve a self help group targeting at homogenous group of clients focusing on one disorder (e.g. for anxiety disorders or AA). It is short term and involves direct didactic instruction.

Formal group therapy:

Exclusion criteria:
Clients who are brain damaged, paranoid, severe narcissistic, hypochondriacal, acutely psychotic or sociopathic.

Inclusion criteria:
1. Long standing psychological problems
2. Require careful selection

Advantages of group therapy:
1) Reduction of cost
2) The context of the group may give members value and a sense of group identity to assist them to cope with current problems in life

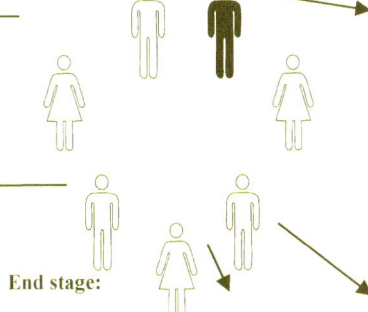
Therapist

Therapeutic process (Yalom's 11 therapeutic factors):

Early stages:

1) **Instillation of hope**: sense of optimism about progress and improvement.

2) **Universality:** one individual's problems also occur in other members and the member is not alone

3) **Information giving:** member will receive information on his or her illness as well as and problems faced by the others

Middle stage:

4) **Altruism:** One member feels better by helping other members and sharing their solutions.

5) **Corrective recapitulation of the primary family:** the group mirrors one's own family and provide a chance for self-exploration of past family conflicts through this process.

6) **Improved social skills** by social learning

7) **Imitative behaviour** by vicarious learning or observation of others.

8) **Interpersonal learning** by corrective experience in social microcosm.

End stage:

9) **Group cohesiveness** occurs after inter-member acceptance and understanding. This will lead to the sense of safety and containment of negative feelings.

10) **Catharsis:** Member feels encouraged and supported by expressing emotionally laden materials.

11) **Existential factors:** after group therrapy, member has more self - understanding and insight in responsibility and capriciousness of existence.

Role of therapist:
1) The therapist can draw heavily upon a group setting to exercise the full range of therapeutic skills while the group therapy reduces the chance of intense transference reactions from the members.
2) The therapist may have particular inclinations and motivations (e.g. past experience of alcohol misuse in an AA groups)
3) The therapist can encourage a combination of individual and group therapy. This will positively integrate these two forms of therapy for maximal benefit.

Couple therapy

Couple therapy involves the therapist (or sometimes two therapists) seeing two clients who are in a relationship. It aims to sort out interpersonal and marital/partnership difficulties. Couple therapy works on issues such as grief in a couple and/or sexual problems. Couple therapy follows one of the following psychotherapeutic models: **CBT** (identifying reinforcement of undesirable behaviour in the couple), **psychodynamic** (understanding one's own emotional needs and how to relate to the needs of the partner), **transactional** (behaviours are analysed in terms of the child, adult and parent within the client and how the client reacts with the partner in the relationship) and **family-based** therapies.

Techniques of couple therapy:

1) **Reciprocity negotiation**: the couple develops ability to express their offers and understand the partner's request. This promotes exchange of positive behaviours and reactions.

2) **Communication training**: This encourages mutual exchange of emotional messages.

3) Structural moves involve the following techniques. **Experiment of disagreement** focuses on a topic where one partner always dominates while the other habitually gives in to avoid disagreement. This exercise will help the passive partner to express an opinion forcefully and the other needs to value the expression. **Role reversal** helps one partner to understand the viewpoints and experiences of the other.

Therapists need to consider the following issues before arranging couple therapy:

1) Therapist-related issues: **Age and life experiences** of the therapist are essential as couples may look for therapists who are married or who have been married. The **gender** of the therapist plays a key role as the therapist may identify with the same sex client. **Culture and religious background** of the therapist in relation to the background of the couple should be considered. Therapists need to monitor their own **countertransference** as otherwise, there is a risk they may experience rescue fantasies or over-identification with the client.

2) Therapy-related issues: The relationship to other **on-going therapies** such as individual psychotherapy should be considered as well as **confidentiality and neutrality** between the couple and therapist.

3) Couple related issues: It requires **motivation** from both partners to attend the session and it is often a challenge to **establish therapeutic alliance** with both partners.

Eye movement and desensitisation reprocessing (EMDR)

EMDR is indicated for post-traumatic stress disorder. EMDR focuses on traumatic experiences and the negative cognitions and affective responses associated with these. The aim is to desensitise the individual to the affective responses. This is accompanied by **bilateral stimulation** by inducing rapid eye movement as the client is asked to follow the regular movement of the therapist's fore finger.

The **procedural phases** of EMDR include:
1) **assessment of target memory of image**;
2) **desensitisation** by holding the target image together with the negative cognition in mind and **bilateral stimulation** continues until the memory has been processed with the chains of association;
3) **installation** of positive images;
4) **scanning** of body to identify any sensations;
5) closure and debriefing on the experience of the session.

Grief counselling

Grief counselling allows client to talk about the loss, to express feelings of sadness, guilt or anger and to understand the course of grieving process. This will allow the client to accept the loss, working though the grief process and adjust one's life without the deceased.

Motivational interviewing

Indications: to enhance motivation to quit substance misuse or other forms of addiction (e.g. internet or gambling).

Techniques:
1) Express empathy of patient's substance abuse by reflective listening
2) Rolls with resistance and understand why patient is reluctant to give up a substance or a habit.
3) Develop discrepancy between one's addiction and goals.
4) Support self-efficacy to quit a drug or bad habit and to provide a menu of treatment options.

MEQ

A 25-year-old male teacher presents with low mood, poor sleep, poor appetite and negative thoughts. He has suffered from depression for 6 months. He works as a teacher. He received good appraisals in the past few years. As a result of depression, he feels that he is useless and he is not competent to be a teacher although his head of department is very supportive of him to seek treatment and get well. He admits that he has problems with his colleagues. You have advised him to take antidepressant but he worries about side effects. He is interested in psychotherapy.

Q.1 Name 3 types of psychotherapies which may be suitable for him.

1. Supportive psychotherapy
2. Problem-solving therapy
3. Cognitive behaviour therapy
4. Interpersonal therapy.

Q.2 He is interested in cognitive behaviour therapy (CBT). He wants to know the specific technique in cognitive therapy which helps him to tackle depression. Your answer is:

Cognitive therapy involves gathering information from dysfunctional thought diary. The therapist uses the diary to identify cognitive bias and negative automatic thoughts. Then the therapist helps the patient to develop capacity to challenge such negative thoughts and cognitive distortions.

Q.3 He wants to know the specific technique in behaviour therapy which helps to tackle depression. Can you give an example?

Activity scheduling: the patient is advised to plan a schedule to engage himself in activities to prevent withdrawal which perpetuates depression.

Q.4 He wants to know the number of sessions required for CBT. Your answer is.

12 to 16 sessions.

Q.5 He worries that he will be busy with work in the near future. How often does he need to seek a psychologist for CBT?

Ideally, he should see a psychologist every week for 3 to 4 months. He can see a psychologist every 2 weeks if he is busy and the duration of CBT will be prolonged to 6 months to one year.

MCQ

1. A 25-year-old man is seeing a male psychologist for dynamic psychotherapy. He feels very angry with the male psychologist as he finds that the male psychologist is exerting his authority like his abusive father. The phenomenon is known as:

A. Acting out
B. Boundary violation
C. Countertransference
D. Transference
E. Resistance.

The answer is D. This phenomenon is known as transference because the patient unconsciously relates the therapist to someone from his past (i.e. his abusive father).

2. A 25-year-old man is seeing a male psychologist for dynamic psychotherapy. The therapist does not want to see the patient because he is quite narcissistic. This patient reminds the therapist previous narcissistic patients who lodged complaints to the Quality Improvement Unit (QIU) of the hospital. The phenomenon is known as:

A. Acting out
B. Boundary violation
C. Countertransference
D. Transference
E. Resistance.

The answer is C. Countertransference involves unconscious emotional needs, wishes or conflicts (i.e. complaints to QIU) arising from the therapist's prior relationship experiences being evoked by the patient during psychotherapy.

3. You are seeing a 50-year-old man presents with cirrhosis as a result of chronic alcohol dependence. During the motivational interviewing, he defends his chronic drinking habit. He needs alcohol to relax himself. He told his wife alcohol is good because he does not need to go out and save money by drinking whole day. Which of the following defence

mechanisms does he manifest?

A. Denial
B. Displacement
C. Rationalisation
D. Splitting
E. Victimisation

The answer is C. He justifies his alcohol dependence with a plausible explanation (e.g. relaxation) after the event (e.g. cirrhosis), rather than examining unacceptable explanation (e.g. saving money by drinking).

EMIS:
Learning Theories / Conditioning

A-	Aversive conditioning
B-	Classical conditioning
C-	Conditioned response
D-	Extinction
E-	Generalization
F-	Operant conditioning
G-	Punishment

Question 1: David has just completed his course of chemotherapy. Usually, he feel nauseated on entering the hospital grounds, but now, he finds that over the period of several months, during which he no longer receive any treatment, the nauseated feeling have diminished – Extinction

Question 2: In order to make sure that Peter would be cooperative with the regular doctor's checkup, his parents buy him a new toy every time he attends the session – operant conditioning

Question 3: There is a new pet dog in the family, who barks quite frequently, without triggers. The owners have decided to apply an aversive stimulus to the dog each time it barks for no reason. – Punishment.

Defense mechanisms in Psychoanalysis

A-	Death wish
B-	Denial
C-	Displacement
D-	Humor
E-	Idealization
F-	Identification
G-	Projective identification
H-	Rationalization
I-	Repression
J-	Reaction formation

Please choose from the above the most appropriate defense mechanisms for each of the clinical vignettes below. Each option could be used once, more than once, or not at all.

Question 1: Patient A has a long-standing alcohol addiction. He told his therapist that he is keen to give up alcohol only if he could find a job which he is able to enjoy – Rationalization

Question 2: An employee of a multinational company has long been very frustrated with his boss, who he thinks is bossy and incompetent. Recently, he has changed his attitude by going the extra mile and try to be helpful to his boss. – Reaction formation

Question 3: Even though they have broken up 2 weeks ago, the women still continues to phone her boyfriend. She continues to talk about him in front of her colleagues as well – Denial

References

Books

[1] Gabbard GO (2005) Major modalities in psychoanalytic/psychodynamic in Gabbard GO, Beck JS, Holmes J (2005). *Oxford Textbook of Psychotherapy*. Oxford: Oxford University Press.

[2] Gelder M, Mayou R, Cowen R (2001) *Shorter Oxford Textbook of Psychiatry*. Oxford: Oxford University Press.

[3] Shemilt J & Naismith J (2007) Psychodynamic theories II (Chapter 3) in Naismith J & Grant S (2007) *Seminars in the psychotherapies*. London: Gaskell.

[4] Gabbard GO (2000) Psychodynamic psychotherapy (3rd ed). Washington: American psychiatric press.

[5] Lackwood K (1999) Psychodynamic psychotherapy (Chapter 7) in Stein S, Hiagh R & Stein J (1999) *Essentials of psychotherapy*. Oxford: Butterworth Heinemann.

[6] Hallstrom C & McClure N (2005) *Your questions answered: depression*. London: Churchill Livingstone.

[7] Hawton K, Salkovskis PM, Kirk J & Clarl DM (2001) *Cognitive behaviour therapy for psychiatric problems: a practical guide*. Oxford: Oxford University Press.

[8] Turner T (2003) *Your questions answered: depression*. London: Churchill Livingstone.

[9] Bateman A & Fonagy P (2004) *Psychotherapy for Borderline Personality Disorder: mentalization based treatment*. Oxford: Oxford University Press.

[10] Stuart S & Robertson M (2003) *Interpersonal psychotherapy: a clinician's guide*. London: Arnold.

[11] Barker P (1998) *Basic family therapy*. Oxford: Blackwell Science.

[12] Gelder M, Mayou R, Cowen P (2001) *Shorter Oxford Textbook of Psychiatry*. Oxford: Oxford University Press.

[13] Hook J (2007) Group psychotherapy in Naismith J & Grant S (2007) *Seminars in the psychotherapies*. London: Gaskell.

[14] Crowe M & Ridley J (2000) *Therapy with couples* (2nd Ed).

London: Blackwell Science.

[15] Schaverien J & Odell-Miller H (2005) The art therapies in in Gabbard GO, Beck JS, Holmes J (2005). *Oxford Textbook of Psychotherapy*. Oxford: Oxford University Press.

[16] Johnstone EC, Cunningham ODG, Lawrie SM, Sharpe M, Freeman CPL. (2004) *Companion to Psychiatric studies*. (7th edition). London: Churchill Livingstone.

[17] Aveline M, Dryden W. (1988_ Group therapy in Britain. Milton Keynes: Open University Press.

[18] Sadock BJ, Sadock VA. (2003) *Kaplan and Sadock's Comprehensive Textbook of Psychiatry*. (9th ed.) Philadelphia (PA): Lippincott Williams and Wilkins.

[19] Barnes B, Ernst S, Hyde K. (1999) An introduction to group work. A group-analytic perspective. Hampshire: Palgrave Macmillan.

Websites

[20] NICE guidelines for depression http://guidance.nice.org.uk/CG90

[21] NICE guidelines for depression in children and young people http://guidance.nice.org.uk/CG28

[22] NICE guidelines for depression in adults with a chronic physical health problem http://guidance.nice.org.uk/CG91

[23] NICE guidelines for anxiety http://guidance.nice.org.uk/CG22

[24] NICE guidelines for obsessive compulsive disorder and body dysmorphic disorder http://guidance.nice.org.uk/CG31

[25] NICE guidelines for post-traumatic stress disorder http://guidance.nice.org.uk/CG26

[26] http://www.annafreudcentre.org/shortcourses.php

[27] NICE guidelines for borderline personality disorder http://guidance.nice.org.uk/CG78

Journals

[28] Palmer RL (2002) Dialectical behaviour therapy for borderline personality disorder. *Advances in Psychiatric Treatment*. **8**: 10-16.

[29] Denman C (2001) Cognitive-analytical therapy. *Advances in Psychiatric Treatment*. **7**:243-256.

[30] Coetzee RH & Regel S (2005) Eye movement desensitisation and reprocessing: an update. *Advances in Psychiatric Treatment*, **11**, 347-354.

Chapter 11: Sleep disorders

Classification	Parasomnia
Epidemiology	REM and non-REM sleep disorder
Assessment of sleep disturbances	Sample MEQ
Dyssomnia	Sample MCQ & EMIs

Introduction

Sleep disorders are highly prevalent in our population. Psychiatrists are often asked to evaluate patients who are having problems with either difficulties initiating sleep or, who are having difficulties with excessive sleep. Sleep disorders could be primary in nature, or arise as a result of an underlying psychiatric condition. Studies have previously demonstrated the implications of sleep deprivation on individuals – poor sleep quality usually leads to impairment in cognitive performance and cause mood disturbances as well.

Classification:

Based on the DSM-5 diagnostic manual, sleep disorders can be classified into:
   a. Dyssomnia
   b. Parasomnia
   c. Sleep disorder related to another mental disorder
   d. Other sleep disorders.

Dyssomnia is defined as a disorder of the quantity or timing of sleep. Thus, it includes both insomnia and also hypersomnia. Insomnia refers to a disturbance in the quantity or the quality of sleep. Hypersomnia refers to excessive sleepiness.

Dyssomnia include the following sleep disorders:
- Breathing related sleep disorder
- Circadian rhythm sleep disorder
- Narcolepsy
- Primary insomnia
- Primary hypersomnia
- Dyssomnia not otherwise specified.

Parasomnia is defined as abnormal behaviours that occur during sleep or during the transition between sleep and wakefulness. Parasomnia includes the following sleep disorders:

- Nightmare disorder
- Sleep terror disorder
- Sleepwalking disorder
- Parasomnia not otherwise specified.

Assessment of sleep disturbances:

Assessment of sleep disorders requires eliciting a detailed medical and psychiatric history from the patient. Apart from having a detailed medical and psychiatric history, it is crucial to perform a more in-depth assessment of the sleep disturbances.

The following questions should be asked during assessment:

Daytime:
1. Do you feel sleepy during the day?
2. Do you take routine naps during the day?
3. Do you find yourself having difficulties with concentration during the day?

Night-time:
4. Could you describe to me a typical night of sleep (in terms of the number of hours you get, the quality of sleep etc.)?
5. Do you find yourself having difficulties with falling asleep?
6. Do you sleep well? Do you find yourself awake during the night? If so, what is the reason? Is it because of poor sleep or you need to go to toilet? (going to toilet twice per night is normal and not considered to be sleep disturbance)
7. Do you find yourself waking up much earlier in the morning?

Cause and course of sleeping problems:
8. Could you tell me how long you have had such sleeping difficulties?
9. What do you think might have precipitated such difficulties?
10. Do you wake up and sleep at the same time during weekends as weekdays?
11. Does your job currently require you to work on shifts or travel frequently?
12. Do you drink caffeinated beverages close to your desired sleeping time?
13. Do you have any other long standing medical problems apart from the difficulties you are experiencing currently with your sleep?

Treatment
14. Did you seek help for your sleep problems? (e.g. GP, psychiatrist, acupuncturist, traditional medicine practitioner)?
15. Are you on any chronic long-term medications to help yourself to sleep (e.g. sleeping pills)? Where do you get those medications?

In some cases, it is also crucial to obtain a sleep history from a sleep partner.

The following questions should be asked:
1. Have you noticed the change in sleep habits of your partner?
2. Have you noticed that your partner has been snoring during his sleep?
3. Does your partner exhibit any abnormal movements (e.g. kicking) during sleep? Were you injured?

Dyssomnias:

a. **Insomnia**
Insomnia is diagnosed when the main problem lies with difficulty in initiating or maintaining sleep. It is important that other physical and mental conditions have been ruled out.

The DSM-5 diagnostic criteria state that there must be sleep difficulties for at least 3 nights per week, for the past 3 month.
These sleep difficulties include
a. Difficulties associated with initiation of sleep
b. Difficulties with maintaining sleep
c. Early morning awakening and inability to return back to sleep
The sleep disturbances must have resulted in marked impairments in terms of functioning.
The DSM-5 specified that the sleep difficulties must have occurred despite there being adequate opportunities for rest.

Epidemiology

Insomnia has been estimated to affect at least 30% of the general population, with women being more affected as compared to men. The incidence rate is higher among the elderly.

Mahendran et al (2007) studied 141 patients seen at the Insomnia Clinic at IMH and found the following:

- 47.5 percent had primary insomnia,
- 52.5 percent had a primary diagnosis of a psychiatric disorder.
- 41.1 percent of those diagnosed with a primary psychiatric disorder had comorbid psychiatric disorders
- 4.3 percent had substance abuse problems.

Aetiology:

1. Intrinsic causes: idiopathic or primary insomnia, sleep apnoea syndrome, periodic limb movement disorder and restless leg syndrome
2. Extrinsic causes: Poor sleep hygiene, sleep disorders due to environmental, adjustment, altitude, and substance-related issues.
3. Circadian rhythm disorders
4. Medical disorders: chronic pain, pulmonary diseases (e.g. COPD), neurological disorders (e.g. Parkinson's disease), endocrine disorders, iron deficiency (restless leg syndrome and sleep apnoea.
5. Psychiatric disorders: generalized anxiety disorder, depression, bipolar affective disorder, chronic pain disorders, posttraumatic stress disorder, anorexia nervosa, somatoform disorder and schizophrenia.

Patients with primary insomnia are usually concerned of getting sufficient sleep for that night. This anticipatory anxiety worsen sleep at night.

Treatment of insomnia:

1. **Non-pharmacological treatment:**

    (a) **Sleep education** – Individuals suffering insomnia needs to be educated with regards to the various stages of sleep, the typical sleep cycle and how the sleep cycle changes with age, in order for them to get an insight into the characteristic of the sleep issue they are seeking treatment for.

    (b) **Sleep hygiene** - A series of steps that are considered to be useful in all patients. The general principles of sleep hygiene include ensuring that the sleep environment is familiar, comfortable, dark, and quiet; ensuring regular bedtime routines, with consistent bedtime and waking up time, and reinforcing going to bed only when tired. Furthermore, patients are advised to avoid factors associated with insomnia such as avoiding overexcitement prior to going to bed, late evening exercises, consuming drinks with caffeine late in the day, excessive smoking and alcohol, excessive daytime sleeping and late meals prior to bedtime.

    (c) **Stimulus control** - Focus on the core principle that an individual should go to sleep only when he or she is tired, and in the event that he or she has difficulties with sleep, he or she should get up and focus on relaxing activity instead prior to returning to bed again.

    (d) **Sleep restriction** – This method can be adopted in order to improve the quality of sleep. Basically, this helps to reduce the total time spent in bed and thus enabling an improvement of the quality of sleep via consolidation. Individuals seeking treatment using this method have to have high intrinsic motivation.

2. **Pharmacological treatment:**

    Pharmacological methods, in particular, the prescription of hypnotics should be considered as the last option for the treatment of insomnia. The underlying cause of insomnia should be carefully worked out, prior to the commencement of treatment using hypnotics. In Singapore, the following non-benzodiazepine hypnotics are available:

    First line:
    1. Hydroxyzine (atarax) 25mg – 50mg ON; it is an anti-histamine; main side effects include drowsiness in the morning,
    2. Mirtazapine (remeron) 15mg ON – a sedative antidepressant and main concern is weight gain.
    3. Agomelatine (valdoxan) 25mg ON – a new antidepressant targeting at the melatonin receptors and regulate sleep and wake cycle. Main side effect includes giddiness.

Second line:
1. Zopiclone (imovane) 7.5 – 15mg ON; half life is 6 hours. Main side effect includes metallic taste.
2. Zolpidem CR (stilnox CR) 10mg ON; half life is 2-3 hours in non-CR form and longer in CR form. Dependence is a possibility.

Benzodiazepine (BZD) should be used only for a short period of time and of not more than 2 weeks. This helps in the prevention of individuals developing dependence to the medications. BZD abuse in Singapore is contributed to by both doctor-shopping behaviour and doctors' prescribing practices (Dong et al 2007). The SMC also regulates the prescription of benzodiazepine by GP and prescription of benzodiazepine should be done by specialists. Common benzodiazepines include:

1. Alprazolam (xanax) 0.25mg – mainly indicated for anxiolytic use; half life is around 11 hours.
2. Clonazepam 0.5mg – mainly indicated for anxiolytic use and REM – movement disorder; half life is around 25 hours.
3. Lorazepam (ativan) 0.5mg ON – intramuscular form is used to calm patient with acute agitation; half – life is around 10 hours.
4. Diazepam (valium) 5mg ON – per rectum use is indicated for epilepsy; half life is around 30 hours.

b. **Hypersomnia and excessive daytime sleepiness**

Individuals suffering form hypersomnia tend to present with recurrent sleep attacks during the day, poor concentration and experience difficulties with transiting from a rested state to full arousal and wakefulness. To fulfil the diagnostic criteria, such episodes must have lasted for several months and must have significant impacts on their physical and psychosocial life.

Hypersomnia tends to affect around 10 - 15% of the general population.

The DSM-5 diagnostic criteria state that there must be excessive sleepiness that have occurred at least 3 times per week, for at least 3 months. The excessive sleepiness is characterized by:
a. Recurrent periods of sleep or lapses back to sleep even within the same day
b. Prolonged sleep episode of more than 9 hours that is not restorative
c. Difficulties associated with being fully awake after abrupt awakening.
The sleep disturbances must have resulted in marked impairments in terms of functioning.

Aetiology:
- Drug effects e.g. long acting benzodiazepine
- Poor sleep routines e.g. playing online game whole night
- Circadian rhythm sleep disorders
- Chronic physical illness
- Frequent parasomnia
- Insufficient night time rest
- Kleine-Levin Syndrome
- Narcolepsy
- Obstructive sleep apnoea
- Psychiatric disorders e.g. melancholic depression

The more prevalent causes of hypersomnia are explored in greater details:

Narcolepsy

The prevalence of narcolepsy amongst the American population has been estimated to be around 5 per 10,000. It affects individuals who are between 10 and 20 years of age. It has been shown to affect both men and women equally.

The DSM-5 Diagnostic criteria states that there must be repeated episodes during which there is a need to fall back into sleep, or napping occasionally within the same day. This must have occurred at least 3 times per week over the past 3 months or so.
In addition, there must be the presence of at least one of the following symptoms:
1. Episodes of sudden bilateral loss of muscular tone with maintained consciousness, that are precipitated by laughter or joking
2. Lack of hypocretin
3. REM sleep latency of less than or equal to 15 minutes

Ateiology:

- HLA-DR2 is the candidate gene for this condition.
- Hypocretin, a hypothalamic neuropeptide transmitter, regulates sleep-wake cycles is involved. The concentrations of hypocretin-1 and hypocretin-2 are reduced in patients suffering from narcolepsy.

Clinical features:

- Cataplexy (which is the sudden and brief episode of paralysis with lost of muscular tone)
- Excessive sleepiness
- Hypnagogic hallucinations (hallucination when falling asleep; hypnopompic (waking up) hallucinations: less common)
- Sleep paralysis

Treatment:

Non-pharmacological treatment: Encourage patients to adopt a regimen of regular naps in the day time.

Pharmacological treatment:

Modanfinil (Provigil) helps to reduce the number of sleep attacks in narcolepsy. It has a better side effect profile than traditional psycho-stimulants.

SSRIs (e.g. fluoxetine) have also been used to help reduce cataplexy. These drugs are indicated for usage in treatment as they have a tendency to help in suppressing REM.

Breathing related sleep disorder

Breathing disturbances that could occur during sleep include apnoea and hypopnoea.

Obstructive sleep apnoea is a common cause of breathing related sleep disorder, which affects at least 4% of the male population. People with sleep apnoea stop breathing at least five times per hour for greater than 10-second period as a result of upper airway obstruction. Middle aged overweight males who snores loudly are typically at risk. Their sleep is fragmented by short arousals following apnoea and having unrefreshed sleep.

Treatment includes weight loss, and at times, the usage of continuous positive pressure ventilation (CPAP) via a face mask at night

Kleine-Levin syndrome

This condition is rare in prevalence, largely affecting male adolescents.

Clinical features include:
- episodes of excessive sleepiness
- increased appetite
- other clinical features might include episodes of sexual disinhibition and other co-morbid psychiatric symptoms
- duration often lasts for days or weeks with long intervals
- free from attacks in between these episodes.

Management:

- Stimulatory SSRIs such as fluoxetine can be used.
- Psycho-stimulants can used. However, such stimulants are usually only effective for a short period of time.

Circadian rhythm sleep disorder

In this condition, the patient's sleep wake schedule is different from the normative sleep wake pattern of other people in the society. This leads to excessive daytime sleepiness and impairments in social or occupational functioning.

There are two types of circadian rhythm sleep disorder:

1. Advanced sleep phase syndrome –early onset of sleep with resultant early morning awakening.

2. Delayed sleep phase syndrome – delayed onset of sleep, usually at 2am. It should be noted that the total number of sleep time is still normal.

Ateiology:

1. Time zone changes
2. Shift work (e.g. security guard)
3. Irregular sleep wake pattern

Management:

1. **Sleep education** – It is important to educate individuals about the stages of sleep, and assist them in establishing good sleep habits. For individuals who are working on shifts, they should be advised to try and have a regular cycle of sleep, and to attempt to nap if necessary to compensate for the absolute number of hours of sleep loss.

2. **Medications** –

    1. Agomelatine or melatonin to resent circadian rhythm.
    2. Hypnotics, such as short acting benzodiazepines.

c. **Parasomnia**

Parasomnia is classified into the following subtypes:

1. Arousal disorders such as confusional arousals, sleep-terrors and also sleep walking
2. Sleep-wake transition disorder such as nocturnal leg cramps, sleep talking and rhythmic movement disorder.
3. Parasomnia associated with REM sleep such as nightmares and sleep paralysis.
4. Other parasomnia includes sleep bruxism, sleep enuresis and also other parasomnia not otherwise stated.

Sleep-walking (aka somnambulism)

Epidemiology

1. This condition is prevalent amongst children and adults.
2. Up to 17% of children are affected, with incidences higher amongst those between 4 to 8 years old.
3. Up to 10% of adults are affected as well.

Clinical features

1. The patient may exhibit complex and automatic behaviours (e.g. wandering without purpose, attempting to dress or undress).
2. These episodes tend to occur in the initial stages of sleep, usually 15-120 minutes after individuals fall asleep.
3. Individuals are usually able to get back to bed and continue on with their sleep after the event has taken place.
4. They usually do not recall the exact incidents that happen. If they become awake during the somnambulism episode, they appear to be disorientated and confused.

Management:

1. Supportive therapy, sleep hygiene, psychoeducation and reassurance are useful.
2. Protective measures such as locking doors and windows or installation of metal bars or frames at window to prevent individuals from accidents (e.g. fall from height).
3. Antidepressants such as imipramine and paroxetine are useful.

Sleep terrors or night terrors

Epidemiology:

Sleep terrors affect an estimated 3% of children and 1% of adults. It is more common in the male gender.

Clinical features:

1. Characterized as the sudden waking up during sleep with loud terrified screaming.
2. Other physiological changes include tachycardia, diaphoresis and mydriasis.
3. Each episode is estimated to last for around 10-15 minutes.
4. Similar to sleep-walking, if the individual is awakened during sleep terror, he or she would be confused and disorientated.
5. The patient is usually unable to recollect in events in details.

Management:

Supportive therapy, sleep hygiene, psychoeducation and reassurance are useful.

Nightmares

A nightmare is defined as an awakening from REM sleep to full consciousness with detailed dream recall ability. It is common especially among children, between the ages of 5 to 6 years old.

No specific treatment is usually required for such disorders. Agents that help to suppress REM sleep, such as tricyclic drugs and SSRIs can be used.

Periodic limb movement disorder

Epidemiology: more common in elderly population, pregnant women and vertebral degenerative disorders.

Aetiology: B12 deficiency, iron deficiency, chronic renal disease and Parkinson's disease.

Clinical features: repeated twitches of the legs in sleep.

Investigations: low B12 and ferritin levels

Treatment: vitamin 12 and iron tablets

Random eye movement (REM) sleep and non random eye movement (non-REM) sleep

1. Sleep is comprised of periods of NREM and REM sleep alternating throughout the night.
2. The deepest stages of NREM sleep occur in the first part of the night.
3. The episodes of REM sleep are longer as the night progresses

Table 1 Compare and contrast clinical features of REM and non-REM sleep

REM sleep	Non-REM sleep
Dreams: • ↑ recall of dreams if woken • ↑ complexity of dreams Sympathetic activity: • ↑ sympathetic activity • ↑ transient runs of conjugate eye movements • ↑ heart rate • ↑ systolic blood pressure • ↑ respiratory rate • ↑ cerebral blood flow Neuromuscular: • ↓ muscle tone • occasional myoclonic jerks • penile erection or increased vaginal blood flow Poililothermic condition: • Thermoregulation stops (No shivering or sweating).	Dreams: • ↓ recall of dreams if woken • ↓ complexity of dreams Parasympathetic activity: • ↓ sympathetic activity • upward ocular deviation with few or no eye movements • ↓ heart rate • ↓ systolic blood pressure • ↓ respiratory rate • ↓ cerebral blood flow Neuromuscular: • abolition of tendon reflexes • no penile erection

Table 2 Compare and contrast disorder associated REM and non-REM sleep

REM sleep related disorders	Non-REM sleep related disorders
1. Nightmares and vivid violent dreams: ↑ in recall of dreaming if awoken from REM sleep. REM behavioural sleep disorder can be treated by clonazepam. 2. Narcolepsy 3. Pickwickian syndrome: compromised ventilation which is	1. Night terrors 2. Sleep walking (somnambulism) 3. Sleep talking: A clinical condition which occurs during sleep and the individual utters words or sounds, without

abruptly worse during REM sleep. 4. Sleep apnoea 5. Sleep paralysis 6. Depressive disorders: short REM latency, ↑REM sleep duration and most antidepressants ↓duration of REM sleep except bupropion. 7. Alcohol misuse: ↑REM sleep duration 8. Seizure: 1) Absence seizure 2) Motor seizure	being aware of it. The words or sounds that are produced usually do not make sense. No treatment is required. 4. Sleep drunkenness: confusion during arousals from sleep. 5. Bedwetting 6. Nocturnal leg cramps 7. Depression: ↑ duration of SWS 8. Old age (> 60 years): ↓ duration of SWS, ↑ latency to sleep onset, more fragmented sleep and ↑ daytime napping

MEQ

You are a GP of the neighbourhood clinic. You are seeing a 40-year-old woman who complains of poor sleep at night. She requests to take a blue tablet from you because her GP prescribed this medication to her in the past. She showed you the package and the drug is called midazolam (dormicum).

Q.1 Would you prescribe midazolam to this patient?
No. Oral midazolam is not indicated as a regular hypnotic because it has high chance to cause dependency. Midazolam is a potent BZD which causes blood brain barrier rapidly.

Q.2 The patient is very upset with the fact that most GPs refuse to prescribe dormicum to her nowadays. She wants to lodge a complaint. What would you tell her about the regulation of prescription of benzodiazepines or sleeping pills?
The Singapore Medical Council advises all GPs to keep a log of the benzodiazepine prescribed and doctors have to justify the reasons for prescription. Investigations will be conducted for unnecessary prescription leading to misuse by patients.

Q.3 Name 3 types of non-addictive medications which you would order instead of dormicum?
1. Antihistamines e.g. hydroxyzine and piriton or
2. Noradrenaline and specific serotonin antidepressants, mirtazapine.
3. Melatonin
4. Antidepressant targeting at the melatonin receptors, agomelatine.

Q.4 She refuses to take any medication except dormicum. Few days later, his daughter brought her to the clinic again and she was unconscious for 10 minutes. What is the most important diagnosis you need to consider?

Benzodiazepine withdrawal seizure.

Q.5 What is your next action?

Send the patient to the nearest Accident and Emergency Department for assessment.

Q.6 Her daughter asks whether admission to a hospital would help her condition. Your answer is:

Yes, she may need detoxification with diazepam in the ward. Diazepam will be slowly titrated from a high dose e.g. 30mg per day to a low dose e.g. 5mg per day over a one to two-week period to prevent further withdrawal seizure.

MCQ

Q.1 Which of the following medications is a non-benzodiazepine hypnotics?

A. Alprazolam
B. Clonazepam
C. Diazepam
D. Lorazepam
E. Zopiclone.

The answer is E.

Q.2 A patient is very disturbed by the sudden death of her mother. She feels very agitated in the day time and needs an anxiolytic to calm her down on a p.r.n. basis for the next two weeks. She has tried hydroxyzine and finds it ineffective. Which of the following oral medications is the most suitable for her case?

A. Alprazolam
B. Benzhexol
C. Clonazepam
D. Diazepam
E. Midazolam.

The answer is A. Option B is an anticholinergic agent and not relevant for this case. Option E is indicated for intravenous or intramuscular use prior to a procedure. Option A has the shortest half-life among the options A, C and D and good for anxiolytic effect in the day time for 2 weeks.

Q. 3 Which of the following is not a REM sleep disorder?

A. Narcolepsy
B. Nightmare
C. Night terror
D. Sleep apnea
E. Sleep paralysis.

The answer is C.

Q. 4 Which of the following is a REM sleep disorder?

A. Night terror
B. Sleep drunkenness
C. Sleep paralysis
D. Sleep walking
E. Sleep talking.

The answer is C.

EMIs

Sleep EEG Waveforms

1. Alpha
2. Beta
3. Delta
4. Gamma
5. Theta

Question 1: The wavelet of EEG that is greater than 13hz in frequency – Beta

EEG waveforms
A. More than 13 hz
B. 8-13
C. 4-8
D. Less than 4hz
E. 7-11
F. Single Sharp Wave
G. Spike wave

Question 1: Alpha – 8 to 13
Question 2: Beta – more than 13
Question 3: Theta – 4 to 8
Question 4: Mu -7 to 11
Question 5: Delta- Less than 4

References

American Psychiatric Association (1994) DSM-IV-TR: Diagnostic and Statistical Manual of Mental Disorders (Diagnostic & Statistical Manual of Mental Disorders) (4th edition). American Psychiatric Association Publishing Inc: Washington.

American Psychiatric Association (2013) DSM-5: Diagnostic and Statistical Manual of Mental Disorders (Diagnostic & Statistical Manual of Mental Disorders) (5th edition). American Psychiatric Association Publishing Inc: Washington.

Dong Y, Winslow M, Chan YH, Subramaniam M, Whelan G. Benzodiazepine dependence in Singapore. Subst Use Misuse. 2007;42(8):1345-52.

Gelder M, Mayou R and Cowen P (2001) *Shorter Oxford Textbook of Psychiatry*. Oxford: Oxford University Press.

Goldbloom DS and Davine J (2010) Psychiatry in primary care. Centre for Addiction and Mental Health, Canada.

Mahendran R, Subramaniam M, Chan YH. Psychiatric morbidity in patients referred to an insomnia clinic. Singapore Med J. 2007 Feb;48(2):163-5.

Sadock J, and Sadock V (2005). *Pocket handbook of Clinical Psychiatry*. Lippincott Williams and Wilkins.

Taylor D, Paton C, Kapur S (2009) *The Maudsley prescribing guideline*.10th edition London: Informa healthcare.

http://science.education.nih.gov/supplements/nih3/sleep/guide/info-sleep.htm [accessed on 29th June 2012]

Chapter 12 Psychosexual medicine

Sexual dysfunctions	Sample MEQ
Abnormal sexual preferences or paraphilia	Sample MCQ
Gender dysphoric disorder	Sample EMIs

Sexual dysfunctions

General criteria for sexual dysfunction:

1) The subject is unable to participate in a sexual relationship as he or she would wish.
2) The dysfunction occurs frequently.
3) The dysfunction has been present for at least 6 months.
4) The dysfunction is not entirely attributable to other psychiatric disorders.

Conditions	Clinical features/ Prevalence / Treatment
Premature ejaculation	There is an inability to delay ejaculation sufficiently during sexual intercourse, Ejaculation occurs very soon after the beginning of intercourse (before or within 1 minute). Ejaculation occurs in the absence of sufficient erection to make intercourse possible **Premature (early) Ejaculation** The DSM-5 states that, over a course of at least 6 months, there must be a recurrent pattern of ejaculation occurring during partnered sexual activity within approximately 1 minute following vaginal penetration and before the individual wishes it to happen. This has to happen at least for almost all occasions of sexual activities. This condition has been estimated to affect around 35-40% of men who are receiving treatment for sexual related disorders amongst the American population. Treatment: 1. Squeeze technique: the female partner grips the penis for a few seconds when the man indicates that he is about to have orgasm. 2. Vaginal containment refers to containing the penis without any movement. 3. SSRIs can delay ejaculation.
Orgasmic dysfunction	**Orgasmic Disorder** The DSM-5 has specified that either of the following symptoms must be experience nearly on all occasions of sexual activity for at least 6 months: a. Delay or infrequent or total absence of orgasm. b. Reduction in the intensity of orgasmic sensations This condition has been estimated to affect around 30% of females and males amongst the American population. **For women:** orgasm does occur in certain situations (e.g. when masturbating or with certain partners). Treatment involves graduated practice in masturbation to orgasm. **For men:** It manifests as lack of or delayed ejaculation. It is not as common as premature ejaculation. Antipsychotics, MAOIs and SSRIs can lead to this problem. It may lead to infertility and yohimbine may assist ejaculation (side effects include anxiety, flu like symptoms, diarrhoea and agitation).

Failure of genital response	**For men:** erection sufficient for intercourse fails to occur when intercourse is attempted. **Erectile dysfunction (ED)** is 2 times more common in men older than 50 years than those younger than 50 years. The estimated prevalence of the disorder amongst the American population has been estimated to be around 8%. The DSM-5 has specified that at least one of the following symptoms must have been experienced nearly on all occasions of sexual activity. The symptoms include: a. Difficulties involving obtaining an erection b. Difficulties involving maintaining an erection till the completion of the sexual activity c. Decrease in erectile rigidity The DSM-5 has specified a duration criteria of at least 6 months. Common causes include anxiety (↑sympathetic tone) DM, venous incompetence, dialysis and spinal cord injury. There are several forms: i) Full erection occurs during the early stages of sexual intercourse but disappears or declines before ejaculation. ii) Erection does occur, but only at times when intercourse is not being considered iii) Partial erection which is insufficient for intercourse but not full erection iv) No penile tumescence occurs at all **Treatment:** sildenafil or vacuum devices. **For women:** There is failure of genital response, experienced as failure of vaginal lubrication and inadequate tumescence of the labia. There are three different forms: i) General: lubrication fails in all relevant circumstances ii) Lubrication may occur initially but fails to persist for long enough to allow comfortable penile entry iii) Situational: lubrication occurs only in some situations (e.g. with one partner but not another, or during masturbation).
Lack or loss of sexual desire	**Sexual Interest / Arousal Disorder** The DSM-5 specified that over a duration of 6 months, an individual has a reduced sexual interest or arousal, as manifested by at least 3 of the following symptoms: a. Decreased interest in sexual activity b. Decreased number of sexual thoughts and fantasies. c. Reduced attempts to initiate sexual activity d. Absence or reduction in sexual activity e. Absence or reduced sexual interest in response to internal or external sexual cues f. Absence or reduced genital or non genital sensations during sexual activity The DSM-5 specified that these symptoms must have persisted for at least a total duration of 6 months. **Cause:** Depression, multiple sclerosis, hyperprolactinaemia, and anticonvulsants (reduction in testosterone level in men).
Sexual aversion and lack of sexual enjoyment	**Treatment:** 1) Treat the underlying psychiatric or medical disorders. 2) Couple therapy.

	3) Cognitive therapy or counseling may be useful. 4) Apomorphine hydrochloride may improve sexual desire. Apomorphine acts centrally by stimulating dopamine release and hence enhance the sexual desire.

Non-organic vaginismus	**Genito-Pelvic Pain / Penetration Disorder**
	The DSM-5 states that there must be recurrent and persistent difficulties involving at least one of the following:
	a. Vaginal penetration during intercourse
	b. Pelvic or vulvovaginal pain during intercourse or during penetration attempts
	c. Fear or anxiety about pain in anticipation of, during, or as a result of vaginal penetration
	d. Tensing and tightening of the pelvic floor muscles during attempted penetration
	The DSM-5 states that these symptoms must be present for a minimum duration of at least 6 months and have caused significant distress to the individual.
	
	Clinical features:
	i) Normal response has never been experienced.
	ii) Vaginismus has developed after a period of relatively normal response:
	   a) when vaginal entry is not attempted, a normal sexual response may occur.
	   b) any attempt at sexual contact leads to generalized fear and efforts to avoid vaginal entry.
	
	Treatment:
	   1) Sensate focus therapy: focus on communication, mutual touching and postpone vaginal intercourse until later stage. This will reduce performance anxiety and explore alternative sexual practices to enhance mutual satisfaction.
	   2) Applying dilators of increasing size to desensitize patients to the painful feelings.

Non organic dyspareunia	**For women:** 1) Pain is experienced at the entry of the vagina, either throughout sexual intercourse or only when deep thrusting of the penis occurs. 2) The disorder is attributable to vaginismus or failure of lubrication. **Treatment:** Treat the underlying disorders such as endometriosis or pelvic infections. **For men:** 1) Pain or discomfort is experienced during sexual response. The timing of pain and the exact location should be carefully recorded. 2) The discomfort is not the result of local physical factors or organic causes.

Usage of sildenafil citrate for treatment of ED:

Penile erection depends on parasympathetic activity and leads to:

1) relaxation of the smooth muscle in the corpora cavernosa;
2) intact pelvic blood supplies to penis;
3) increase in arterial inflow and reduction of venous outflow from penis;
4) the build up of blood pressure in penis close to the systolic blood pressure.

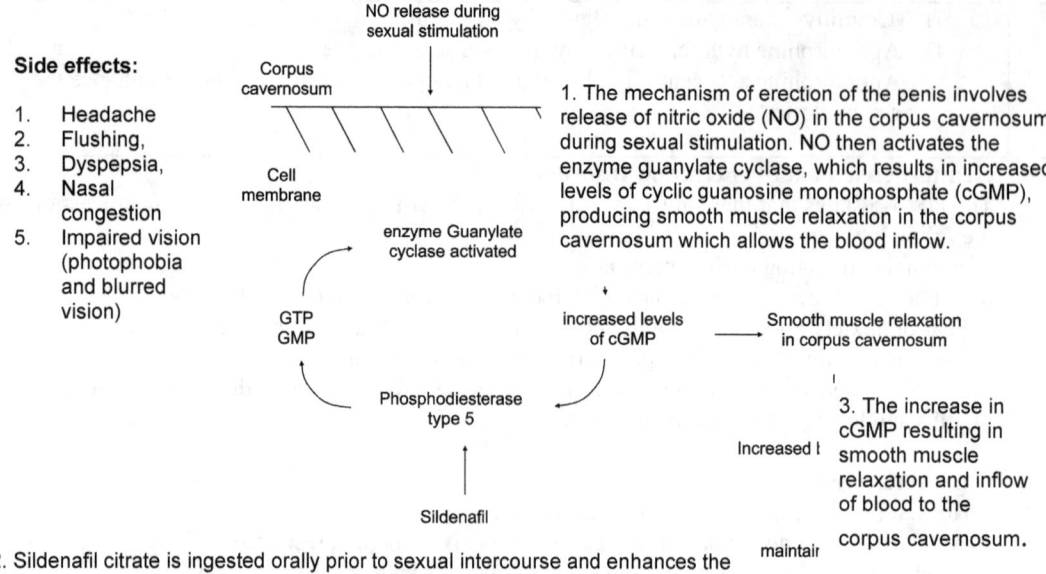

Side effects:
1. Headache
2. Flushing,
3. Dyspepsia,
4. Nasal congestion
5. Impaired vision (photophobia and blurred vision)

1. The mechanism of erection of the penis involves release of nitric oxide (NO) in the corpus cavernosum during sexual stimulation. NO then activates the enzyme guanylate cyclase, which results in increased levels of cyclic guanosine monophosphate (cGMP), producing smooth muscle relaxation in the corpus cavernosum which allows the blood inflow.

3. The increase in cGMP resulting in smooth muscle relaxation and inflow of blood to the corpus cavernosum.

2. Sildenafil citrate is ingested orally prior to sexual intercourse and enhances the effect of nitric oxide (NO) by inhibiting phosphodiesterase type 5 (PDE5), which is responsible for degradation of cGMP in the corpus cavernosum.

Other medications to treat ED

1. Vardenfil (Levitra) is a PDE-5 inhibitor that is used for treatment of erectile dysfunction. The mechanism of action is similar to that of sildenafil. The most common side effect is nausea. The usual starting dose is 10mg, and it is recommended for usage 1-2 hours prior to the commencement of sexual activity.

2. Tadalafil (Cialis) also works via the inhibition of PDE-5. However, it has a longer half-life of 17.5 hours as compared to sildenafil (4-5 hours) and vardenfil (4-5 hours). The recommended starting dose is 10mg, taken as needed prior to sexual activity. Common side effects include headache, indigestion, back pain, muscle aches, facial flushing and rhinitis.

Abnormal sexual preferences and paraphilia

Paraphilia	The person experiences recurrent intense sexual urges and fantasies involving unusual objects or activities with a duration of at least 6 months. He or she either acts on the urges or would be markedly distressed by the urges.
Fetishism	The DSM-5 diagnostic criteria states that over a time duration of at least 6 months, there must be recurrent and intense sexual arousal from the usage of non-living objects or highly specific focus on non-genital body parts. These objects are not limited to articles of clothing used in cross-dressing, or devices that are designed for the purpose of tactile genital stimulation 1. The fetish refers to a non-living object (e.g. rubber garments, underclothes and high heeled shoes). The fetish has become the most important source of sexual stimulation or is essential for satisfactory sexual response. 2. More common in men, 20% are homosexuals. It may be due to classical conditioning or temporal lobe dysfunction. Amphetamine and cocaine misuse are associated with fetishism. Onset is common during adolescence. 3. Treatment: behaviour therapy. 4. Prognosis: Most fetishism will diminish after developing a satisfying heterosexual relationship. The prognosis is worse in single man or those with forensic history related to fetishism.
Fetishistic transvestism	**Transvestic Disorder** The DSM-5 diagnostic criteria specified that there must be recurrent and intense sexual arousal from cross-dressing, for a period of at least 6 months. 1. The individual wears articles of clothing of the opposite sex in order to create the appearance and feeling of being a member of the opposite sex. 2. The cross-dressing is closely associated with sexual arousal. Once orgasm occurs and sexual arousal declines, there is a strong desire to remove clothing. 3. It is more common in men and most males are heterosexual. It may be associated with classical conditioning and temporal lobe dysfunction. 4. No specific treatment has been identified. 5. Prognosis: the condition becomes less severe as sexual drive decline.
Exhibitionism	**Exhibitionistic disorder** The DSM-5 states that there must be the presence of recurrent and intense sexual arousal from exposure of one's genitals to an unsuspecting person, over a period of at least 6 months.
Voyeurism	**Voyeuristic disorder** The DSM-5 states that there must be the presence of recurrent and intense sexual arousal from observing an unsuspecting person who is naked, in the process of disrobing or engaging in sexual activity over a total duration of 6 months. The DSM-5 specifies that the diagnosis could only be made if the individual is at least 18 years of age. 1. The voyeur is usually a man who is shy, socially awkward with girls and has difficulty with normal sexual expression. 2. Treatment: no specific treatment is recommended.
Paedophilia	**Pedophilic Disorder** The DSM-5 diagnostic criteria states that over a duration of at least 6 months, the individual must have recurrent, intense sexually arousing fantasies, urges or behaviours that involves sexual activity with a prepubescent child, or any children younger than the

	age of 13 years old. The DSM-5 has specified that the individual must be at least 16 years of age and must be at least 5 years older than the child or children. 1.
Sadomasochism	**Sexual Masochism Disorder** The DSM-5 states that over a total duration of at least 6 months, the individual must have recurrent and intense sexual arousal from the act of being humiliated, beaten, bound, or otherwise made to suffer. **Sexual Sadism Disorder** The DSM-5 states that over a total duration of at least 6 months, the individual must have recurrent and intense sexual arousal from the physical or psychological suffering of another person.
Other types of paraphilia	1. **Auto-erotic asphyxia**: Induction of cerebral anoxia to heighten sexual arousal. 2. **Frotteurism**: Rubbing male genitalia against another person. 3. **Coprophilia**: Sexual arousal is induced by watching the act of defecation. 4. **Coprophagia**: Sexual arousal is induced follows the eating of faeces. 5. **Sexual urethism**: Sexual arousal is obtained by stimulation of urethra in women. 6. **Urophilia**: Sexual arousal is obtained by watching the act of urination.

Gender Dysphoria disorder

Gender Dysphoria
In Children
The DSM-5 states that over a duration of at least 6 months, there must be marked incongruence between one's expressed gender and assigned gender, which is manifested by at least 6 of the following:
a. Persistent and strong wish to be of the other gender
b. For boys, a persistent preference for cross dressing in female attire; and for girls, a persistent preference for wearing only typical masculine clothing.
c. Strong preference for cross-gender roles during play
d. Strong preference to engage in activities typically done by the other gender
e. Strong desire to be with playmates of the other gender
f. For boys, rejection of typical masculine toys and for girls, rejection of typical feminine toys
g. Strong dislike of one's sexual anatomy
h. Strong wish that primary and secondary sexual characteristics would match one's experienced gender.

In adults
The DSM-5 states that over a duration of at least 6 months, there must be marked incongruence between one's expressed gender and assigned gender, which is manifested by at least 2 of the following:
a. Marked incongruence between one's experienced gender and primary and secondary characteristics
b. Strong desire to get rid of one's primary and/or secondary sex characteristics
c. Strong desire for the primary and/or secondary characteristic of the other gender
d. Strong desire to be of the other gender
e. Strong desire to be treated as the other gender
f. Strong conviction of having typical feelings and reactions of the other gender

Clinically, it is still important to differentiate between:

Transsexualism	1. The individual desires to live and be accepted as a member of the opposite sex, usually accompanied by the wish to make his or her body as congruent as possible with the preferred sex through surgery and hormonal treatment. 2. Duration required for persistent transsexual identity for at least 2 years. 3. Exclusion: schizophrenia or chromosome abnormality.

	4. Treatment: Sex-reassignment surgery.	
Dual-role transvestism	1. The individual wears clothes of the opposite sex in order to experience temporarily membership of the opposite sex. There is no sexual motivation for the cross-dressing. 2. The individual has no desire for a permanent change of gender. 3. Age of onset: puberty and the behaviour is concealed without guilt. 4. Treatment: No specific treatment is required but allowing the person to ventilate their feelings may help to reduce the frequency of cross-dressing. 5. Prognosis: Some of them become fetishists	

Gender identity disorder of childhood	For girls:
	1. The child shows persistent and intense distress about being a girl, and has a stated desire to be a boy.
	2. Either of the following must be present:
	    a. Persistent marked aversion to feminine clothing and insistence on wearing stereotypical masculine clothing, e.g. boys' underwear and other accessories
	    b. Persistent rejection of female anatomical structures, as evidenced an assertion that she wants to have a penis; ii) refuse to urinate in a sitting position; iii) assertion that she does not want to grow breasts or have menses.
	3. The girl has not yet reached puberty.
	4. Duration of symptoms: at least 6 months.
	
	For boys, the symptoms are very similar to girls. The pre-pubertal child shows persistent and intense distress about being a boy, and has a desire to be a girl. The child prefers female activities and clothing and rejects male anatomical structure. Duration of symptoms is 6 months.	

MEQ

You are the GP of the neighbourhood. A 50-year-old man complains of erectile dysfunction for 6 months. You have been treating his diabetes. He also suffers from major depression and takes a SSRI, fluoxetine on a daily basis. His relationship with his wife is poor as a result of financial problems.

Q.1 List 4 possible causes of his erectile dysfunction.

1. Primary erectile dysfunction
2. Sexual dysfunction associated with depression
3. Side effects of fluoxetine
4. Organic causes e.g. diabetes
5. Poor marital relationship.

Q.2 What question would you ask to differentiate organic from psychogenic erectile dysfunction?

The presence of early morning erection differentiates organic from psychogenic erectile dysfunction.

Q.3 How do you differentiate erectile dysfunction as part of the depressive symptoms from antidepressant induced erectile dysfunction?

The history taking process would differentiate these two conditions. Erectile dysfunction associated with depression occurs prior to the initiation of antidepressant. Antidepressant induced erectile dysfunction occurs after initiation of antidepressant. The severity of erectile dysfunction may be dose-related and improved after stopping the antidepressant.

Q.4 After careful assessment, the erectile dysfunction is related to antidepressant. List 4 treatment advices.

1. Reduce the dose and frequency of antidepressant.
2. Change antidepressant from fluoxetine to mirtazapine or buporion with less sexual dysfunction.
3. Consider CBT as treatment of depression and stop antidepressant.
4. Drug holiday, stop antidepressant on weekends or during a holiday when sexual activity takes place.
5. Consider sildenafil to treat erectile dysfunction.
6. Refer the patient for sensate focus sex therapy with his wife.

MCQ

Q.1 Which of the following is the most likely diagnosis for a male national serviceman who describes persistent and intense distress about assigned sex, together with an insistence that he is of the female gender?

A. Dual-role transvsestism
B. Egodystonic sexual orientation
C. Sexual maturation disorder
D. Transexualism
E. Transvestism

The answer is D.
This patient suffers from transexualism. Transvestism refers to cross-dressing with no intention to change gender.

Q.2 Which of the following is not an established effect of sildenafil?

A. Flushing
B. Hypertension
C. Nasal congestion
D. Priapism
E. Vision disturbance.

The answer is E. Sildenafil causes smooth muscle relaxation and causes hypotension rather than hypertension.

Q.3 Which of the following is not recommended in sensate focus sex therapy?

A. Adopt a behaviour therapy approach.
B. Aim to reduce performance anxiety.
C. Focus on communication in the beginning.
D. Focus on mutual stimulation in the beginning.
E. Focus on vaginal intercourse in the beginning.

The answer is E. Sensate focus therapy suggests focusing on vaginal intercourse at a later stage and not in the beginning.

EMIS

A Exhibitonism
B Fetishism
C Paedophilia
D Sexual masochism
E Sexual sadism
F Transvestic fetishism
G Voyeurism
H Frotteurism
I Kelptomania
J Pyromania

1. Use of inanimate objects to achieve arousal – B
2. Touching and rubbing against non consenting individual – H
3. Tendency to observe unsuspecting persons naked or engaged in some sexual activity – G
4. Tendency to wear clothes of the opposite sex for sexual arousal – F
5. Sex with prepubescent child - C

References

Books

[1] Bancroft J. H. J.(1999) *Human sexuality and problems (2nd ed)* Edinburgh: Churchill Livingstone.

[2] World Health Organisation (1992) *The ICD-10 Classification of Mental and Behavioural Disorders: Clinical Descriptions and Diagnostic Guidelines.* Geneva: World Health Organisation.

[3] Taylor D, Paton C, Kapur S (2009) *The Maudsley prescribing guideline.* 10th edition London: Informa healthcare.

[4] Gelder M, Mayou R and Cowen P (2001) *Shorter Oxford Textbook of Psychiatry.* Oxford: Oxford University Press.

[5] Green R, and Fleming D (1991). Transsexual surgery. Follow-up: status in the 1990s. In Bancroft, C. Davis, and D. Weinstein. Annual review of sex search Society for Scientific Study of Sex, Society for Scientific Study of Sex : Mt Vernon, IA.

[6] Sadock J, and Sadock V (2005). *Pocket handbook of Clinical Psychiatry.* Lippincott Williams and Wilkins.

[7] American Psychiatric Association (2013) *DSM-5: Diagnostic and Statistical Manual of Mental Disorders (Diagnostic & Statistical Manual of Mental Disorders)* (5th edition). Washington: American Psychiatric Association Publishing Inc.

Website

[7] **The Open University and BBC on family and child development**
http://open2.net/healtheducation/family_childdevelopment/2005/extractone.html

[8] Rx List Drug Information
http://www.rxlist.com/script/main/hp.asp

Chapter 13 Somatoform, dissociative and conversion disorders

Somatoform disorder	Body dysmorphic disorder
Somatoform autonomic dysfunction	Dissociative and conversion disorders
Persistent somatoform pain disorder	Factitious disorders and malingering
Hypochondriacal disorder	MEQ and MCQ and EMIs

Changes for Somatic symptoms and related disorders in DSM-5
Somatoform disorders are now referred to as somatic symptom and related disorders.
The current DSM-5 now reduces the number of disorders and subcategories to avoid overlap. Diagnosis of somatization disorder, hypochondriasis, pain disorder and undifferentiated somatoform has been removed.

Somatic Symptom Disorder
The DSM-5 diagnostic criteria states that there is excessive preoccupations and persistent anxiety about one or more bodily symptoms, such as excessive amount of time and energy are devoted to these bodily symptoms that have led to significant impairments in life. The DSM-5 specified time duration of at least 6 months

Nevertheless, it is still important clinically to know about somatoform disorder, somatoform autonomic dysfunction and persistent somatoform pain disorder.

Somatoform Disorder

Diagnostic criteria for somatoform disorder (ICD-10)
1. There must be a history of at least 2 years complaints of multiple physical symptoms that cannot be explained by any detectable physical disorders.
2. There must also be chronic preoccupation with the symptoms that caused persistent distress and has led the patient to seek repeated consultations or investigations with GPs or specialists.
3. There is persistent refusal to accept medical reassurance that there is no adequate physical cause for the symptoms.
4. There must be a total of six or more symptoms from the following list, with symptoms occurring in at least two separate body systems:

Figure 1 **Summary of ICD-10 diagnostic criteria**

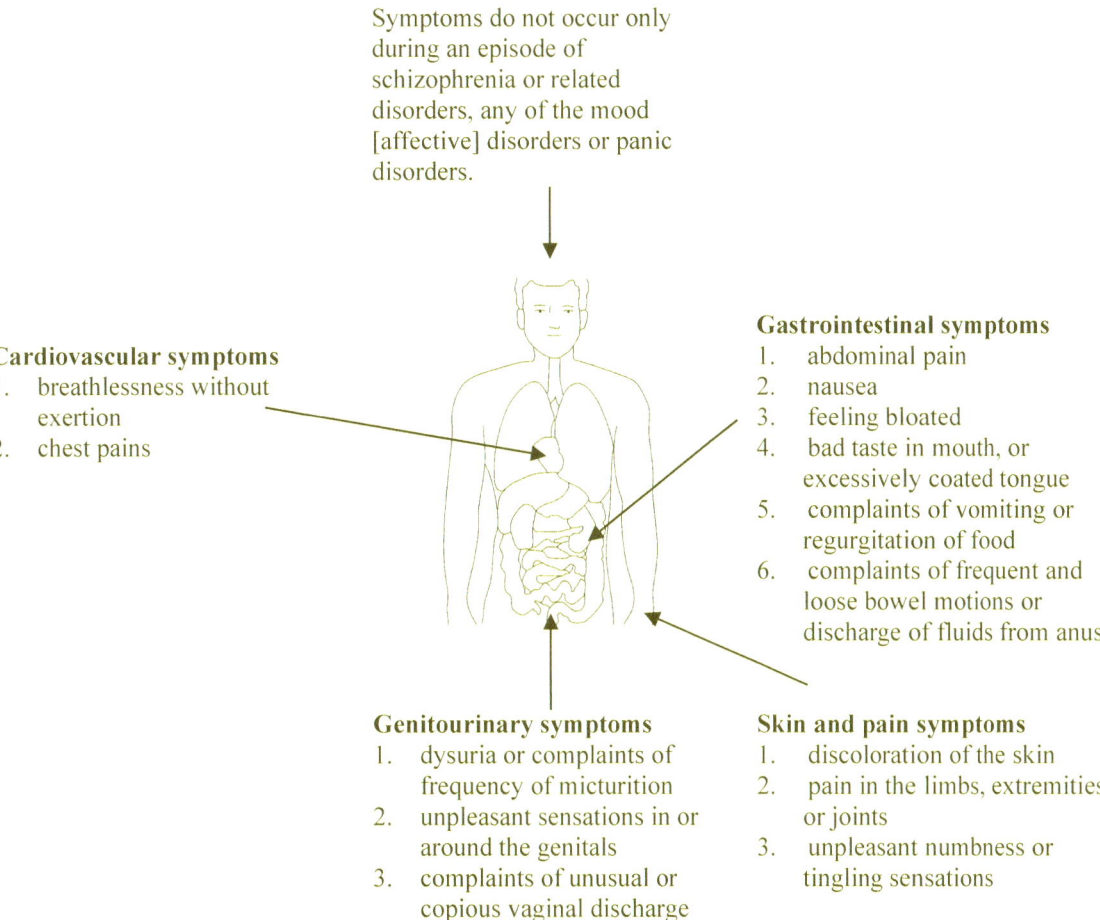

Previously, in DSM-IV-TR, the diagnostic criteria for somatoform disorder include age of onset below 30 years, 4 pain symptoms, 2 gastrointestinal symptoms, 1 sexual problem and 1 pseudo-neurological symptom after exclusion of other potential differential causes.

Epidemiology

Western countries

1. The estimated prevalence of somatisation disorder has been estimated to affect around 0.2-2% of women and around 0.2% of men in the United States.
2. There is an increased prevalence of such disorders among women. The female to male ratio has been estimated to be around 5:1.
3. Studies have also demonstrated that somatisation disorders are especially prevalent among patients who have had little education and are part of a lower socioeconomic class.
4. Patients diagnosed with somatisation disorder tend to have other psychiatric comorbidity such as antisocial personality disorder.

Singapore

Yap et al (2004) reported that depression and somatisation are related to the self-report of muscle pain in Singapore patients. In addition, severe somatisation may be associated with an increase in jaw disability.

Aetiology

1. **Genetics:** Genetic linkages have been proposed as a result of the increase in prevalence (10-20%) of somatisation disorder among family members.

2. **Learning theory:** a person learns somatisation from a parental figure.

3. **Psychodynamic theory:** somatisation symptoms are repressed impulses.

4. **Other psychological factor**: Alexithymia (lack of language to describe emotion and describe emotion in terms of physical complaints).

5. **Psychosocial factors include** dysfunctional family relationship, a past history of abuse in the family, low social economic status and education.

Differential diagnosis

Other medical disorders should be excluded.

How is somatisation disorder different from medical illness?

1. Somatisation disorder does not involve any laboratory abnormality.
2. Patients with somatisation disorder present wide range of symptoms and the symptoms may not correspond to current anatomical and physiological knowledge.
3. Somatisation disorder has an early onset and chronic course.
4. Somatisation disorder may follow a traumatic event.
5. Somatisation disorder is associated with antisocial personality trait (based on US findings).

Other psychiatric differential diagnosis:

1. Depression
2. Anxiety
3. Body dysmorphic disorder
4. Substance abuse
5. Psychotic disorder
6. Personality disorder
7. Adjustment disorder or grief
8. Factitious disorder.

Management

The management of somatisation disorder involves the following principles:

1. **Regular appointments:** Patients diagnosed with somatisation disorder should have a fixed, regular physician whom they can seek help from and be seen on a regular basis.
2. **Avoid unnecessary investigation** but physical examination should be performed if new complaints arise.
3. **Offer empathy** to the sufferings experienced by the patient.
4. **Pharmacotherapy for somatisation disorder**: SSRI for somatisation disorder; SNRI for somatoform pain disorder.
5. **Psychotherapy**: self-help techniques, supportive psychotherapy, cognitive behaviour therapy (e.g. challenge cognitive distortions, activity scheduling to enhance physical activities).
6. **Avoid Polypharmacy:** review the list of unnecessary medications and advise patients to avoid medications which cause dependency (e.g. opioid analgesics).
7. **Increase in social and occupational functioning**: refer to job agencies (e.g. Community Development Council for job referral).

Somatoform autonomic dysfunction

ICD-10 criteria:
1. There must be symptoms of autonomic arousal that are attributed by the patient to a physical disorder of one or more of the following systems or organs:
    1. Heart and cardiovascular system
    2. Upper gastrointestinal tract (oesophagus and stomach).
    3. Lower gastrointestinal tract
    4. Respiratory system
    5. Genitourinary system.

Figure 2   Summary of somatoform autonomic dysfunction

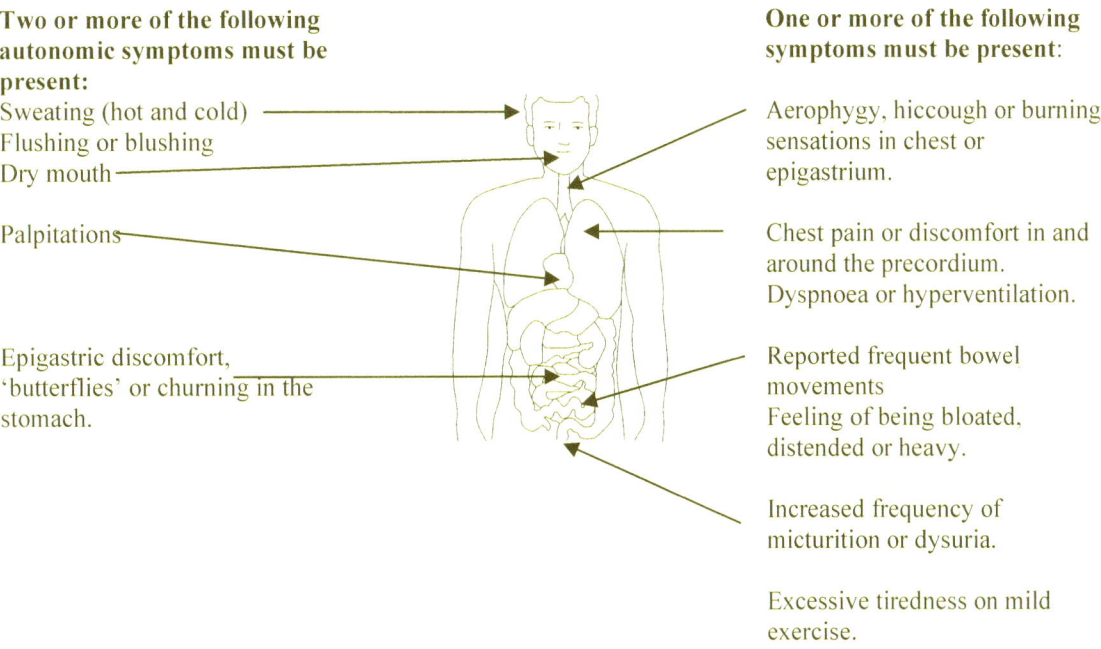

Two or more of the following autonomic symptoms must be present:
Sweating (hot and cold)
Flushing or blushing
Dry mouth
Palpitations
Epigastric discomfort, 'butterflies' or churning in the stomach.

One or more of the following symptoms must be present:
Aerophygy, hiccough or burning sensations in chest or epigastrium.
Chest pain or discomfort in and around the precordium.
Dyspnoea or hyperventilation.
Reported frequent bowel movements
Feeling of being bloated, distended or heavy.
Increased frequency of micturition or dysuria.
Excessive tiredness on mild exercise.

Persistent somatoform pain disorder

Definition: There is persistent severe and distressing pain (for at least 6 months and continuously on most days), in any part of the body, which cannot be explained adequately by a physiological process or a physical disorder.

Epidemiology: Somoatoform pain disorder is more common in women (F:M = 2:1) and occurs in older patients (40s to 50s).

The relationship between somatoform pain disorder and depression:

Biological theory:
1. Increase in cortisol levels
2. Decrease in immune functions

Psychological theory:
1. Learned helplessness as no relief of chronic pain
2. Role transition as a result of pain
3. Secondary gain: attention from family members

Acute pain (< 6 months) is associated with anxiety disorder. Chronic pain (> 6 months) is associated with depressive disorder.

Management
1. **Cognitive therapy**: Ask the patient to review thoughts and beliefs with regards to chronic pain.
2. **Behaviour therapy**: Involves enhancement of pre-existing activity levels and encouraging positive coping behaviours.
3. **Family support:** Educate the partner or spouse on how to respond to pain.
4. **Problem solving techniques**
5. **Antidepressants** (e.g. SNRI) may reduce depression and anxiety associated with pain.

Hypochondriacal disorder / Illness Anxiety Disorder (DSM-5)

Epidemiology

Western countries:

1. The estimated prevalence has been estimated to be around 4-6 percent.
2. Men and women are equally affected.
3. The disorder commonly affects individuals who are between the ages of 20 to 30-year old.

There is no epidemiology figure in Singapore.

Aetiology

1. **Misinterpretation and amplification**: patients with hypochondriacal disorder tend to have lower threshold for physical discomfort and misinterpret and attribute their bodily symptoms into a condition that is more sinister.

2. **Social learning and sick role:** the sick role offers secondary advantage to allow the individual to be free from certain obligations and their usual duties.

3. **Variant of other mental disorders** e.g. several depressive disorder.

Diagnostic criteria (DSM-5)

Illness Anxiety Disorder
There has been excessive preoccupation with illness for a minimum of at least 6 months. These include symptoms such as somatic symptoms, which even if present are mild; pervasive anxiety about health; and excessive health related behaviours such as repeatedly checking for signs of illness and/or maladaptive avoidance such as avoiding doctors appointments and hospitals.

Differential diagnosis

Other medical disorders should be excluded.

Differences between somatoform disorder and hypochondriasis:

1. Patients with hypochondriasis are concerned of having a particular disease or disorder.
2. Patients with somatoform disorder are concerned with the number of symptoms they have.

Differences between body dysmorphic disorder (BDD) and hypochodriasis:

1. BDD involves a concern of body defect. Hypochondriasis involves concerns of having a serious illness.
2. BDD is associated with an increase in social phobia and compulsive checking.
3. Patients with BDD are interested in surgical correction while patients with hypochondriasis are interested in diagnostic workup.

Management

1. The general approach is similar to the management of somatoform disorder.
2. **Cognitive therapy**: involves reattribution and developing alternative explanations of concerns of serious illness. Cognitive restructuring can modify dysfunctional assumptions.
3. **Behaviour therapy**: involves self-monitoring of worries, negative thoughts and illness related behaviours. It also involves exposure and response prevention and reducing repeated reassurance – seeking behaviours.
4. **Pharmacotherapy**: Psychotropic drugs (e.g. SSRIs) are usually used to treat comorbidity such as depression.

Prognostic factors

Good prognosis factors include:

1) High social economic status
2) Acute onset
3) Treatment responsive (depression and anxiety)
4) No underlying personality disorder
5) No underlying medical problem.

Body dysmorphic disorder (BDD)

Epidemiology

Western countries
1. The most common age of onset is between 15 – 30 years.
2. BDD is more common in women.
3. More likely to be single and unmarried.

Singapore
Hsu et al (2009) reported a prevalence of 29.4% in an aesthetics centre warrants systematic screening for BDD.

Aetiology

1. Family history of OCD or BDD
2. Reduction of serotonin
3. Psychodynamic theory: displacement of unresolved conflict to a body part
4. Childhood: rejection of others as a result of minor defect in the body.
5. Social factors: disharmonious family and culture which emphasises on beauty.

Body Dysmorphic Disorder
The DSM-5 specified that individuals with this disorder tend to have excessive preoccupation with perceived defects or flaws in physical appearance, to the extent that individuals have performed repetitive behaviours (eg. repeated checking, excessive grooming or seeking reassurance) in response to the perceived defects. The preoccupation has caused much

distress and has affected the individual's level of functioning.
Subtypes include
a. With good or fair insight
b. With poor insight
c. With absent insight or delusional beliefs

Clinical features

Patients with BDD have an imagined defect and leads to significant distress.

3 common behavioural problems:
1. Social avoidance: 30%
2. Suicide: 20% and self-injury: 70-80%
3. Checking rituals on body defect.

Common reactions to the imagined defects:
1. Camouflage: 90%
2. Minor check: 90%
3. Compulsion: 90%
4. Skin pick: 30%.

Common body sites by frequencies:
1. Hair: 63%
2. Nose: 50%
3. Skin: 50%
4. Eye: 30%
5. Face: 20%
6. Breast < 10%
7. < 5%: Neck, forehead and facial muscle.

BDD is different from OCD in the following aspects:

1. Less likely to have social phobia in OCD
2. Less likely to attempt suicide in OCD
3. Less likely to misuse substance in OCD
4. More insight in patients with OCD
5. Better interpersonal relationship in OCD.

BDD is similar to OCD in the following aspects:

1. Obsessions and compulsions are common in both disorder
2. SSRI is the treatment for both disorders.
3. Both disorders cause significant distress.

Comorbidity:

1. Social phobia: 38%
2. Substance: 36%
3. Suicide: 30%
4. OCD: 30%
5. Depression: 20%

Management

1. Risk assessment is important because patient may perform self-operation especially in those who are rejected by various doctors for further treatment.

2. SSRIs are indicated for treatment of BDD. 50% of patients with BDD respond to SSRIs.
3. Second generation antipsychotics (e.g. risperidone) may be useful for patients with BDD exhibiting psychotic features.
4. Psychotherapy: CBT is useful. Cognitive therapy provides cognitive restructuring and behaviour therapy prevents response such as frequent mirror checking on imagined defects.
5.

Dissociative and conversion disorders

Conversion disorder

Conversion disorder is defined as an illness associated with either deficits in motor or sensory function as a result of internal psychiatric conflict or secondary gain. Common symptoms include sudden paralysis of one limb or loss of vision after a traumatic event or conflict.

Epidemiology

Western countries

1. Conversion disorder is more prevalent among women as compared to men (F: M = 2:1).
2. There is a relationship between conversion disorders and low socio-economic status.
3. Comorbidity: depressive disorder (50%) and anxiety disorder (40%)

Aetiology

1. **Biological factors:** decrease in metabolism of dominant hemisphere; increase in metabolism of non-dominant hemisphere and increase in cortisol arousal.
2. **Psychoanalytic theory**: conversion disorder is a result of the repression of underlying unconscious conflicts, with the subsequent conversion of anxiety symptoms into a physical symptom.
3. **Learning theory:** a person may acquire these dysfunctional learned behaviours (e.g. pretend to have a limb or blindness) in childhood to cope with difficult situation.

Research findings in Singapore

Teo and Choong (2008) studied 13 Singapore children with conversion disorder. Paralysis was the most common neurological symptom. Patients presented with multiple, complex conversion symptoms and other neurological symptoms such as seizures and headache. Majority of them eventually had good outcome in terms of academic grades and social functioning.

Clinical features

Table 1 Summary of clinical features of conversion disorder

Motor symptoms	Sensory symptoms	5 salient features
1. Paralysis 2. Mutism 3. Gait disturbance (astasia –abasia is a common disturbance seen in conversion disorder. Patient with this gait is noted to be ataxic, staggering while walking and with irregular arm movements.) 4. Pseudo seizure 5. Anaesthesia or paresthesia	1. Blindness 2. Tunnel vision	1. Not congruent with anatomical and physiological knowledge 2. No structural change or damage 3. Vary with level of stress 4. Fluctuation in the course 5. Inconsistent symptoms.

Conversion Disorder

The DSM-5 diagnostic criteria specified that there ought to be at least 1 symptom of altered voluntary motor or sensory function, whose etiology cannot be accounted for by existing neurological or medical conditions. The altered function has resulted in marked impairments for the individuals.

The DSM-5 further sub-specifies in accordance to the symptomatology:

a. With paralysis or weakness
b. With abnormalities in movements
c. With swallowing symptoms
d. With epileptic seizures
e. With anesthesia or sensory loss
f. With special sensory symptoms
g. With mixed symptoms

How is pseudo-seizure different from a genuine seizure?

Table 2 Compare and contrast pseudo-seizure and a genuine seizure

Pseudo-seizure	Genuine seizure
No history of seizure	History of seizure
Anxiety and absence of aura	Aura
Induced by stress	Not induced by stress
Conscious during seizure	Unconscious during seizure
Occurs in front of witnesses to seek attention	May have not witness
Asymmetrical body movements during seizure	Symmetrical body movements during seizure
No delirium	Post-ictal confusion
No increase in prolactin	Raised prolactin after seizure if blood is taken within 30 minutes
No injury as a result of seizure	Injury as a result of seizure
No incontinence	Incontinence after seizure
Normal EEG	Abnormal EEG

Management

Psychotherapy such as insight-oriented supportive therapy and behavior therapy are helpful. SSRI is indicated if there is comorbidity of depression and anxiety.

Prognosis

Good prognostic factors include:
1. High intelligence

2. Good premorbid function
3. Acute onset
4. Clear stressor as a precipitant

65% of patients have complete recovery after an episode of conversion disorder.

DSM-5 Changes for Dissociative disorders
1. Derealization is now included together with what was called depersonalization disorder and is now termed as depersonalization/derealization disorder.
2. Dissociative fugue is now a specifier of dissociative amnesia rather than a separate diagnosis.

Nevertheless, it is still important clinically to know about:

Dissociative Disorder

Epidemiology

Western countries

1. Dissociative amnesia has a prevalence of 6% in the general population.
2. The prevalence of depersonalisation has been estimated to affect as high as 19% of the general population. It is more common in women with a F:M ratio = 4:1.
3. Dissociative identity disorder tends to affect women with a F:M ratio = 5:1.

Singapore

Ng BY (2004) studied dissociative trance state in Singapore. His findings are summarised as follows:

Common stressors for the trance include:
1. Problems with military life (38%)
2. Conflicts over religious and cultural issues (38%)
3. Domestic disharmony and marital woes (24%).

Positive predictors for trance include:
1. Conflicts over religious and cultural issues
2. Prior exposure to trance states
3. Being a spiritual healer or his/her assistant.

Aetiology

1. Dissociative amnesia: as a result of severe psychological trauma, the patient temporarily and unconsciously shut downt the memory of all events in life.
2. Depersonalization and derealisation: traumatic stress or fatigue (e.g. post-call in doctors)
3. Dissociative identity disorder – maltreatment and severe childhood trauma are predisposing factors.

Table 3   Summary of diagnostic criteria of dissociative disorders

Disorders	DSM-5 diagnostic criteria
Dissociative amnesia	**Dissociative Amnesia** The DSM-5 diagnostic criteria states that individual with dissociative amnesia would have marked difficulties with recall of important information, usually pertaining to traumatic or stressful memories, that cannot be explained by normal forgetting.
Dissociative fugue	1. The individual undertakes an unexpected yet organized journey away from home or from the ordinary places. Self-care is largely maintained. 2. There is amnesia (partial or complete) for the journey, which also meets criterion 2 for

	dissociative amnesia. 3. It also includes confusion about personality identity and assumption of a new identity.\ (Now a specifier condition instead in DSM-5)
Dissociative stupor	1. There is profound diminution or absence of voluntary movements and speech and normal responsiveness to light, noise and touch. 2. Normal muscle tone, static posture and breathing are maintained.

Trance and possession disorders	1. **Trance.** There is temporary alteration of the state of consciousness, shown by any two of:
	  A. loss of the usual sense of personal identity;
	  B. narrowing of awareness of immediate surroundings and selective focusing on environmental stimuli;
	  C. limitation of movements, postures and speech to repetition of a small repertoire.
	
	2. **Possession disorder.** The individual is convinced that he or she has been taken over by a spirit, power, deity or other person.

Dissociative identity disorder or multiple personality disorder	**Dissociative identity disorder** The DSM-5 diagnostic criteria state that dissociative identity disorder is characterized by the presence of 2 or more distinct personality states. The disruption in identity causes the following signs and symptoms: a. Discontinuity in sense of self b. Discontinuity in sense of agency c. Changes in affect, behaviour, consciousness, memory, perception, cognition and sensory-motor functioning. The DSM-5 states that there must be repetitive gaps in memory, that cannot be explained by ordinary forgetting. These symptoms cannot be explained as part of a cultural or religious practice.
Other dissociative syndrome	**Ganser's syndrome:** give approximate answers, psychogenic physical symptoms, hallucinations and apparent clouding of consciousness.

Differential diagnosis

1. DDX for dissociative amnesia include ordinary forgetfulness, organic disorders like dementia and delirium, post-traumatic amnesia, substance related amnesia and also transient global amnesia.
2. DDX for depersonalization include medication-related side effects, substance abuse, post traumatic stress disorder, schizophrenia or other dissociative disorders.

Management

Dissociative amnesia
1. Supportive psychotherapy in the initial stage. Consider CBT when the patient recovers from the amnesia.
2. Cognitive therapy helps to identify the specific cognitive distortions, especially so in the case of trauma related cognitive distortions.
3. Pharmacological agents that have been previously used include sodium amobarbital or diazepam in a process called abreaction to help patient to facilitate recall traumatic events in a semi-conscious state but this process is seldom performed nowadays.

Depersonalization disorder
1. Pharmacotherapy: SSRI may be helpful in frequent depersonalisation disorder and comorbid depressive disorder.

Dissociative identity disorder
1. Supportive psychotherapy in the initial stage. Consider CBT when the patient is stable.
2. CBT is useful in dealing with the multiple cognitive distortions that the patient has.
3. Antidepressants reduce depression and enable the mood to be stabilised.

Factitious disorders

Epidemiology

1. The prevalence in western countries is around 0.8%.
2. Factitious disorder is more common in women aged between 20 – 40 years.
3. Patients came from a background of medical –related fields e.g. medical laboratory technician, paramedics, allied health workers.

Classification

There are four types of factitious disorder:
1) Predominantly psychiatric signs and symptoms
2) Predominantly physical signs and symptoms
3) Combined physical and psychiatric signs and symptoms.
4) Munchausen syndrome (aka hospital addiction syndrome) or Munchausen syndrome by proxy (hospital addiction imposed on a child by his or her parent) are classified under factitious disorder. Munchausen syndrome by proxy is a form of childhood abuse.

Diagnostic criteria (DSM-5)

Factitious Disorder Imposed on Self
The DSM-5 diagnostic criteria state that an individual must have presented himself as injured, impaired or ill to others, without the intention of gaining obvious external rewards.

Factitious Disorder Imposed on another
(previously by proxy)
The DSM-5 diagnostic criteria states that an individual must have presented another individual to others as injured, impaired or ill, without the intention of gaining obvious external rewards

Additional clinical features

1. Centre of attention
2. Dependent on others
3. Atypical and vague symptoms
4. Feeling of worthlessness
5. Long history with multiple AED visits.
6. Involves pathological lying.

Management

1) Treat psychiatric comorbidity such as depression and anxiety.
2) SSRIs can reduce impulsivity.
3) Supportive psychotherapy to explore alternative behaviour to avoid frequency admission to hospital.
4) Family therapy is indicated for Munchausen syndrome by proxy.

Malingering

Diagnostic criteria (DSM-5)

1. Intentional production or feigning of physical or psychological signs or symptoms
2. Motivation is a result of external incentive (e.g. making a false claim of insurance).

Table 4   Compare and contrast malingering and conversion disorder

Criteria	Malingering	Conversion disorder
Attitude to mental health professionals	Suspicious, uncooperative, resentful, aloof, secretive and unfriendly.	Cooperative, and described as appealing, clinging and dependent.
History	Tend to give every detail of the accident or symptoms.	A vague account of the symptoms.
Attitude to physical examination and further investigations	Try to avoid physical examination and investigations to confirm the diagnosis.	Allows physical examination and investigations.
Attitude towards employment	Refuse employment.	More likely to accept employment as compared to a malingerer.

MEQ

You are working as a GP in the neighbourhood. A 30-year-old woman requests a referral letter from you to the National Skin Centre for hair transplant. She has seen private dermatologists locally and aboard. She does not like her scalp and firmly believes that she has baldness but no dermatologist agrees with her. Physical examination reveals that she has normal amount of hair and there is no sign of baldness. She tends to count the number of hair loss when she washes her hair. She is very upset because she feels that she is losing more and more hair. She has frequently applied a '101' solution to her head to grow more hair. She frequently looks into the mirror to check her hair. She has good past health and not taking medication on a regular basis.

Q.1 Will you refer her to the National Skin Centre for hair transplant immediately after seeing this patient?
No

Q.2 What is your diagnosis?
Body dysmorphic disorder (BDD)

Q.3 The patient believes that she suffers from obsessive compulsive disorder. How is your diagnosis in Q.2 different from obsessive compulsive disorder? Name 3 differences.
1. Patients with BDD tend to have more social phobia.
2. Patients with BDD are more likely to attempt suicide.
3. Patients with BDD are more likely to misuse drugs.
4. Patients with BDD have less insight.
5. Patients with BDD have poor interpersonal relationship

Q.4 Name 1 question you would ask in the risk assessment.
1. Do you have plan to perform the hair transplant on your own if doctors at National Skin Centre refuse to offer hair transplant? (assess self operation).
2. Do you have thoughts of ending your life? (assess suicide).

Q.5 Name one pharmacological agent which is useful to treat her condition.

SSRI e.g. fluoxetine or flvoxamine.

Q.6 Name one psychological intervention which helps to reduce repetitive checking into the mirror?

Exposure and response prevention: ask patient to set a time limit (e.g. 3 hours without looking into the mirror).

MCQ

Q.1 A 35-year-old unemployed man believes that he suffers from AIDS. He needs to go to a private hospital for monthly HIV test despite negative results. What is the diagnosis?

A. Conversion disorder
B. Dissociative disorder
C. Factitious disorder
D. Hypochondriacal disorder
E. Somatisation disorder.

The answer is D. This man suffers from hypochondriacal disorder because he believes he has a serious illness (e.g. AIDS) without any laboratory evidence.

Q.2 A 30-year-old unemployed man comes to the Accident and Emergency Department (AED) today and requests for admission. He mentions to the AED doctor that he would jump down from a building if he is not allowed to be admitted. Mental state examination reveals a cheerful man without any psychiatric sign. He has been admitted to various hospitals for 30 times in the past five years. There is no clear diagnosis and he seems to enjoy staying in the hospital. His mother has similar behaviour and her parents have financial difficulty because she used all money by admitting herself to various private hospitals. What is the diagnosis?

A. Conversion disorder
B. Dissociative disorder
C. Factitious disorder
D. Hypochondriacal disorder
E. Somatisation disorder

The answer is C. This man suffers from Munchausen syndrome and seems to run in his family.

Q.3 A 22-year-old national serviceman is admitted to the ward because he cannot remember his name, his current vocation and his personal information. He is brought in by his camp medical officer urgently. He has recently broken up from his girl friend and he is very affected by this event. What is the diagnosis?

A. Conversion disorder
B. Dissociative disorder
C. Early-onset dementia
D. Hypochondriacal disorder
E. Somatisation disorder.

The answer is B. This man suffers from dissociative amnesia.

EMIS:

A – As if phenomenon
B – If not phenomenon
C – What if phenomenon
D – What next phenomenon
E – Why me phenomenon

Depersonlization is also described as – as if phenomenon

A – Circumstantial thinking
B – Confabulation
C – Over inclusive thinking
D – Perseveration
E – Tangentiality

Suggestibility is a prominent feature of – Confabulation

Somatoform and dissociative disorder
A – Dissociative amnesia
B – Dissociativce fugue
C – Somatisation disorder
D – Possession disorder
E – Multiple personality disorder
F – Hypochondriacal disorder
G – Ganser syndrome
H – Munchasen syndrome

1. 50 year old women presented with a 6 years history of multiple physical symptoms not attributed to any physical causes. She has been seeking repeated consultations from her GP and various other specialists – Somatisation disorder
2. 40 year old male finds himself 20 miles away from his home for no apparent reason – Dissocative fugue
3. 40 year old prisoner who is awaiting his court trial keeps fiving repeated wrong answers to questions, which are nonetheless in the right ballpark – Ganser syndrome
4. 40 year old women is constantly pre-occupied that she has breast cancer despite all the necessary investigations showing that it is unlikely – Hypochondriacal disorder

References

American Psychiatric Association (1994) *DSM-IV-TR: Diagnostic and Statistical Manual of Mental Disorders (Diagnostic & Statistical Manual of Mental Disorders) (4th edition)*. Washington: American Psychiatric Association Publishing Inc.

American Psychiatric Association (2013) *DSM-5: Diagnostic and Statistical Manual of Mental Disorders (Diagnostic & Statistical Manual of Mental Disorders) (5th edition)*. Washington: American Psychiatric Association Publishing Inc.

Gelder M, Mayou R and Cowen P (2006) *Shorter Oxford Textbook of Psychiatry (5th ed)*. Oxford: Oxford University Press.

Hsu C, Ali Juma H, Goh CL. Prevalence of body dysmorphic features in patients undergoing cosmetic procedures at the National Skin Centre, Singapore. Dermatology. 2009;219(4):295-8.

Johnstone EC, Cunningham ODG, Lawrie SM, Sharpe M, Freeman CPL. (2004) *Companion to Psychiatric studies*. (7th edition). London: Churchill Livingstone.

Ng BY, Chan YH. Psychosocial stressors that precipitate dissociative trance disorder in Singapore. Aust N Z J Psychiatry. 2004 Jun;38(6):426-32

Sadock BJ, Sadock VA. (2003) *Kaplan and Sadock's Comprehensive Textbook of Psychiatry*. (9th ed.) Philadelphia (PA): Lippincott Williams and Wilkins.

Semple D, Smyth R, Burns J, Darjee R, McIntosh A (2005) *Oxford Handbook of Psychiatry*. Oxford: Oxford University Press.

Teo WY, Choong CT. Neurological presentations of conversion disorders in a group of Singapore children. Pediatr Int. 2008 Aug;50(4):533-6.

Trimble M (2004) *Somatoform disorders – A medicolegal guide*. Cambridge: Cambridge University Press.

World Health Organisation (1992) *ICD-10: The ICD-10 Classification of Mental and Behavioural Disorders: Clinical Descriptions and Diagnostic Guidelines*. Geneva: World Health Organisation.

Yap AU, Chua EK, Tan KB, Chan YH. Relationships between depression/somatization and self-reports of pain and disability. J Orofac Pain. 2004 Summer;18(3):220-5.

Chapter 14 Consultation liaison psychiatry and women mental health

An introduction to consultation liaison psychiatry	Psycho-oncology
Delirium / acute confusional state	Neuropsychiatry
Common psychiatric issues in consultation liaison psychiatry	Women mental health
Cardiovascular disease and psychiatry	Antenatal psychiatric disorders
Liver impairment and psychiatry	Postpartum psychiatric disorders
Renal impairment and psychiatry	Premenstrual syndrome
Endocrinology and psychiatry	Revision MEQ, MCQ and EMIs

An introduction to consultation liaison psychiatry

The two models of working with general hospital patients are:
1. **Consultation** – with individual patients,; patients are referred by medical or surgical teams for a psychiatric opinion.
2. **Liaison** - a psychiatrist becomes an integrated member of a medical or surgical team, and develops a collaborative working relationship.

Development of consultation liaison psychiatry in Singapore

The National University Hospital (NUH) is the first hospital in Singapore to establish consultation liaison psychiatry service in 2002.

The need for consultation - liaison psychiatry in a hospital setting:

(a) **Frequency of mental health problems:**
- 20-40% of all general hospital outpatients and inpatients have some degree of psychological illnesses.
- 1/3 of patients in outpatient clinics have medically unexplained symptoms.
- 25% of male medical inpatients consume alcohol at a hazardous level in western countries

(b) **Recognition of problems:**
- Less than half of cases of mental health problems in medical inpatients are recognised.
- Of these, less than 10% are referred for psychiatric assessment and follow-up based on findings from western countries.
- A liaison psychiatry service can contribute to the early detection and treatment of mental health problems.

(c) **Effective delivery of mental health care:**
Patients with chronic physical illnesses or whose psychiatric disorder is closely linked to inpatient treatment. In such cases, it is often more convenient to refer them to the mental health service in the hospital.

(d) **Effective coverage for A&E**
- 2-5% of patients attending A&E have primary psychiatric problems.
- 20-30% have significant psychiatric problems co-existing with physical disorders.

Delirium / Acute confusional state

Definition

1. Delirium is a common neuropsychiatric complication that often cause confusion in elderly patients after a major operation.
2. Disturbed sleep and wake cycle is the most common symptom reported by patients suffering from delirium.

Epidemiology

Western countries

1. Delirium occurs in 10 – 30% of hospitalised medically ill patients with higher rates in elderly (10 – 40%).
2. General Surgery (10-15% delirious); cardiothoracic surgery (30%), Hip operation (50% delirious), elderly (older than 65 years) in ICU (70% delirious), palliative care (88% delirious).

Singapore

1. Merchant et al (2005) reported that post-operative confusion is a common complication after hip fracture surgery. Predictors for such complication include being of female gender and pre-fracture mobility.
2. Delirium also occurs in young children. Bong et al (2009) reported that the incidence of emergency delirium is approximately 10% in a population of healthy, un-premedicated Singaporean children undergoing day surgery at KK Hospital. Young age, poor compliance at induction, lack of intraoperative fentanyl use and rapid time to awakening were predictive risk factors for emergency delirium.

Aetiology

The treatment team needs to explore underlying treatable medical causes including systemic infections, drug intoxication and withdrawal (steroid, benzodiazepine, alcohol), endocrine disorders and vitamin deficiencies (Addison's disease, thyroid disease, vitamin B12 insufficiency), Wilson's disease (autosomal recessive inheritance, Kayser Fleischer's ring on slit light lamp exam), head trauma, intracranial neoplasm, CNS infections, hypotension, metabolic encephalopathies (hyper or hyponatraemia, uraemia), haematological disorders (severe anaemia, coagulopathy), and seizure disorders. The treatment team also needs to look for changes in the patient's environment and recent alteration in medication. For example, antibiotics and naproxen can cause delirium.

Risk factors

The following are risk factors for developing delirium:
- Old age
- Polypharmacy (e.g. steroids)
- Anaemia
- Electrolyte disturbance

- History of substance misuse.

Pathophysiology

Figure 1   The risk factors and pathophysiology of delirium

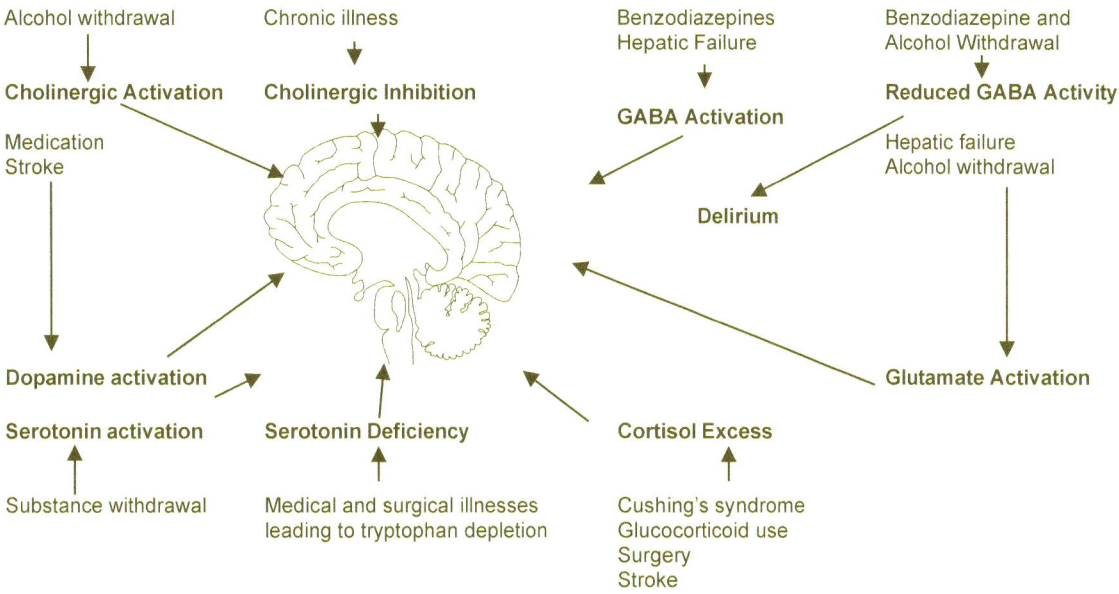

Diagnostic criteria

Delirium
The DSM-5 diagnostic criteria states that there must be:
1. Changes in attention and awareness
2. These changes or disturbances have developed over a short duration of time (usually characterized as within hours to few days) and has represented a change from baseline attention and awareness.
These disturbances are noted to be fluctuating during the course of a day.
3. In addition, there must be changes with regards to cognition (memory deficits, disorientated, language, visuospatial ability or perception)
4. There is evidence from clinical histories, physical examination and also biochemical investigations that the disturbances have arose due to physiological consequences of a medical condition, substance intoxication or withdrawal or due to multiple etiologies.

The DSM-5 has specified several etiologies, which include:
a. Substance intoxication delirium
b. Substance withdrawal delirium
c. Medication induced delirium
d. Delirium due to another medical condition
e. Delirium due to multiple etiologies

The DSM-5 has also specified that acute delirium usually last for a few hours or days, whilst persistent delirium last for weeks or months.

There are several other subtypes, which include:
a. Hyperactive - individuals have a hyperactive level of psychomotor activity that may be accompanied by mood lability, agitation, and refusal to cooperate with medical care.
b. Hypoactive - Usually accompanied by marked sluggishness and lethargy that approaches stupor
c. Mixed level of activity - Individuals have normal level of psychomotor activity even though attention and awareness are disturbed.

Type of delirium:

Delirium may present in either a hyperactive or hypoactive form.
1. Hyperactive delirium is easier to identify and frequently presents with anxiety, agitation, or overt psychotic symptoms.
2. Hypoactive delirium is more difficult to diagnose and may present with depression-like symptoms of hyper-somnolence and social withdrawal. Both forms of delirium respond to psychiatric treatment.

Management

The primary objective in treating delirium is to identify and treat the underlying aetiology. The treatment team needs to monitor vital signs, fluid input and output, and oxygenation. The treatment team should perform regular laboratory investigations (30% of patient have negative investigation results and cannot find the underlying cause), discontinue unnecessary medications, rehydrate the patient if necessary and identify sources of pain.

Generalized slowing is a common EEG finding.

Non-pharmacological treatment

Advise nursing staff to place the patient in a room near the nursing station. The room should be quiet and provide a calm environment with day and night lighting. Minimise transfers and encourage the presence of family members. Chong et al (2011) proposed a new model of delirium care in Tan Tock Seng Hospital. The following are recommendations for non-pharmacological treatment:

1. No mechanical restraints
2. Thrice daily patient orientation via reality orientation board
3. Thrice daily cognitive stimulation
4. Early mobilization
5. Provision of visual or hearing aids
6. Rehydration via oral feeding
7. Sleep enhancement via warm milk, relaxation tapes or music
8. Bright light therapy from 6 to 10 pm
9. Minimise immobilisation equipment such as IV drips or Foley's catheter
10. Encourage family members to come daily.

Pharmacological treatment

The secondary objective is to manage the symptoms associated with delirium, including possible agitation, confusion, or psychotic thought content or process. The mainstay of pharmacological treatment of delirium involves the use of pharmacological agents including antipsychotic medications. Delirious patients usually respond to daily doses of antipsychotic drugs equivalent to 0.5-5.0 mg/day of haloperidol. To avoid extrapyramidal side effects, clinicians should use the lowest effective dose or second generation antipsychotics such as risperidone and olanzapine but be wary that the second generation antipsychotics may increase the risk of cerebrovascular accident.

Once the patient becomes less confused, the treatment team can slowly reduce the dose of the antipsychotics. The treatment team should stop the antipsychotics 7- 10 days after the delirious symptoms resolve.

Avoid benzodiazepine and anticholinergic agents which make patient more confused. Pain management is important in treatment of delirious patient.

Prognosis

- The course of delirium ranges from less than 1 week to 2 months with a typical course of 10 – 12 days. Chong et al (2011) recommended repeated assessments undertaken of comorbidity scoring, duration and severity of delirium, cognitive, functional measures at baseline, 6 months and 12 months later.
- Mortality: 6 – 18% (Risk is increased by 2 times as compared to normal controls)
- Persistent cognitive impairment: 60%.
- Dementia: the risk is increased by 3 times after delirium.

Table 1. Common problems seen in consultation liaison psychiatry

Disorders	Descriptions
Adjustment disorder	Adjustment disorder usually occurs in approximately 25% of medical patients. Adjustment disorder is usually related to all aspects of physical illnesses or its treatment e.g. after receiving the diagnosis of serious illnesses and loss of physical health.
Anxiety and depressive disorder	Anxiety and depression are twice as common in medical patients as in the general population. They are particularly common among inpatients who have physical illnesses that affect the brain, such as a stroke. They may also occur more often in patients who have painful, chronic or life-threatening illnesses, such as heart disease or rheumatic arthritis. Singapore findings: • Zhang et al (2011) found that the prevalence of depression in COPD is 24.6%. • Ho et al (2011) found that the 26% of RA patients at NUH percent presented with anxiety, 15% with depression and 11% with both. • Mak et al (2011) found that the frequency and level of anxiety were significantly higher in SLE patients than patients with gout, RA and healthy controls in NUH.

Somatic symptoms of depression	Somatic symptoms in medically ill patients are poor indicators of depression. Symptoms such as fatigue, weight los, pain and insomnia may be a result of physical illnesses rather than depressive disorders. Hence, doctors should pay attention to mood (e.g. fearfulness) and cognitive symptoms (pessimism, hopelessness, cognitive errors such as overgeneralisation) when assessing depression in medical patients.
Capacity assessment	Psychiatrist is consulted to assess the capacity of a patient suffering from schizophrenia and chronic renal failure because she refuses renal dialysis.

Cardiovascular diseases and psychiatry

1. Prolonged QT interval caused by antipsychotics may lead to potentially lethal arrhythmia.
2. Sertraline is the antidepressant recommended for patients with myocardial infarction.
3. Mirtazapine is the antidepressant recommended for patients taking warfarin.
4. Avoid rapid escalation of dose in elderly at risk of cardiac pathology.

Table 2. Recommendations from the Maudsley's guidelines on the choice of psychotropic medication for people with atrial fibrillation

	Antipsychotics	Antidepressants	Mood stabilisers	Hypnotics	Dementia
Recommended	Aripiprazole	Mirtazapine as AF patients often take NSAIDs and warfarin.	Lithium Valproate	Benzodiazepines	Rivastigmine
Not recommended	Clozapine Olanzapine Paliperidone	Tricyclic antidepressants (TCAs)	Nil	Nil	Avoid other acetylcholinesterase inhibitors in paroxysmal AF.

Figure 2   The pathophysiology between cardiovascular medicine and psychiatry

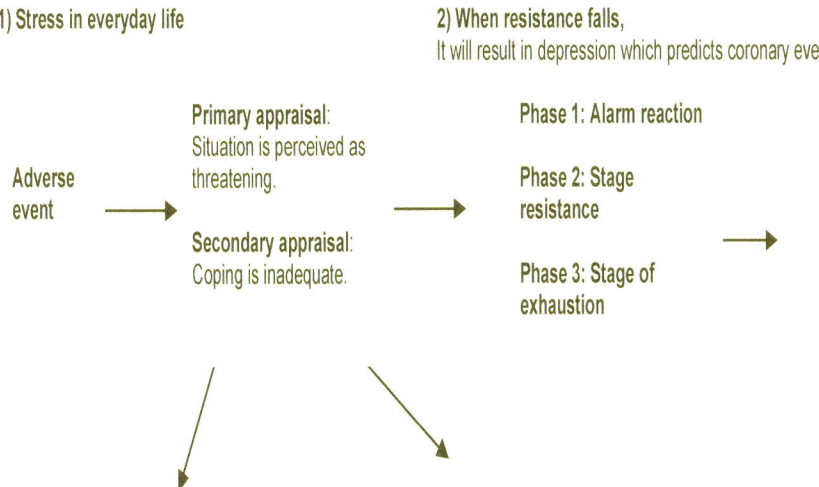

1) Stress in everyday life

Adverse event →
- **Primary appraisal:** Situation is perceived as threatening.
- **Secondary appraisal:** Coping is inadequate.

2) When resistance falls, it will result in depression which predicts coronary events.

- Phase 1: Alarm reaction
- Phase 2: Stage resistance
- Phase 3: Stage of exhaustion

3) Stimulation of sympathetic system during stress

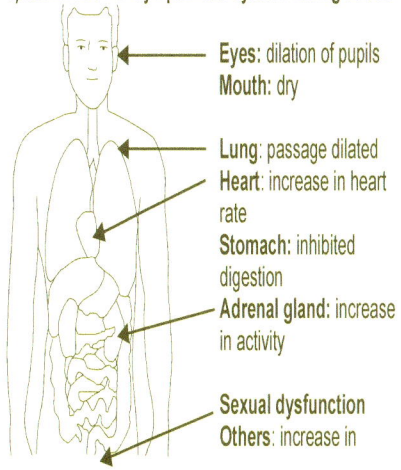

- **Eyes:** dilation of pupils
- **Mouth:** dry
- **Lung:** passage dilated
- **Heart:** increase in heart rate
- **Stomach:** inhibited digestion
- **Adrenal gland:** increase in activity
- **Sexual dysfunction**
- **Others:** increase in

Coronary artery disease (CAD) and stress
The mental stress of ordinary life is the most common precipitant of myocardial ischaemia in patients with CAD. Type A behaviour (aggressive, impatient, hostile) can increase the incidence of recurrent myocardial infarction (MI) and cardiac death in previous MI. 20% of people with acute MI suffer from depressive disorder.

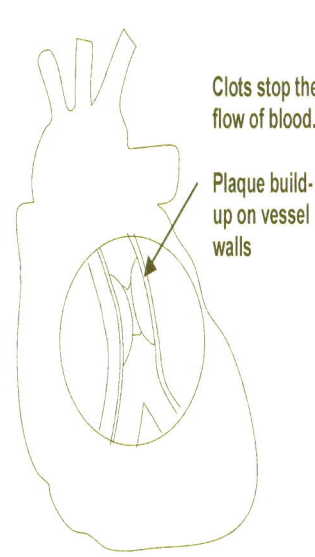

Clots stop the flow of blood.

Plaque build-up on vessel walls

Hypertension and stress
Psychosocial stress increases adrenaline and noradrenaline levels, which subsequently increases peripheral vascular resistance, resulting in hypertension and hypertrophy of the heart.

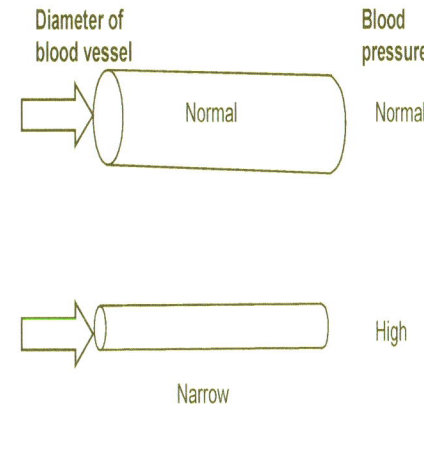

Diameter of blood vessel — Blood pressure
Normal — Normal
Narrow — High

Ventricular arrhythmia and stress

Both the brain and the peripheral sympathetic nervous systems are implicated as causes of stress-induced arrhythmias. An acute emotional trigger, often provoking anger, is an immediate precipitant of arrhythmia in patients with a relatively chronic state of helplessness. Helplessness is an underlying sense of entrapment without possible escape. Arrhythmia can lead to sudden cardiac death.

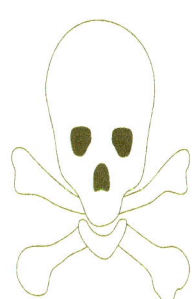

Liver impairment

Most psychotropic drugs are extensively metabolised by the liver. Liver disease is likely to lead to an impairment of drug metabolism and increased drug plasma levels. As a rule of thumb, the staring dose of psychotropic medications should be lowered in patients with liver impairment.

Table 3 Recommendations from the Maudsley's guidelines on the prescription for people with liver diseases

Types	Antipsychotics	Antidepressants	Mood stabilisers	Hypnotics
Indicated	Haloperidol: low dose Sulpiride: no dosage reduction if renal function is normal	Paroxetine Citalopram	Lithium	Lorazepam Zopiclone
Contraindicated	Avoid antipsychotics which have extensive hepatic metabolism such as chlorpromazine which gives rise to anticholinergics side effects.	Avoid TCA which gives rise to constipation Avoid fluoxetine with long half live and accumulate of metabolites	Avoid carbamazepine as it induces hepatic metabolism Avoid valproate as it is highly protein bound and metabolised by liver.	Avoid diazepam which has long half life.

Renal impairment

Advanced uraemia causes lethargy, asterixis, myoclonus, deterioration in total intelligence, impairment in working memory. Dialysis can improve uraemic encephalopathy. Sexual dysfunction and impaired quality of life continue during dialysis.

Renal impairment is important for drugs or active metabolites which are dependent on the kidney for elimination (e.g. lithium). Furthermore, renal function declines with age.

Table 4 Recommendations from the Maudsley's guidelines on the choice of psychotropic medication for people with renal failure

Types	Antipsychotics	Antidepressants	Mood stabilisers	Hypnotics
Indicated	**First-generation antipsychotic:** haloperidol 2-6 mg/day **Second-generation antipsychotic:** olanzapine 5 mg/day	Escitalopram Sertraline	Valproate, Carbamazepine Lamotrigine	Lorazepam Zopiclone
Contraindicated	Sulpiride (excreted by kidney)	Tricyclic antidepressant (because of anticholinergic effects)	Lithium	Diazepam (long half-life)

Table 5 Summary of electrolyte disturbances and psychiatric manifestations

	Hypo ↓	**Hyper ↑**
Sodium	**Causes**: Antidepressants, Lithium, carbamazepine, diuretics, SIADH, compulsive water drinking **Neuropsychiatric features**: anorexia, fatigue, headache confusion and convulsion (if Na < 115)	**Causes**: dehydration, diarrhoea, CNS lesions, severe burns. **Neuropsychiatric features**: reduced consciousness (Na > 160), confusion, stupor and coma if sodium is very high.
Potassium	**Causes**: vomiting, nausea (in bulimia nervosa), laxative addiction, diuretics, hypomagnesaemia, renal tubular acidosis and Cushing's disease **Neuropsychiatric features**: muscular weakness, lethargy, drowsiness	**Causes**: excessive consumption of NSAIDS, or potassium supplements, acute and chronic renal impairment, tumour lysis syndrome **Neuropsychiatric features**: weakness, sensory perception deficits, delirium
Calcium	**Causes**: hypoparathyroidism, phenytoin, secondary hyperparathyroidism **Neuropsychiatric features**: weakness, depression, delirium, seizure	**Causes**: hyperparathyroidism, malignancy, chronic renal failure, dehydration, **Neuropsychiatric features**: anorexia, vomiting, nausea; Late stage: drowsiness, depression, delirium, seizure
Magnesium	**Causes**: starvation, chronic alcoholism, acute intermittent porphyria **Neuropsychiatric features**: weakness, depression, delirium, seizure, hallucination	**Causes**: renal failure, magnesium containing oral medication. **Neuropsychiatric features**: drowsiness, weakness, coma
	Metabolic Acidosis	**Metabolic alkalosis**
Acid – base balance	**Causes**: starvation, alcohol, diabetic ketoacidosis, renal failure, severe sepsis **Psychiatric features**: fatigue, anorexia, hyperventilation, laboured breathing (Kussmaul respiration), coma, convulsions.	**Causes**: vomiting, chloride depletion **Psychiatric features**: irritability, hypoventilation, lethargy, confusion pH> 7.55.

Endocrinology

Hyperthyroidism

Fifty percent of patients present with psychiatric symptoms. Anxiety and depression are common. Depressive symptoms are not linearly related to thyroxine levels. Delirium occurs in thyroid crisis (3-4%). Psychosis is rare in hyperthyroidism (1%).

Hypothyroidism

Twenty percent of patients present with psychiatric symptoms. Fatigue accompanied by mental and physical slowing is a central psychiatric feature. Depression and anxiety are common. It is a must to assess cognitive impairment. Myxoedema causes psychotic symptoms (paranoid delusions, auditory or visual hallucinations) and delirium.

Cushing syndrome

Fifty to eighty percent of patients suffer from depression with moderate to severe symptoms. Depression will resolve after the control of hypercortisolism. Suicide has been reported in 3 to 10% of cases. Cognitive impairment such as amnesia and attentional deficits are common.

Addison disease

1. 90% of patients with adrenal disorders present with psychiatric symptoms.
2. Memory impairment is common.
3. Depression, anxiety and paranoia tend to have a fluctuating course with symptom free intervals.
4. Psychosis in 20% of patients.
5. Fatigue, weakness and apathy are common in early stage.
6. Adrenal crisis may lead to delirium.

Hyperparathyroidism

1) Disturbance of mood and drive: more prevalent. Depression may progress to psychosis and suicide.
2) Delirium is caused by high calcium levels or parathyroid crisis.
3) Cognitive impairment: impaired attention, mental slowing and impaired memory.
4) Psychosis (5-20%): mainly persecutory delusions and hallucinations.

Surgery

Table 6 Recommendations from the Maudsley's guidelines on the choice of psychotropic medication for people going for surgery

	Antipsychotics	Antidepressants	Mood stabilisers	Hypnotics
Recommendations	Most antipsychotics are probably safe to continue.	SSRIs should be stopped on the day of surgery because it may increase the bleeding time.	Sodium valproate Carbamazepine	Most benzodiazepines are safe to continue
Potential complications with surgery	Clozapine may delay recovery from anaesthesia.	MAOIs (irreversible) have fatal interaction with pethidine. Combination between MAOIs(irreversible) and sympathomimetic agents result in hypertensive crisis. TCA: α blockade leads to hypotension, prolong QTc, lower seizure threshold and require careful selection for anaesthesia agents.	Lithium is recommended to be discontinued before major surgery because it affects renal function and cause electrolyte disturbances.	Benzodiazepines lead to reduced requirements for induction and maintenance of anaesthesia.

Psycho-oncology

Epidemiology

25% of people with cancer suffer from depressive disorder.

Theoretical background of psycho-oncology

Figure 2   Pathophysiology between cancer and psychiatric symptoms

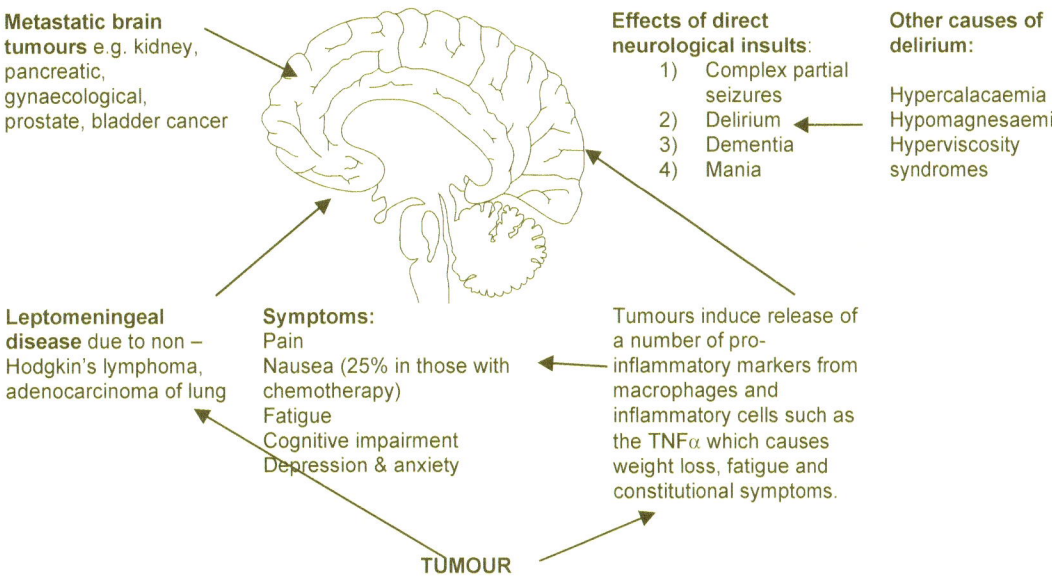

Metastatic brain tumours e.g. kidney, pancreatic, gynaecological, prostate, bladder cancer

Effects of direct neurological insults:
1) Complex partial seizures
2) Delirium
3) Dementia
4) Mania

Other causes of delirium:

Hypercalacaemia
Hypomagnesaemia
Hyperviscosity syndromes

Leptomeningeal disease due to non – Hodgkin's lymphoma, adenocarcinoma of lung

Symptoms:
Pain
Nausea (25% in those with chemotherapy)
Fatigue
Cognitive impairment
Depression & anxiety

Tumours induce release of a number of pro-inflammatory markers from macrophages and inflammatory cells such as the TNFα which causes weight loss, fatigue and constitutional symptoms.

TUMOUR

Neuropsychiatric side effects of chemotherapeutic agents

- Cognitive impairment/dementia due to leukoencephalopathy (Cytosine arabinoside, methotrexate)
- Acute confusion (most agents)
- Psychiatric Disorders
    - Manic and depressive symptoms (Steroid, interferon causing depression)
    - Psychosis (Procarbazine)
    - Personality change (Cytosine arabinoside)
- Fatigue (Fluorouracil, interleukin, interferon)
- Seizures (Vincristine, vinblastine, alkylating agents)
- Anorexia (most agents)
- Neuropathies and sexual dysfunction (especially breast, ovary, uterus, cervical cancer)
- Cataracts (Steroids)
- Anticholinergic effects: (antiemetic agents)

Counselling a cancer patient

OSCE station

Name: Mr. A   Age: 40 years old   Occupation: unemployed

You are the medical resident receiving training in hepatology. Mr. A, a hepatitis C carrier and an intravenous drug abuser, complains of severe weight and appetite losses, progressive lethargy, yellowing of the eyes and skin and abdominal distension. He consults a hepatologist who finds that he has deep jaundice and gross ascites. A CT scan of the abdomen reveals multiple liver masses and peritoneal deposits. Ascites fluid analysis shows malignant cells, accompanied by a very high serum α-fetoprotein protein level. The diagnosis of advanced hepatocellular carcinoma is made and Mr. A is informed that he has a very limited life expectancy.

Tasks: To address the end of life issues and his concerns

Table 7   Approach to counsel Mr. A

Possible questions raised by a cancer patient	Suggested approach
Why am I so unlucky to get this cancer?	Establish rapport and express empathically that you are sorry to hear what has happened.
Was it because I used drugs? Am I a bad person?	For issues of guilt, encourage the patient to avoid blaming himself. You can encourage him by saying that those who do not use drugs can also develop liver cancer e.g. people who get hepatitis B from birth.
I hate looking myself in the mirror. I look thin and my skin looks yellowish. What's wrong with me?	Need to explore his perception of his body image and look for possible jaundice.
The gastroenterologist says there is no cure for hepatitis C. Wouldn't it be better to give up?	Ask patient to think about the positive aspects of his life to look forward to and encourage him to fight the illness. Explore his view on his own death. Does he have any fear?
I am very worried that I will die soon. Will I die in severe pain? Can you just ask my doctor in charge to give me an injection and kill me? I don't want to suffer.	Address his suffering: Is he willing to ask for more pain control? Address the diversity of experiences of pain. Explain that euthanasia is illegal in Singapore and not an option. Explore his reasons for asking about this and discuss alternatives such as enhancing his pain management.
My mother doesn't know I have cancer. How should I tell her? She will be very sad. I am worried my family do not want me as I am a burden to them.	Explore his relationship with his family and his concerns around informing his mother of his diagnosis. Also inform him of the risks of hiding his illness from other family members. Address his concern as a burden and his concern of being abandoned.
Is there God? I have committed crimes and used drugs. Will I be forgiven?	Explore spiritual issues and his religious faith. Does he feel guilty about his past. Explore the need to be forgiven and who should forgive him e.g. God, family or friends.
Doctor, you are the psychiatrist. I need your emotional support. Will you stay with me until the day I die?	Explain boundary and your schedule in an empathetic way: you will visit him regularly. Get other friends or caregivers involved to reach a conjoint effort.

Table 8 Summary of neuropsychiatric consequences in HIV infection and common neurological disorders

	Cognitive impairment	Psychosis	Depression	Mania	Anxiety related disorder	Other disorders
HIV infection Neurological and neuropsychiatric sequelae develop in more than 50% of people of advanced HIV disease.	**Course:** Singapore (Chan et al 2012): HIV-associated cognitive disorder (HAND) occurs in 22.7% of HIV patients. Older patients with less education and severe illness are at highest risk of HAND. Delayed recall is most commonly affected. Visuospatial dysfunction is most strongly associated with prevalent HAND. **Mild cognitive impairment** in the early stage: 70-80% Fluctuation in the first 2 years **Associated dementia complex** in the later stage: 30%	**Symptoms:** Paranoia is common. **Treatment:** Olanzapine / risperidone can be used but need to monitor for metabolic syndrome.	**Frequency:** 30-50% **Symptoms:** Apathy Amotivation Anergia Cognitive impairment. **Treatment:** Use SSRIs such as escitalopram with least drug interaction. **Suicide risk:** increase by 30 times.	**Symptoms:** Secondary mania is common. **Treatment:** Valproate is well tolerated. Olanzapine / risperidone is indicated for impulsivity, agitation, disinhibition.	**Symptoms:** Anxiety is common. **Treatment:** For anxiety: use SSRI such as escitalopram. For mixed anxiety and depression: use venlafaxine but need to monitor blood pressure if dose > 300mg/day Non-benzodiazepine: Hydroxyzine (antihistamine) is recommended.	**Neuropathic pain:** treatment involves TCA but make sure patient has no tendency to take an overdose of TCA. **Insomnia:** treatment include sleep hygiene, lorazepam (long acting without active metabolites) for 2 weeks or hydroxyzine (antihistamine) **Sexual dysfunction** **Eating disorders** **Chronic pain:** 80% **Delirium:** 30%.
Cerebro-vascular accidents CVA	30% of stroke patients show severe cognitive impairment.	Complex auditory and visual hallucinations called peduncular hallucinosis occur in people with infarct in pons and midbrain.	30% suffer from post-stroke depression. Singapore data (Loong et al 1995): 55% suffer from post-stroke depression. Right-sided stroke is	Possible in people with CVA, especially left-sided stroke.	Common in people with CVA.	6% develop seizures. 50% die within 3 years.

			associated with depression.			
Parkinson disease (PKD)	**Frequency:** 25-40% dementia **4 types of cognitive impairments in PKD:** 1) Generalised global dementia 2) Focal and specific cognitive deficits 3) Drug-induced 4) Depression-related.	**Frequency:** 30% have hallucinations (usually visual hallucinations) **Risk factors:** -Greater age -Longer duration of illness - Coexisting cognitive impairment - Depression - Sleep disturbance - Poor visual acuity.	**Frequency:** 40% of people with PKD develop depression follow diagnosis; 1-2% of patients with PKD develop severe depression Diagnosis is difficult to make because of the flatten affect. Hence, biological symptoms are given more weight in diagnosing depression.	Uncommon	Often mixed with depression.	**Personality change** include being more introverted, over-controlled and exhibit anhedonia.
Epilepsy	**Epileptic dementia:** rare in epilepsy patients.	**Frequency:** 3% in epilepsy patients. **Schizophrenia:** The risk is 2-3 times more likely as compared to the general population. **Other forms of psychosis:** post-ictal, brief inter-ictal and chronic inter-ictal.	**Frequency:** 9 – 22% of epilepsy patients. The risk is 17 times more likely when compared to the general population. Patients are 25 times more likely to attempt suicide when compared to the general population.	Uncommon in people with epilepsy.	Often mixed with depression	Pathological aggression: 4-50%. Crime offences: 3 times more likely to commit crime when compared to the general population
Head injury	The important predictor of cognitive impairment is the duration of post traumatic amnesia.	**Frequency:** Schizophrenia 2.5% Paranoid psychosis 2%.	Depression and anxiety are common. Frequency of psychotic depression is 1%.	Frequency of secondary mania is 9%.	Anxiety is common.	Personality change is common after frontal lobe injury. 50% of patients with head injury develop post concussion syndrome (cognitive impairment, irritability).

Women mental health

Antenatal psychiatric disorders

Self-limiting minor depressive illness and generalised anxiety are common in the first trimester. Most studies of the aetiology of antenatal depression have found a link between psychosocial problems (e.g. marital conflict and a lack of support) and antenatal psychiatric disorders. Specific risk factors include previous termination of pregnancy, ambivalence towards the pregnancy and feeling anxious about pregnancy.

Bipolar disorder:

1. Risk of relapse is high in the first 90 days after delivery.
2. Planning before pregnancy is important.
3. The choice of psychotropic medication is based on the risk to benefit ratio of each drug (e.g. past treatment response and risk of teratogenic effect) and a conjoint decision is made between a psychiatrist and his or her patient.

Table 9 Maudsley's guidelines for use of psychotropic medications in pregnant women with bipolar disorder

Mother and foetus	Mania	Bipolar depression	Not recommended
The risk of relapse is high if medication is stopped abruptly. **Lithium:** Incidence of Ebstein's anomaly is between **0.05 to 0.1%** (after maternal exposure to lithium in the first trimester). **Valproate:** Incidence of foetal birth defect (mainly neural tube defects) is **1 in 100**. **Carbamazepine:** Incidence of foetal birth defect is **3 in 1000**.	☑ Mood stabilising antipsychotics: haloperidol, olanzapine. Olanzapine increases risk of gestational diabetes. ECT is indicated if antipsychotic fails.	☑ CBT for moderate bipolar depression Fluoxetine has the most data on safety and indicated for severe bipolar depression especially for those patients who have very few previous manic episodes.	☒ Valproate is the most teratogenic mood stabiliser. Lamotrigine requires further evaluation and it is not routinely prescribed in pregnancy. It causes oral cleft (9 in 1000) and Stevens–Johnson syndrome in infant.

☑ = recommended   ☒ = not recommended

Depressive disorder

Table 10   Maudsley's guidelines for use of psychotropic medications in pregnant women with depressive disorder

Mother and foetus	TCA	SSRI	Not recommended
Patients who are at risk of relapse or developing moderate to severe depression should be treated with antidepressants. ECT is safe for pregnant women. SSRI causes pulmonary hypertension (after 20 weeks) in newborn. The neonates may experience withdrawal (agitation and irritability) especially with paroxetine and venlafaxine.	☑ Amitriptyline and imipramine TCAs have been used for many years without causing teratogenic effects.	☑ Fluoxetine has the most safety data. ☒ Paroxetine: foetal heart defects in the 1st trimester and more dangerous than other SSRIs	☒ Moclobemide Reboxetine Venlafaxine Bupropion Mirtazapine These drugs should be avoided due to lack of safety data.

☑ = recommended   ☒ = not recommended

Schizophrenia

Table 11 Maudsley's guidelines for use of psychotropic medications in pregnant women with schizophrenia

Mother and foetus	Antipsychotics and other treatment	Not recommended
Low risk of relapse if medication is continued with good social support. Antipsychotic discontinuation syndrome occurs in the neonate and mixed breast/bottle feeding can minimise withdrawal.	☑ Haloperidol Chlorpromazine Trifluoperazine Olanzapine (Gestation DM and weight gain)	☒ Depot antipsychotics Anticholinergic drugs. Clozapine (agranulocytosis in foetus)

☑ = recommended   ☒ = not recommended

Postpartum psychiatric disorders

Epidemiology:

Table 12   Compare and contrast three types of post-partum disorder.

	Postnatal blues	Postnatal depression	Puerperal psychosis
Onset	3-5 days	2-4 weeks	1-6 weeks
Prevalence	50%	10-15%	0.2%

Duration	2-3 days	4-6 weeks (1 year if untreated)	6-12 weeks

1. Although the rate of postnatal depression is similar to the non-partum rates, the risk for developing severe depression is 5 times greater than the lifetime risk. Social deprivation is associated with higher risks.
2. The peak period for admission as a result of psychiatric problems usually occurs 3 weeks after delivery.

Singapore

1. The prevalence rates of postnatal depression (PND) as high as at least 7%. Chen et al (2011) proposed that there is a need for early detection and intervention of postpartum mental illness amongst Singaporean mothers.
2. Chen and Lau (2008) reported that suicide and psychiatric causes are not significant causes of maternal mortality in Singapore. There was only one identified maternal death among 589 female suicides aged 15-45 years, occurring in a teenager within the first month postpartum.
3. Chee et al (2008) reported that women who had brought their infants for three or more non-routine visits to the infant's doctor had a significantly higher prevalence of depression (32.6%) than those with fewer visits (13.6%).

Postnatal blues

Aetiology: no proven hormonal differences between those with and without postnatal blues, although the temporal relationship to childbirth and the frequency of the postnatal blues make biological factors likely to be an aetiological factor.

Risk factors include:
1. Premenstrual tension
2. Previous gynaecological problems
3. Higher neuroticism scores
4. Marked change in salivary progesterone
5. Not shown to be associated with birth complications, caesarean section or bottle feeding.
6. Greater risk of subsequent postnatal depression, particularly if severe blue.

Clinical features include anxiety, depression, tearfulness, emotional lability (highs and lows), irritability, apparent confusion and mild hypochondriasis.

Management

1. Postnatal blue usually resolves spontaneously.
2. Medications is not necessary.
3. Support and reassurance is important.

Postnatal depression (PND)

Aetiology:

Strong predictors of postnatal depression include:
1. Past history of psychopathology
2. Psychological disturbance during pregnancy
3. Poor marital relationship
4. Poor social support
5. Stressful life events
6. Low social economic status
7. Obstetric complications

8. Other factors include old mother, marital and family conflict, substance abuse, previous pregnancy loss, ambivalence about the current pregnancy, and frequent antenatal admissions for obstetric problems.

Singapore

Chee et al (2005) reported that a negative 'confinement' experience is a significant risk factor for postnatal depression, and is not universally welcomed by Singaporean women.

Clinical features

Many of the features of depression will occur as a result of becoming a new mother and are less useful in detecting postnatal depression. Common symptoms include:
1. Irritability
2. Tearfulness
3. Poor sleep
4. Tiredness.

Mothers may not complain of low mood, but may present the following:
1. Feeling inadequate as a mother
2. Loss of confidence in mothering
3. Anxieties about the baby's health
4. Concerns that the baby is malformed
5. Reluctance to feed or handle the baby
6. Thoughts of harming the baby.

Management:

Chen et al (2011) developed a PND intervention programme in KK hospital. The programme consists of two phases:

1. Postpartum women were screened with the Edinburgh Postnatal Depression Scale (EPDS) and provided appropriate care plans.
2. Individualised clinical intervention using a case management multidisciplinary team model.
3. The PND intervention programme achieved 78% reduction in the EPDS symptom and around 70% had improved health quality scores.
4. Antidepressant is indicated. For the choice of antidepressant, please refer to Figure 3.

Prognosis

1. 1 in 3 recurrence of severe postnatal depression in subsequent childbirth.
2. Recurrence is higher among women when PND is the first episode of depressive illness in a life time.
3. Without treatment, 30% of women remain depressed at the end of 1 year.

Puerperal psychosis

Aetiology:

Risk factors:
1. Past psychiatric history e.g. schizophrenia, mania
2. Past history of postpartum psychosis
3. Family history of psychiatric illness
4. Timing of onset implicates the role of reproductive hormones and raise in the level of dopamine. Progesterone concentration in the blood falls about 1000-fold in the days after childbirth. Oestrogen falls to a lesser extent. Oestrogen increases serotonin levels and it has anti-dopaminergic or anti-psychosis effect.

Clinical Features:
1. Onset is usually sudden.

2. Symptoms include confusion, disorientation lability of mood, manic symptoms, sleep disturbance and mild confusion.
3. Delusions and hallucinations may involve baby and family.
4. Suicide and infanticide thoughts may be present and must be explored during assessment.
5. The clinical picture may resemble schizophrenia, affective disorders (80%) or an organic illness such as delirium.

Treatment:
1. Admission to psychiatric ward prevents harm to baby and patient.
2. Antipsychotic drug is indicated (see Figure 3)
3. ECT is an option if antipsychotic drug fails.

Figure 3   Breastfeeding and psychotropic medication

☑ = recommended   ☒ = not recommended

Bipolar disorder:

☑ Valproate can be used but advise mother to ensure adequate contraception to prevent unexpected pregnancy.

Depressive disorder:
☑ Paroxetine and sertraline.

Schizophrenia:
☑ Sulpiride and olanzapine

Anxiety and insomnia:
☑ Lorazepam for anxiety and zolpidem for insomnia. Advise mother not to sleep with her baby to prevent accident when mother is over-sedated.

Substance abuse:
☑ Methadone is compatible with breastfeeding but dose has to be kept as a minimum.

It is suggested that psychotropic drugs should be taken immediately after breast feeding, before the infant's longest sleep period, to avoid feeding during peak milk levels.

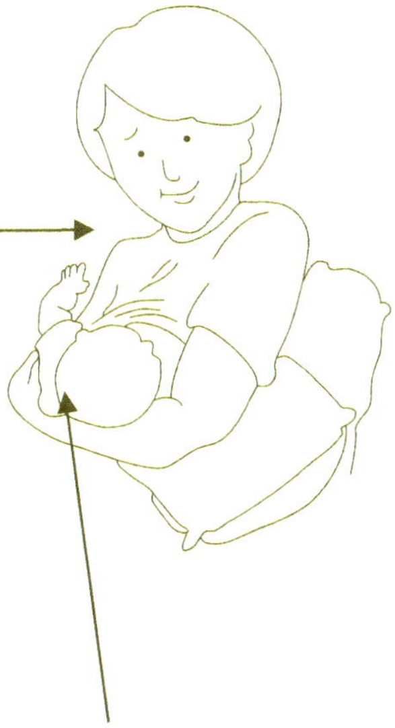

Premature infants, infants with renal, hepatic, cardiac and neurological impairments are at a greater risk from exposure to psychotropic drug. Hence, use the lowest effective dose and avoid poly-pharmacy.

Bipolar disorder:
☒ Lithium (concentration in breast milk is 50% of serum concentration).

Depressive disorder
☒ Escitalopram ad fluoxetine are present in breast milk at relatively high levels.

☒ Mirtazapine, venlafaxine

Schizophrenia:
☒ Clozapine, aripiprazole, quetiapine, risperidone and depot antipsychotic (infants may show EPSE).

Prognosis:
1. 20% of patients have relapse in subsequent childbirth.
2. 20% of patients develop bipolar disorder in later life.

Substance misuse and pregnancy

Foetal alcohol syndrome (FAS):

Clinical features of FAS include:

a. Pre- or postnatal growth deficiency
b. CNS disorders – including developmental delay, intellectual impairment, structural abnormalities
c. Facial abnormalities including short palpebral fissues, thin upper lip, flattened mid face, indistinct philtrum.

Tobacco: intra-uterine growth retardation, low birth weight and developmental delay in the child.

Cocaine: low birth weight, dose-dependent relationship with brain circumference and brain weight and motor abnormalities in infants.

Opiates: no evidence of increase in congenital defects with methadone and methadone is a safe option for pregnant women but need to consult a psychiatrist. Up to 90% of infants born to opiate dependent mothers show signs of withdrawal.

Premenstrual syndrome (PMS)

PMS occurs from the day of ovulation to the onset of menstruation.

Epidemiology of PMS:

1. 40% of women experience symptoms of PMS and 5% meet the criteria of PMS with impairment in functioning.
2. 30-40% of women with PMS have depressive disorder.

Clinical features of PMS:

1) Marked depression, 2) Anxiety, 3) Anger, 4) Affective lability, 5) Reduction in Interest, 6) Difficulty in concentrating, 7) Lethargy, 8) Lack of energy, 9) Overeating, 10) Hypersomnia or insomnia, 11) Feeling overwhelmed, 12) Other symptoms such as breast tenderness, headache, muscle pain.

The DSM-IV-TR criteria requires the presence of 5 out of the above 11 symptoms with marked social impairment.

Management

Non-pharmacological: special diets (increase in complex carbohydrates and dietary fibre to 20-40g/day and reduced intake of refined sugar and salt), exercise, cognitive therapy and relaxation.

Pharmacological: SSRIs are effective and well tolerated in patients suffering from PMS.

Revision MEQ

You are the medical resident receiving geriatric training. A 80-year-old woman is admitted for treatment of urinary tract infection (UTI). In the past few days, she has become confused and disorientated. She sees ghosts in the ward at night. She has good past health and there is no past psychiatric history. Her son is very concerned and wants to speak to your urgently.

Q.1 What is your provisional diagnosis?
Delirium or acute confusional state.

Q.2 A clinical fellow in your team disagrees with your diagnosis. He thinks this patient has nothing wrong because he spoke to the patient this morning and patient was alert with normal orientation. What is your explanation?
Delirium is fluctuating in course. Hence, the patient can be normal in the morning and becomes confused at night.

Q.3 What is the most likely cause of delirium in this case if the patient does not have other medical illness and all laboratory results are normal except abnormal Urine FEME and culture?
Urinary tract infection.

Q.4 If the patient has mistaken the curtain as a ghost at night, what is this phenomenon known as?
Illusion

Q.5 The clinical fellow suggests to give the patient benzodiazepine to help the patient to sleep better at night. What is your recommendation?
You should inform the clinical fellow that benzodiazepine causes confusion in frail elderly.

The treatment team should consider prescribing an antipsychotic e.g. haloperidol to treat delirium.

Q.6 Her son wants to know how he could help to manage the delirium of his mother. What is your recommendation?

You should advise the son to do the following:
1. Frequent re-orientation to his mother.
2. Frequent visit.
3. Bring a hearing or visual aid if his mother has sensory deficit.

Revision MCQ

1. You are a GP working in the heartland. A 70-year-old woman suffers from depression and you started fluoxetine one month ago. Her daughter calls you and say that her mother is very confused and admitted to a general hospital. The doctor says that there is an electrolyte abnormality. Her daughter wants to know the relationship between recent antidepressant use and electrolyte abnormality. Which of the following electrolytes is most likely to be involved?
A. Calcium
B. Magnesium
C. Phosphate
D. Potassium
E. Sodium.
The answer is E. Antidepressant use is associated with hyponatraemia in old people.

2. A 30-year-old woman with bipolar disorder is 12-week pregnant. She has been taking her lithium every night. Which of the following abnormality may occur if she continues to take lithium?
A. Dandy-Walker syndrome
B. Erb's palsy
C. Ebstein's anomaly
D. Foetal lithium syndrome
E. Neural tube defect.
The answer is C. The mother has 1 in 1000 chance to develop Ebstein's anomaly if she continues to take lithium. This congenital condition is characterised by apical displacement tricuspid valve leaflets, leading to part of the right ventricle becoming part of the right atrium. It is associated with atrial septal defect. Neural tube defect is associated with antenatal valproate and carbamazepine use.

3. You are the resident in obstetrics. A 25-year-old mother coming back for postnatal follow-up complains of low mood for 2 months after delivery. Which of the following

questionnaires is the most suitable scale to screen for post-natal depression in Singapore?
A. Beck depression inventory
B. Edinburgh post-natal depression scale
C. Glasgow post-natal depression scale
D. Hospital depression and anxiety scale
E. London post-natal depression scale.

The answer is B. Option A and D are validated questionnaires but not most suitable in post-natal depression.

EMIS:

Disorders seen in women:
A – Menarchy
B – Menopause
C – Postnatal blues
D – Premenstrual syndrome
E – Postnatal depression
F – Postpartum psychosis
G – Late onset paraphrenia
H – Briquet's syndrome
I – Pseudocyesis

Question 1: The onset of this disorder is within 1-2 weeks postpartum – Postpartum psychosis

Question 2: Occurs in around 10% of women postpartum – Postnatal depression

Question 3: Usually affects up to 40% of women of reproductive age – Premenstrual syndrome

Question 4: Usually peaks 3-4 days postpartum – Postnatal blues

Question 5: Firmly believes that oneself if pregnant and developed objective pregnancy signs in the absence of pregnancy – Pseudocyesis.

References

American Psychiatric Association (1994) *DSM-IV-TR: Diagnostic and Statistical Manual of Mental Disorders (Diagnostic & Statistical Manual of Mental Disorders) (4th edition).* Washington: American Psychiatric Association Publishing Inc.

American Psychiatric Association (2013) *DSM-5: Diagnostic and Statistical Manual of Mental Disorders (Diagnostic & Statistical Manual of Mental Disorders) (5th edition).* Washington: American Psychiatric Association Publishing Inc

Bentley P. (2008) *Memorizing medicine. A revision guide.* London: Royal Society of Medicine Press.

Bird J & Rogers D (2007) *Seminars in general adult psychiatry.* London: Gaskell.

Bong CL, Ng AS. Evaluation of emergence delirium in Asian children using the Pediatric Anesthesia Emergence Delirium Scale. Paediatr Anaesth. 2009 Jun;19(6):593-600.

Chan LG, Kandiah N, Chua A. HIV-associated neurocognitive disorders (HAND) in a South Asian population - contextual application of the 2007 criteria. BMJ Open. 2012 Feb 13;2(1):e000662. Print 2012.

Chen H, Wang J, Ch'ng YC, Mingoo R, Lee T, Ong J. Identifying mothers with postpartum depression early: integrating perinatal mental health care into the obstetric setting. ISRN Obstet Gynecol. 2011;2011:309189. Epub 2011 Sep 15.

Chen YH, Lau G. Maternal deaths from suicide in Singapore. Singapore Med J. 2008 Sep;49(9):694-7.

Chee CY, Lee DT, Chong YS, Tan LK, Ng TP, Fones CS. Confinement and other psychosocial factors in perinatal depression: a transcultural study in Singapore. J Affect Disord. 2005 Dec;89(1-3):157-66

Chee CY, Chong YS, Ng TP, Lee DT, Tan LK, Fones CS. The association between maternal depression and frequent non-routine visits to the infant's doctor--a cohort study. J Affect Disord. 2008 Apr;107(1-3):247-53. Epub 2007 Sep 14.

Chong MS, Chan MP, Kang J, Han HC, Ding YY, Tan TL. A new model of delirium care in the acute geriatric setting: geriatric monitoring unit. BMC Geriatr. 2011 Aug 13;11:41.

Citron K, Brouillette MJ, Beckett A (2005) *HIV and Psychiatry – A training and resource manual.* Cambridge: Cambridge University Press:

Flacker JM, Lipsitz LA. (1999) Neural mechanisms of delirium: current hypotheses and evolving concepts. Journal of Gerontology, *Biological Science and Medical Science*, 54,B239-46.

Friedman M, Thoreson CE, Gill H, et al. (1987) Alteration of type A behavior and its effect on cardiac recurrences in post-myocardial infarction patients: summary results of the Recurrent Coronary Prevention Project. *American Heart Journal,*114, 483-490.

Gurnell M. (2001) *Medical Masterclass-Endocrinology.* London: Blackwell Science.

Ho RC, Fu EH, Chua AN, Cheak AA, Mak A. Clinical and psychosocial factors associated with depression and anxiety in Singaporean patients with rheumatoid arthritis. Int J Rheum Dis. 2011 Feb;14(1):37-47.

Kerrihard TN & Breitbart General issues in hospital HIV psychiatry in Citron K, Brouillette MJ, Beckett A (2005) *HIV and Psychiatry – A training and resource manual*. Cambridge: Cambridge University Press.

Lishman, W. A. (1998) Organic Psychiatry. *The Psychological Consequences of Cerebral Disorders (3rd ed)*. Oxford: Blackwell Scientific.

Loong CK, Kenneth NK, Paulin ST. Post-stroke depression: outcome following rehabilitation. Aust N Z J Psychiatry. 1995 Dec;29(4):609-14.

Lown B (1987) Sudden cardiac death: biobehavioral perspective. *Circulation,* **76**,186-196.

Mak A, Tang CS, Chan MF, Cheak AA, Ho RC. Damage accrual, cumulative glucocorticoid dose and depression predict anxiety in patients with systemic lupus erythematosus. Clin Rheumatol. 2011 Jun;30(6):795-803.

Merchant RA, Lui KL, Ismail NH, Wong HP, Sitoh YY. The relationship between postoperative complications and outcomes after hip fracture surgery. Ann Acad Med Singapore. 2005 Mar;34(2):163-8.

Olumuyiwa JO, Akim AK (2005) *Patient Management Problems in Psychiatry*. London: Churchill Livingstone.

Puri B, Hall A, Ho RC (2013) Revision Notes in Psychiatry (3rd ed). Hodder Arnold: London.

Wise MG, Rundell JR. (2002) *Textbook of consultation – liaison psychiatry*. Washington: American Psychiatric Publishing Inc.

Rozanski A, Bairey CN, Krantz DS, et al. (1988) Mental stress and the induction of silent myocardial ischemia in patients with coronary artery disease. *New England Journal of Medicine*, **318**, 1005-1012.

Sadock BJ, Sadock VA. (2003) *Kaplan and Sadock's Comprehensive Textbook of Psychiatry*. (9th ed.) Philadelphia (PA): Lippincott Williams and Wilkins.

Taylor D, Paton C, Kapur S (2009) *The Maudsley prescribing guideline*. 10th edition London: Informa healthcare.

Zhang MW, Ho RC, Cheung MW, Fu E, Mak A. Prevalence of depressive symptoms in patients with chronic obstructive pulmonary disease: a systematic review, meta-analysis and meta-regression. Gen Hosp Psychiatry. 2011 May-Jun;33(3):217-23. Epub 2011 Apr 27.

Chapter 15 Psychogeriatrics

Normal aging and local research findings	Lewy body dementia (LBD)
Overview of dementia	Reversible causes of dementia
Alzheimer's disease (AD)	Other psychiatric disorders in elderly
Vascular dementia (VaD)	Revision MCQ
Frontotemporal dementia (FTD)	Revision MEQ & EMIs

Normal Aging and local research findings

Singapore
Choo et al (1990) predicted that Singapore is undergoing a rapid transition into an ageing society. This is a result of dramatic fall in the birth rate combining with a fall in infant and early childhood mortality rate as well as an improvement in the life expectancy.

Associate Professor Ng Tze Pin and his colleagues from Department of Psychological Medicine, NUS performed the Singapore Longitudinal Ageing Studies (SLAS) and the following is a summary of the interesting research findings.
1. The results of the SLAS study suggested that continued work involvement or volunteerism provides opportunities for social interaction and engagement and may be associated with enhanced mental well-being (Schwingel et al, 2009).
2. Successful aging was determined by female gender, ≥6 years of education, better housing, religious or spiritual beliefs, physical activities and exercise, and low or no nutritional risk. (Ng et al, 2009)
3. APOE-E4 allele (not E2) was significantly enhanced the risk of cognitive decline associated with depressive symptoms (Niti et al, 2009).
4. Metabolic syndrome was associated with increased risk of cognitive decline in Chinese older adults (Ho et al, 2008).
5. Daily omega-3 PUFA supplement consumption was independently associated with less cognitive decline in elderly Chinese (Gao et al, 2011).
6. Tea consumption was associated with better cognitive performance in community-living Chinese older adults (Feng et al 2010).
7. Statin use <u>was not</u> associated with depressive symptom scores in Singapore elderly (Feng et al 2010).

Figure 1 Summary of biological changes in normal aging

The weight and volume of the brain:
↓5% by the age of 70,
↓10% by the age of 80
↓20% by the age of 90
↑ in ventricular size and subarachnoid space

Gastrointestinal tract:
1. ↓ rate of gastric emptying.
2. ↓ secretion of gastric acid
3. ↓ absorption of drugs and slower onset of action
4. ↑ in gastric pH.

Other pharmacokinetic changes:
1. ↓in body mass and body fat.
2. ↓ albumin
3. ↑ drug concentration, ↑ level of free drugs and ↑ $t_{1/2}$.

Neuropathological changes in normal aging:
1. ↑ in astrocytes and microglia.
2. ↓ in oligodentrocytes.

Neuropsychological functions:
1. ↓ in performance IQ is more rapid than verbal IQ
2. ↓ in problem solving ability, working memory, long term memory and psychomotor function.

Sleep:
1. ↓ SWS sleep
2. ↓ in α and β waves in EEG.

Kidneys:
↓ in renal function (35% by age 65 and 50% by age 80) and this leads to accumulation of drugs mainly excreted by the kidneys (e.g. lithium and sulpiride).

Oxidative damage by free radicals compromise the ability to produce ATP. This affects the ability to meet energy requirements.

Overview of dementia

Dementia is a condition which involves progressive and cognitive deficits. Dementia usually affects memory first (except frontotemporal lobe dementia), with subsequent progression to cause dysphasia, agnosia, apraxia, diminished ability with executive function and eventually personality disintegration.

It is crucial to differentiate between dementia and delirium.

Major neurocognitive disorder
The DSM-5 diagnostic criteria states that:
1. There must be significant cognitive decline from a previous level of performance in one or more of the following cognitive domains - attention, executive function, learning and memory, language, perceptual motor and social cognition.
2. These deficits affect independence in performing everyday activities.

MIld neurocognitive disorder
The DSM-5 diagnostic criteria states that:
1. There must be **modest** cognitive decline from a previous level of performance in one or more of the following cognitive domains - attention, executive function, learning and memory, language, perceptual motor and social cognition.
2. These deficits do **not** affect independence in performing everyday activities.

Delirium
The DSM-5 diagnostic criteria states that there must be:
1. Changes in attention and awareness
2. These changes or disturbances develop over a short duration of time (usually characterized as within hours to few days) and represent a change from baseline attention and awareness. These disturbances tend to fluctuate during the course of a day.
3. Additional changes with regards to cognition (memory deficits, disorientated, language, visuospatial ability or perception)
4. There is evidence from clinical history, physical examination and biochemical investigations that the disturbances are due to physiological consequences of a medical condition, substance intoxication or withdrawal or due to multiple etiologies.

The DSM-5 specifies several etiologies, which include:
a. Substance intoxication delirium
b. Substance withdrawal delirium
c. Medication induced delirium
d. Delirium due to another medical condition
e. Delirium due to multiple etiologies

The DSM-5 specifies that acute delirium usually last for a few hours or days, whilst persistent delirium lasts for weeks or months.

There are several other subtypes, which include:
a. Hyperactive - individuals have a hyperactive level of psychomotor activity, that may be accompanied by mood lability, agitation, and refusal to cooperate with medical care.
b. Hypoactive - individuals have marked sluggishness and lethargy that approaches stupor

c. Mixed level of activity - Individuals have normal level of psychomotor activity even though attention and awareness are disturbed. This also includes those with rapidly fluctuating activity level.

Table 1. Compare and contrast between dementia and delirium.

	Dementia	Delirium
Course of illness	Insidious in onset, static or progressive in nature, typically occurs over months to years. Less fluctuation in symptoms.	Sudden in onset, occurs over hours to days, associated with fluctuation with lucid spells and sun-downing (i.e. symptoms getting worse at night).
Cognitive symptoms:		
Speech quality	Dysarthria.	Mute or normal.
Use of language	Anomia, aphasia (lack of speech).	Incoherent and illogical
Attention	Normal or slow response.	Reduced ability to focus or obvious shift in attention.
Orientation	Normal except in advanced dementia	Disorientated to time, place and person.
Memory	Difficulty in recall, encoding and consolidation.	Difficulty in encoding associated with recent memory loss after the onset of delirium.
Other psychiatric features		
Affect or mood features	Depressed or abulic (reduction in social drive).	Usually dysphoric, typically labile and rarely manic.
Psychosis	<ul><li>Hallucination is uncommon (e.g. Alzheimer's disease).</li><li>Lewy body dementia is associated with visual hallucinations.</li><li>Delusion of theft is common (e.g. accusing the maid stealing his item).</li></ul>	<ul><li>Visual hallucinations (40-70%) involving distortions and illusions.</li><li>Delusions (40-70%) are usually transient, fragmentary and persecutory.</li></ul>

Dementia is classified as cortical and subcortical dementia.

Table 2 Compare and contrast cortical and subcortical dementia.

	Subcortical dementia	Cortical dementia
Definition	Dementia affects neuroanatomical structures (e.g. nuclei) beneath the cerebral cortex.	Dementia mainly affects the cerebral cortex.
Examples	1. Parkinson disease. 2. Huntington disease. 3. Progressive supranuclear palsy. 4. AIDS dementia complex. 5. Multiple sclerosis. 6. Wilson's disease. 7. Binswanger's disease. 8. Normal pressure hydrocephalus. 9. Vascular dementia and Lewy body dementia (LBD) (can be a mixture of cortical and subcortical dementia).	1. Alzheimer's disease 2. Fronto-temporal lobe dementia (FTD) or Pick's disease.
Course of illness		
Onset	Insidious	Insidious
Duration	Months to years	Months to years
Course	Static or progressive	Progressive
Neuropsychological symptoms		
Speech quality	Dysarthria or mute	Progressive mutism in FTD.
Use of language	Normal or anomia	Aphasia (Nominal aphasia: word finding difficulty).
Attention	Normal or slow response	Normal or mild impairment
Memory	Difficulty in recall are prominent	Difficulty in encoding and consolidation
Cognition	Slow cognition with impairment in spatial orientation, visual discrimination and angle perception.	Acalculia, agnosia (autoprosopagnosia, apperceptual and colour agnosia) and apraxia
Other psychiatric features		
Affect	Depressed or abulic (organic albulia refers to reduction in social drive.)	Apathy and depression are common in AD. Disinhibition is common in FTD.
Psychosis	May be present	May be present. In AD, 30% of patients have delusions and 15% of patients have hallucinations.

Epidemiology

Western countries

Figure 2 Percentage contributed by different dementia subtypes

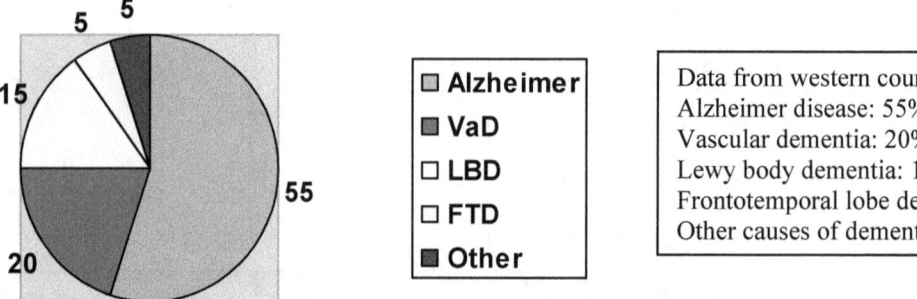

Data from western countries:
Alzheimer disease: 55%
Vascular dementia: 20%
Lewy body dementia: 15%
Frontotemporal lobe dementia: 5%
Other causes of dementia: 5%

1) As a result of the aging population, the prevalence of dementia will increase by 3 times in the next 50 years worldwide.
2) Dementia is rare before the age of 60 years. The risk doubles every 5 years after the age of 65 until the age of 85. By the age of 75, 12% of elderly suffer from moderate-to-severe dementia. By the age of 80 and 90, 20% and 40% of elderly suffer from moderate-to-severe dementia respectively.
3) Common causes of late onset dementia include Alzheimer's disease (AD), vascular dementia (VaD) and Lewy body dementia (LBD).
4) Common causes of early onset dementia include AD, VaD and frontotemporal dementia (FTD).
5) VaD is a common cause of early and late onset of dementia. VaD is more common in Japan and China.

Singapore

1. Sahadevan et al (2008) studied 14800 old people in Singapore and found that the overall age- and race-standardized dementia prevalence was 1.26%. Hence, the dementia prevalence of Singapore is lower than Western countries.
2. Prevalence (in 5-year age bands) was 0.08% (50-54 years), 0.08% (55-59 years), 0.44% (60-64 years), 1.16% (65-69 years), 1.84% (70-74 years), 3.26% (75-79 years), 8.35% (80-84 years), and 16.42% ($\geq$85 years).
3. Interesting data on ethnic differences were found. Sahadevan et al (2008) concluded that Malays had twice the risk for AD as compared to Chinese. Indians had more than twice the risk for AD and VaD than Chinese.
4. Ampil et al (2005) studied a total of 357 patients at National Neuroscience Institute (NNI). 190 (53.2%) suffered from VaD and 167 (46.8%) suffered from AD.
5. VaD was more common among Chinese and Malays and AD was more common in Indians and Eurasians.
6. Factors that may contribute to the observed ethnic variability in dementia include differential frequency of the ApoE-e4 allele, frequency of vascular risk factors, lifestyle choices, and cultural attitudes toward health care utilization.

Caregiver stress:

There were several studies from Singapore on caregiver stress related to the care of dementia patients.
1) Tan et al (2005) reported that neuropsychiatric symptoms were common among dementia patients and were positively correlated with caregiver distress. Family caregivers were significantly more distressed than professional caregivers over the delusion, agitation, depression and aberrant motor domains.
2) Tew et al (2010) reported that most Singaporean caregivers (85.7%) preferred care of dementia patients at home and only 14.3% chose institutionalisation. Four factors were associated with choice of nursing home: caregiver working, no domestic maid, lower caregiver gain and behavioural problems associated with dementia.

Assessment of dementia in OSCE exam:

> A 70-year-old woman is brought by her son because she has become more forgetful.
>
> Task: take a history to establish the diagnosis of dementia.

In clinical practice, dementia patients are often brought by concerned family members rather than complaining memory loss themselves. Dementia patients may not have insight or are in denial of memory loss.

1. Onset of memory loss: gradual or sudden.
2. Extent of memory impairment: recent memory (more likely to be impaired) or long term memory (e.g. childhood history may not be affected).
3. Reactions to memory loss: confabulation (covering the memory loss by making up an answer), denial or catastrophic reaction (anger when being challenged of memory problems).
4. Extent of cognitive impairment: judgement, decision making, problem solving.
5. Explore aetiology: e.g. family history of AD, history of stroke, history of Parkinson's disease (e.g. resting tremor, shuffling gait, masked face) and history of multiple head injury.
6. Explore possible causes of reversible dementia: e.g. normal pressure hydrocephalus (gait abnormalities, urinary incontinence), dietary habits (vitamin B12 deficiency) and thyroid disorder.
7. Assess mood status: history of depression and possibility of pseudodementia as a result of depression (patient tends to give don't know answer). Assess sleep pattern and appetite.
8. Assess common psychotic features e.g. delusion of theft (accusing the domestic helper stealing an item because the patient cannot find), auditory or visual hallucination.
9. Assess behaviour problems: e.g. violent (e.g. attacking domestic helpers), disinhibition, and wandering behaviour.
10. Assess risk: e.g. risk of having a fire or flooding at home as patient may forget to switch off stove or water tap, risk of fall, risk of financial exploitation, risk of self harm or suicide, risk of violence.
11. Assess activities of daily living (ADL): there are two types of ADL - basic ADL which include bathing, feeding and toileting by oneself; instrumental ADL which include withdrawing money from the bank, shopping and using public transport.
12. Assess coping by patient (e.g. memory aids, reminders).
13. Assess coping by caregiver and strain on caregiver.
14. Explore past medical history and chronic medical treatment.
15. Assess education level and past occupation.

Cognitive assessment

For mini-mental state examination (MMSE) and frontal lobe assessment, readers can refer to the website which shows the videos made by the authors. The cut-off for MMSE is 24 out of 30 in a Singaporean who has 'O' level education. The score range for mild dementia is between 20 – 24. The score range for moderate dementia is between 10 – 19 and the score range for severe dementia is between 0 – 9. MMSE score is affected by education level and there is a lack of frontal lobe assessment. There is no score for frontal lobe assessment and it is based on an overall impression and judgement by the assessor. Readers are reminded that the possibility of performing MMSE and frontal lobe assessment is low in the undergraduate OSCE exam in Singapore. If a candidate wants to perform the complete MMSE or frontal lobe assessment in an undergraduate OSCE exam in Singapore, please read the instructions carefully and make sure you do not misinterpret the task.

The Montreal Cognitive Assessment is a validated questionnaire to screen for dementia in Singapore and it also assesses frontal lobe function. (http://www.mocatest.org/). MoCA has both Chinese and English versions. If the score is less than 26, the subject has cognitive impairment.

For general practitioners, family doctors and other specialists, clock face drawing test is an easy method to screen for dementia without further training.

Figure 3 Clock face drawing test and cognitive impairment

Clock face drawing (instruction: 10 minutes past 11 o'clock)	Cognitive function and possible neuroanatomical lesion
	**Normal** Even spacing between numbers (intact parietal lobe function) Hour arm and minute arm are correctly placed (intact prefrontal cortex function)
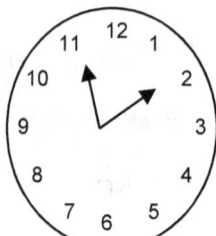	**Subtle abnormalities in cognitive function** Even spacing between numbers (intact parietal lobe function) It is not very clear whether the patient knows which is an hour arm and which a minute arm is. (possible impairment in prefrontal cortex)
	**Abnormal cognitive function** Even spacing between numbers (intact non-dominant parietal lobe function) Wrong place for hour and minute arms (impaired dominant parietal cortex and prefrontal cortex functions)
	**Grossly abnormal cognitive function** Uneven space between numbers (impaired non-dominant parietal lobe function) No hour or minute arm (Impaired dominant parietal cortex and prefrontal cortex functions)

Assessment of activities of daily living (ADL)

Basic ADL: mnemonics: "DEATH". <u>D</u>ress, <u>E</u>at, <u>A</u>mbulate, <u>T</u>ransfer/<u>T</u>oilet, <u>H</u>ygiene
Advanced ADL: mnemonics: "SHAFT": shopping, housekeeping, accounting, food preparation/meds, telephone/transportation.

Differential diagnosis of dementia

1. Other irreversible causes of dementia (e.g. VaD, LBD, FTD)
2. Reversible causes of dementia (e.g. normal pressure hydrocephalus, vitamin B12 deficiency, neurosyphilis, hypothyroidism)
3. Mild cognitive impairment (mild memory impairment without functional or occupational decline)

4. Delirium
5. Depression and pseudodementia (e.g. giving 'don't know' answer, poor concentration in depression results in poor registration of information)
6. Amnestic disorder (e.g. Korsakoff psychosis as a result of chronic alcohol misuse, heavy metal poisoning)
7. Underlying mental retardation or intellectual disability
8. Late onset psychosis (resembles behavioural problems associated with dementia)
9. Worried-well syndrome (an anxious person believes that he or she has dementia).

Investigations:

1. Full blood count
2. Liver function test
3. Renal function test
4. Thyroid function test
5. Calcium panel
6. Syphilis screen or VDRL
7. Vitamin B12 and folate
8. EEG
9. Chest X-ray
10. CT / MRI brain scan (important for age of onset < 60 years, focal neurological sign and rapid progression of dementia).

Alzheimer's disease (AD)

Table 3 Compare and contrast risk and protective factors for AD

Risk factors associated with AD:	Protective factors against AD:
- Old age - Down syndrome - APO e4/e4 alleles: (↑ the risk of AD) - Family history of Down syndrome and vascular risk factors - Head injury and ↑ risk of forming neurofibrillary tangles - Aluminium exposure.	- High education level - APO e2 alleles (people of oriental origin have lower prevalence of APO e4 alleles) - Consumption of fish - Bilingualism and late retirement may delay onset of AD.

Neuropathology of AD

β-amyloid (Aβ) cascade hypothesis

Aβ deposition is caused by mutations in:
Amyloid Precursor Protein (APP) gene on Ch 21 (25% of early onset AD)
Presenilin 1 (PS 1) gene on Chromosome 14 (Presenilin is implicated in β-amyloid). Chromosome 14 accounts for 75% of early onset AD
Presenilin 2 (PS2) gene on Chromosome 1.

Aβ deposition is predisposed by:
Apolipoprotein E (Apo E e4) allele on Chromosome19.

NFTs are intracellular but amyloid deposits are extracellular.
NFT is composed of paired helical filaments with ubiquinated or phosphorylated tau protein. Tau protein links neurofilaments and microtubles. In elderly, NFTs are confined to cells in hippocampus and entorhinal cortex but also found in amygdala, neocortex, locus coeruleus and raphe nuclei.

Amyloid filaments →

Aβ deposition in extracellular plaques and blood vessels leading to amyloid angiopathy. These senile plaques are found in the neocortex (layer 3 and 4), amygdala, hippocampus and entorhinal cortex

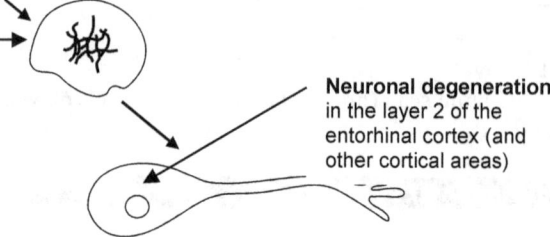

Neuronal degeneration in the layer 2 of the entorhinal cortex (and other cortical areas)

Neurochemical changes in AD:
1. ↓ in acetylcholine in the nucleus basalis of Meynett
2. ↓ in dopamine beta-hydroxylase
3. ↓ in dopamine
4. ↓ in noradrenaline and 5HT in the cortex.

Major or mild neurocognitive disorder due to Alzheimer's disease
The DSM-5 criteria are met for either major or mild neurocognitive disorder.
There is insidious onset and gradual progression of the impairments in one or more cognitive domains (at least two domains must be impaired for major neurocognitive disorder).
Criteria are met for either probable or possible Alzheimer's disease as follows:
a. Evidence of a causative Alzheimer's disease genetic mutation from family history or genetic testing
b. Three of the following must all be present: Clear evidence of memory decline; steady progression and gradual decline; no evidence of mixed etiology

Other clinical features:

1. **Psychotic symptoms**: paranoid delusions (15%), auditory or visual hallucinations (10-15%; visual hallucination is more common than auditory hallucination).
2. **Behavioural disturbances**: aggression, wandering, explosive temper, sexual disinhibition, searching behaviour.
3. **Personality changes**: exaggeration of premorbid personality.
4. **Orientation**: If disorientation occurs in advanced dementia, it is more common for disorientation in time than for place.
5. **Neurological features**: extrapyramidal features (60%), epilepsy (75%), reduction of REM sleep, frequent nocturnal waking periods and shortened sleep periods.

Management for AD

NICE guidelines from the UK recommend the following:

Mild AD:

- Offer patients the chance to participate in a structured group cognitive stimulation programme irrespective of the status of prescriping acetylcholinesterase inhibitors.

Moderate AD

- Consider the acetylcholinesterase inhibitors (AChEIs) including donepezil, galantamine and rivastigmine.
- The least expensive drug should be chosen taking into account of the daily dose and the price per dose.
- Consider an alternative acetylcholinesterase inhibitor if adverse event or drug interaction occurs.
- MMSE, functional and behavioural assessment should be performed every 6 months.
- Treatment should be continued if either the MMSE score remains at or above 10 points or global, functional and behavioural assessment indicate beneficial effects.

Severe AD:
- Memantine is used in moderately severe to severe AD in well designed clinical setting.

Acetylcholinesterase inhibitors (AChEIs) aims at increase the levels of acetylcholine (Ach) and improves cognition for patients with AD.

Table 4 A comparison of acetylcholinesterase inhibitors (AChEIs)

	Donepezil	Rivastigmine	Galantamine
Indications	AD, VaD	AD, LBD (improve hallucinations and delusions)	AD
Specific for CNS	Selective	Selective	Selective
Reversibility	Reversible	Pseudo-irreversible	Reversible
Enzymes inhibited	AChE	AChE and BChE	AChE
Plasma half-life	70 hours	10 hours	6 hours
Frequency of administration	Once per day	Twice per day	Twice per day
Daily dose	5-10mg/day	3-6mg/day	8-12mg/day
Organ of elimination	Liver	Kidney	Both liver and kidney
Price in SGD (2012)	5mg tablet ($ 5.99)	3mg tablet ($ 2.97) 9.5mg/24 hour patch ($5.2)	8mg tablet ($4.7)
Administration	Oral tablets only	Oral tablets and patches (24 hour)	Oral tablets only

AChE = acetylcholinsterase; BChE = butyrylcholinesterase

Common side effects of AChEI:

Common side effects:

In general, AChEIs are safe and well-tolerated. Along with useful effects, AChEIs may cause unwanted side-effects but not every patient experiences them. Some of these side – effects may improve as the human body adjusts to the new medication. The common side- effects affect less than 1 in 10 people who take these drugs. The common side-effects include:

- Diarrhoea (excessive cholinergic effects)
- Difficulty in sleeping (excessive cholinergic effects)
- Dizziness
- Feeling agitated
- Headache
- Loss of appetite
- Muscle cramps (excessive cholinergic effects)
- Tiredness.

Severe but uncommon side effects include (< 1 in 100):
- Bleeding in gastrointestinal tract
- Reduction in heart rate.

Memantine (10mg tablet = $3.67) is not an acetylcholinesterase inhibitors and works on the glutaminergic and NMDA (neuroexcitatory) receptors as an antagonist to improve cognitive function. Common side effects include:
- Agitation
- Confusion
- Drowsiness
- Giddiness
- Nausea.

Uncommon side effects of memantine include:
- Vomiting
- Increased libido
- Hallucinations
- Hypertonia.

Vascular dementia (VaD)

Aetiology:

1. Hypertension is the most significant risk factor for VaD (contributing to 50% of VaD).
2. Metabolic syndrome also plays a key role.
3. Polycythaemia, low levels of high density lipoprotein, homocystinuria and sickle cell anaemia can cause VaD.

Clinical features

VaD has unpredictable course, more rapid and stepwise deterioration. VaD demonstrates more impairment in attention and executive function as compared to AD.

Table 5 NINDS-AIREN diagnostic criteria of VaD

A relationship between dementia and cerebrovascular accidents (CVA) manifested or inferred by the presence of one or more of the following:	Clinical features consistent with the diagnosis of probable VaD:
1. Amnesia and cognitive impairment in at least 1 domain with resultant disability. 2. Focal sign and image findings 3. Onset of dementia within 3 months following a recognised stroke. 4. Abrupt deterioration in cognitive functions (fluctuating and stepwise).	1. Early presence of a gait disturbance. 2. History of unsteadiness and frequent unprovoked falls. 3. Early urinary symptoms not explained by urologic disease 4. Pseudobulbar palsy. 5. Mood changes or abulia.

Major or Mild Vascular Neurocognitive disorder
The DSM-5 criteria are met for either major or mild neurocognitive disorder.
The clinical features must be consistent with an underlying vascular etiology, as suggested by the following:
a. Onset of cognitive deficits that are related temporally to one or more cerebrovascular events
b. Evidence of decline in complex attention and frontal executive function

Investigations

1. CT and MRI may show infarcts, lacunes and leukoaraiosis.
2. SPECT and PET scans may show patchy hypoperfusion.

Management
1. The NICE guidelines (UK) currently do not recommend the use of AChEIs or memantine for cognitive decline in VaD except in properly constructed clinical studies. Donepezil may be beneficial but not licensed in VaD.
2. Treatment of underlying risk factors such as hypertension and diabetes is important.

Prognosis
1. Prognosis of VaD is worse than that of AD. The mean survival of AD is 6 years and the mean survival of VaD is only 3 years.
2.

Fronto-temporal lobe dementia (FTD)

Compare and contrast AD and FTD:
- Patients with FTD have younger age of onset (age of onset: 45-65 years; F:M = 2:1), more severe apathy, disinhibition, reduction in speech output, loss of insight and coarsening of social behaviour as compared to patients with AD. Primitive reflexes such as grasp, pour, palm mental reflexes reappear.
- Patients with AD have more impairment in calculation, constructions and lower Mini-Mental State Examination (MMSE) scores and higher prevalence of depression (20%) compared to FTD.
- Both DAT and FTD have insidious onset.

Pathology: frontotemporal atrophy, swollen achromatic neurons (balloon cells) and presence of pick bodies.

Table 6   Signs and symptoms of FTD:

Frontal-temporal lobe features:	Affective features:	Supporting features:
1. Loss of executive functions: loss of interest, preservation, disinhibition (jocularity and hypersexuality), inflexibility and impulsiveness, lack of personal and social awareness. 2. Primitive reflex and stereotypies (compulsion without obsessions) and strange eating habits. 3. Progressive reduction in speech, poor verbal fluency and echolalia. 4. Preserved visuospatial ability.	1. Depression 2. Anxiety 3. Hypochondriasis 4. Emotionally unconcerned.	1. Age of onset < 65 years. 2. Family history of FTD 3. Bulbar palsy 4. Akinesia, rigidity, and tremor 5. Early incontinence.

Major or mild Frontotemporal Neurocognitive disorder
The DSM-5 criteria are met for either major or mild neurocognitive disorder.
There has been gradual onset and progression of impairments.
There must be the presence of:
1. 3 or more behavioural symptoms such as behavioural disinhibition, apathy, loss of sympathy or empathy, perseverative behaviour and hyperorality

2. Language difficulties such as marked decline in language ability, especially so for speech production, word finding, object naming, grammar or word comprehension.

Investigations for FTD:

1. Psychometry shows characteristic impairments in executive function, verbal fluency and agnosia.
2. Structural imaging may not show characteristic lesion in early stage and functional imaging shows anterior hypoperfusion.

Management:
1. No active pharmacological intervention is indicated for cognitive impairment. SSRI is indicated for non-cognitive feature.
2. Psychosocial interventions may be useful.

Lewy body dementia (LBD)

Epidemiology

- LBD is the third most common cause of late onset dementia.
- LBD commonly affects men more than women.

Pathology

- Neuroanatomical areas affected include the hippocampus, temporal lobes and neocortex.
- Cholinergic deficit is much more pronounced in LBD as compared to AD.

Signs and symptoms of LBD

Cognitive symptoms include enduring and progressive cognitive impairment with emphasis on impairments in consciousness, alertness and attention. Cognition is fluctuating and short term memory is not affected in early stage. LBD has less episodic amnesia, more executive dysfunction and more apraxia as compared to AD.

Parkinson features include slowness, muscle stiffness, trembling of the limbs, a tendency to shuffle when walking (mild Parkinsonism: 70%, no Parkinsonism: 25%), loss of facial expression, changes in the strength and tone of the voice.

Common non-cognitive features include apathy, depression, hallucinations (complex visual hallucination: 80%, auditory hallucination: 25%) and delusions (paranoid delusions: 65%).

Other features include neuroleptic sensitivity (60%), falls, syncope and spontaneous loss of consciousness.

Major or Mild Neurocognitive Disorder with Lewy Bodies
The DSM-5 criteria are met for either major or mild neurocognitive disorder.
There has been gradual onset and progression of impairments.
The DSM-5 specified that the core diagnostic features include:
1. Fluctuating levels of cognition with marked variations in attention and alertness
2. Visual hallucinations (recurrent) that are characterized as being well formed and detailed
3. Features of parkinsonism, with onset subsequent to the development of cognitive decline
Additional supportive diagnostic criteria:
1. Presence of rapid eye movement sleep behaviour disorder
2. Severe neuroleptic sensitivity

Investigations:

- There is little atrophy in early stages of the illness in imaging with sparing of medial temporal lobe.
- 90% of patients with LBD have EEG abnormalities.

Management:

1. Do not use antipsychotic drugs for mild-to-moderate non cognitive symptoms in LBD because of the risk of severe adverse reactions (e.g. EPSE). If patients with LBD need antipsychotics, consider quetiapine.
2. Consider AChEIs for people with LBD who have non-cognitive symptoms causing significant distress or leading to challenging behaviour. Rivastigmine has the best research evidence for improvement of cognitive functions in LBD. AChEIs may improve cognitive symptoms, delusions and hallucinations.

Reversible causes of dementia

1. **Normal pressure hydrocephalus:** triad of cognitive impairment, gait disturbance and urinary incontinence. Aetiology includes subarachnoid haemorrhage, trauma and meningitis. The CT or MRI brain reveals ventricular enlargement without significant cortical atrophy. Treatment is ventriculoperitoneal shunt.
2. **Subdural haematoma:** Recent memory loss, morning headaches, fits, diplopia, hyperreflexia, and extensor plantar responses. Treatment is surgical evacuation.
3. **Reversible causes picked up by laboratory tests: vitamin B_{12} deficiency** (memory impairment in 75% of patients. One common cause is pernicious anaemia associated with diffuse and focal degenerations), **folate deficiency**, **thiamine (vitamin B_1) deficiency** (horizontal nystagmus and lateral rectus palsy, ataxia and memory or cognitive deficits in 10% of patients), **neurosyphilis** (Argyll Robertson pupils (60%), dysarthria (80%), spasticity (50%), tabes dorsalis (20%) and depression (30%)) **hypothyroidism** and **primary hypoparathyroidism** (slow, insidious onset).
4. **Hepatic encephalopathy:** symptoms include impaired cognition, a flapping tremor (asterixis) and a decreased level of consciousness. The treatment is neomycin and lactulose.
5. **Heavy metal poisoning:** e.g. aluminium, thallium (causing hair loss) and mercury.

Other psychogeriatric disorders

Late onset schizophrenia (paraphrenia)

Epidemiology: Prevalence is less than 1% (17-24/100,000) and it is more common in elderly women.

Very early onset schizophrenia occurs after 60-year-old. Late onset schizophrenia occurs between 40 to 59-year-old.

Risk factors: Sensory impairment (e.g. deafness), social isolation, paranoid or schizoid personality (45%).

Clinical features: Onset is insidious. In 20% of late onset schizophrenia, delusion is the only symptom. The most common delusion is persecutory (90%). Auditory hallucination is the most common hallucination (75%). Visual hallucinations (60%) and early cognitive impairment are recognised features.

Compared to adult schizophrenia: Affective flattening, negative symptoms and presence of all first rank symptoms (<30%) are less common in elderly.

Imaging: Larger ventricles on the CT brain scan

DDX: Delusional disorder, dementia and delirium.

Treatment: The decision on using either the first generation (e.g. haloperidol) or second generation (e.g. risperidone) antipsychotics is always under debate. According to the Maudsley Guidelines [UK], the initial warnings of increasing risk of CVA in elderly apply to both the first and second generation antipsychotics. The first generation antipsychotics are also associated with increased mortality. Doctors should discuss with the patient and seek their preference on the choice of antipsychotics.

Prognosis: Psychotic symptoms usually respond to antipsychotics although duration of treatment may need to be indefinite.

Late onset depressive disorder

Epidemiology: Prevalence in the community is 3%; attending GPs (30%), elderly in residential care (40%), elderly as medical inpatients (45%). Mild depression is more common in female (F:M = 2:1) but severe depression have equal sex ratio. Less than 10% of depression emerge in old age.

Aetiology and risk factors: *3Cs: Cancer, Cardiovascular diseases, Central nervous system diseases.* Living alone and presence of dementia are risk factors. 10-20% of widows suffering from grief and require treatment. Marriage is a protective factor against depression in old age.

Salient features: Psychomotor retardation/agitation (30%), depressive delusions (poverty and nihilistic) are common. Paranoia involves derogatory and obscene auditory hallucinations, complaints of nervousness and irritability are also common Late onset depression is associated with deep white matter changes and enlargement of ventricles.

Vascular depression: Association with apathy, psychomotor retardation, impaired executive function and mortality. Anterior infarct is more common than posterior infarct. Family history of depression increases the risk of vascular depression.

Compared to adult depression: Late onset depression is not associated with family history of depression.

Comorbidity: cognitive impairment in depressed patients (70%). Some elderly suffers from pseudodementia with difficulties in concentration and lack of motivation in the background of depression.

Instruments: Geriatric Depression Scale (GDS score > 11 in GDS -30 or 5 in GDS-15) indicates depression)

Treatment: Antidepressant: SSRIs are better tolerated than TCAs. SSRIs increase the risk of GI or other bleeding, particularly in elderly taking NSAIDs or warfarin. Elderly are prone to develop hyponatraemia. Elderly have poorer treatment response and take longer to respond (6-8 weeks) in general.

Psychotherapy is useful and elderly need shorter session. ECT is effective in 80% of severe depression especially for those with anxiety and agitation. Cardiac pacemaker is not an absolute contraindication. 10% of elderly depression is resistant to conventional treatment.

Prognosis: Organic brain disorder and chronic depression are associated with poor prognosis. Long term prognosis is not favourable for 40% of cases.

Suicide

Epidemiology: Suicide is common in elderly. In Singapore, Tan and Wong (2008) found that 53.8% of elderly verbalized thoughts of wanting to kill themselves. Men were three times more likely to report suicidal thoughts. The association between depression severity and suicidal ideations is not strongly supported. 95% of elderly suicide is not feigned or manipulative. Parasuicide is rare in elderly with equal sex incidence. Deliberate self harm (DSH) in over-65s only accounts for 5% of all DSH.

Risk factors: male gender, elderly with low cholesterol, cancer, CVA, epilepsy, multiple sclerosis, social isolation and first year of bereavement. The risk factors for suicide and DSH in elderly are similar.

Salient features: Suicide may take place within the first few hours of admission and within weeks after discharge.

Suicide is less common among those staying in residential care and people have obsessive compulsive personality.

Management of DSH and suicide in elderly: All acts of DSH in people over the age of 65 years should be taken as evidence of suicide intent until proven otherwise. Psychiatrists need to rule out depression, cognitive impairment and poor health.

Late onset bipolar disorder

Epidemiology: The prevalence of bipolar disorder is less than 0.1% in old people older than 65 years. The F:M ratio is 2:1.

Aetiology: The influence from genetics is less as compared to adults. It is more common for depression to switch to mania in elderly. Organic factors such as cerebral insult are a common cause.

Salient features: More irritable and more likely to develop toxic effect if patient is treated by lithium as compared to adult patients; more mixed presentation and paranoid ideation; less euphoria and less hyperactivity.

Treatment: An atypical antipsychotics (with less weight gain) is considered as first line. Consider lithium as second line agents for female patients and valproate for male patients.

Maintenance level of lithium for elderly: 0.4 – 0.6 mmol/L.

Late onset anxiety disorder

Epidemiology: The overall prevalence is between 1- 10%. Phobic disorder is more prevalent in the community samples compared to the hospital samples. The prevalence is listed as follows: social phobia: 1%, agoraphobia: 2-7%, generalised anxiety: 4% and simple phobia: 4%.

Onset is usually in young adulthood.

Aetiology and risk factor: After physical illness and accident, old people often loose their confidence to go out. Adverse life events and loneliness are also risk factors for late onset anxiety disorder.

Salient features: Fear of crowd or public transport is common. Open space phobia is well recognised in elderly. Compared to adults, panic disorder is less common

Treatment: In elderly, SSRIs are better tolerated than TCAs. SSRIs increase the risk of GI or other bleeding, particularly in elderly taking NSAIDs or warfarin. Elderly is prone to develop hyponatraemia. Elderly have poor treatment response and take longer to respond (6-8 weeks). Psychotherapy (e.g. exposure therapy) is useful to overcome agoraphobia and elderly need shorter session as compared to adults.

Substance abuse in elderly

Epidemiology: Alcohol is commonly used in old people. The M:F ratio in problem drinkers is 2-6:1 in western countries. Men started drinking in adulthood and continue into old age.

Causes: It is often precipitated by sudden access to excess time and money after retirement. Higher social class is a risk factor for late onset alcohol abuse. Genetic factor plays a less significant role in the late onset alcohol dependence. In the elderly, there is reduction in body mass and total body water. The same amount of alcohol consumed during adulthood may lead to higher peak blood alcohol levels in old age.

Comorbidity: Depression, anxiety and cognitive impairment.

Treatment: Similar to treatment in adults, it involves motivational enhancement, detoxification and joining Alcoholic anonymous (AA). Naltrexone is safe in elderly but disulfiram should be avoided because it causes cardiac and hepatic adverse effects.

Prognosis: Late onset alcohol dependence usually resolves without formal treatment.

Benzodiazepine dependence: The prevalence is 10% and more common in women. Risk factors include anxiety disorder, depression and personality disorder. 60% of elderly are successful in abstinence after detoxification.

Other substances commonly misused by elderly in Singapore: cough mixture, laxatives and analgesics.

Revision MEQ

You are a GP working in the heartland neighbourhood. A 57-year-old postman complains of poor memory and worries that he suffers from dementia. He mentions that he forgot to bring his wallet when he went for shopping one day. He continues to work without occupational impairment. There is no psychiatric history or family history of mental illness. He has good past health. Mental state examination reveals a middle-aged man who is anxious. He does not show any psychotic feature. He consults you to establish a diagnosis of dementia and anti-dementia treatment.

Q.1 Is this 57-year-old man likely to suffer from dementia?

No, he is unlikely to suffer from dementia.

Q.2 List 5 clinical features in this vignette which suggest that he is unlikely to suffer from dementia.

1. He is younger than 60 years.
2. There is only one episode.
3. There is no family history of dementia.
4. There is no past history of medical illness such as stroke and related risk factor.
5. He has good occupational function.

Q.3 List two differential diagnosis for the subjective memory impairment.

1. Worried-well phenomenon: anxious people worry that they have dementia.
2. Pseudodementia: poor attention/concentration as a result of depression.

Q.4 He is still concerned and requests a cognitive assessment. Name two assessments which you would perform to screen for dementia.

1. Mini-mental state examination (MMSE)
2. Montreal Cognitive Assessment (MoCA)
3. Clock-face drawing test.

Q.5 He wants to have laboratory investigations to rule out reversible causes of dementia. List four investigations:

1. Full blood count
2. Vitamin B12 and folate
3. Liver function test
4. Syphilis screen (VDRL)
5. Thyroid function test (TFT).

Revision MCQs

Q.1 An 80-year-old man is brought in by family for poor memory. The history and MMSE assessment suggest that he suffers from dementia. The family wants him to have a CT brain scan but the patient is not very keen. Which of the following is not an indication for a CT brain scan?

A. Focal neurological sign
B. Gait disturbance
C. Slow progression of dementia
D. Recent head injury
E. Use of anticoagulant (e.g. warfarin).

The answer is C. Rapid progression of dementia is an indication for CT brain scan. Slow progression suggests that the patient may suffer from Alzheimer's disease and there is not a strong indication for CT brain scan.

Q.2 A 75-year-old woman is brought in by her daughter because she has been seeing multiple GPs to obtain medications. She seems to be dependent on various medications and her daughter has no clue of what she is taking. Her daughter wants to find out the pattern of substance misuse in elderly. Which of the following drugs is least likely to be abused by old people in Singapore?

A. Analgesics
B. Benzodiazepines
C. Diuretics
D. Cough mixture
E. Laxatives.

The answer is C. Diuretics is used by young people with anorexia nervosa to loose weight but it is uncommonly to be abused by elderly in Singapore.

Q.3 A 70-year-old woman presents with mania after left cerebrovascular accident. As compared to adult bipolar patient, she is more likely to:

A. Require higher dose of lithium
B. Demonstrate irritability
C. Demonstrate reckless behaviour
D. Have high sexual drive
E. Have grandiose delusions.

The answer is B. Late onset bipolar disorder is associated with less grandiosity, less violent or reckless behaviour and less likely to be associated with concomitant misuse of recreational drugs.

EMIs

Dementia

A- Alzheimer's dementia
B- Delirium
C- Herpes simplex encephalitis
D- HIV dementia
E- Huntington's disease
F- Multi infarct dementia
G- Organic hallcinosis
H- Organic psychotic disorder
I- Picks disease
J- Post encephalitic syndrome

Please choose the most appropriate diagnosis for each of the under-mentioned clinical vignettes.

Question 1: A 80 year old lady presents to the clinic with focal neurological signs, associated with memory impairments. Her family gave a history suggestive of a progressive step wise deterioration from her baseline. – Multi infarct dementia / vascular dementia (usually has associated vascular risk factors like DM, HTN, and IHD)

Question 2: A 80 year-old male presents to the accident and emergency department with significant impairment in consciousness. He was not orientated, and was noted to be not able to hold his attention, even for short spans of time. There is also marked disturbances in his sleep wake cycle. – Delirium

Question 3: A 70 year old male presented to the old age clinic with a history of progressive, gradual deterioration of cognition, without any focal neurological signs – Alzheimers disease

Dementia

A – Stepwise deterioration
B – Insidious onset and gradual progression
C- Emotional incongruity
D – Motor features of Parkinson's disease
E – Personality change and behavioural disorder
F – History of transient ischemia attacks
G – Myoclonus
H – Visual hallucinations
I – Auditory hallucinations
J – Fluctuating levels of consciousness
K – Striking insight loss

Chose the sailent features for each of the following:
Question 1: Lewy body dementia - D h   J
Question 2: Frontotemporal dementia – B E K
Question 3: Vascular dementia – A F
Question 4: Neurological feature associated with rapidly evolving fatal dementia – G

References

American Psychiatric Association (1994) *DSM-IV-TR: Diagnostic and Statistical Manual of Mental Disorders (Diagnostic & Statistical Manual of Mental Disorders)* (4th edition). Washington: American Psychiatric Association Publishing Inc.

American Psychiatric Association (2013) *DSM-5: Diagnostic and Statistical Manual of Mental Disorders (Diagnostic & Statistical Manual of Mental Disorders)* (5th edition). Washington: American Psychiatric Association Publishing Inc.

Ampil ER, Fook-Chong S, Sodagar SN, Chen CP, Auchus AP. Ethnic variability in dementia: results from Singapore. Alzheimer Dis Assoc Disord. 2005 Oct-Dec;19(4):184-5.

Agronin ME, Maletta GJ (2006) *Principles and practice of geriatric psychiatry*. Philadelphia: Lippincott William and Wilkins.

Areosa SA, Sherriff F, McShane R (2005). Areosa Sastre, Almudena. ed. "Memantine for dementia". *Cochrane Database Syst Rev* (3): CD003154

Choo PW, Lee KS, Owen RE, Jayaratnam FJ. Singapore--an ageing society. Singapore Med J. 1990 Oct;31(5):486-8.

Cummings JL (1986) Subcortical dementia. Neuropsychology, neuropsychiatry and pathophysiology. *British Journal of Psychiatry*, **149**: 682-697.

Feng L, Gwee X, Kua EH, Ng TP. Cognitive function and tea consumption in community dwelling older Chinese in Singapore. J Nutr Health Aging. 2010 Jun;14(6):433-8.

Feng L, Yap KB, Kua EH, Ng TP Statin use and depressive symptoms in a prospective study of community-living older persons. Pharmacoepidemiol Drug Saf. 2010 Sep;19(9):942-8.

Gao Q, Niti M, Feng L, Yap KB, Ng TP. Omega-3 polyunsaturated fatty acid supplements and cognitive decline: Singapore Longitudinal Aging Studies. J Nutr Health Aging. 2011 Jan;15(1):32-5.

Gelder M, Mayou R and Cowen P (2001) *Shorter Oxford Textbook of Psychiatry*. Oxford: Oxford University Press.

Graham DI, Nicoll JAR and Bone I (2006) *Introduction to Neuropathology (3rd edition)*. London: Hodder Arnold.

Ho RC, Niti M, Yap KB, Kua EH, Ng TP. Metabolic syndrome and cognitive decline in chinese older adults: results from the singapore longitudinal ageing studies. Am J Geriatr Psychiatry. 2008 Jun;16(6):519-22.

Johnstone EC, Cunningham ODG, Lawrie SM, Sharpe M, Freeman CPL. (2004) *Companion to Psychiatric studies*. (7th edition). London: Churchill Livingstone.

Kay R (2002) *Casebook of Neurology*. Hong Kong: Lippincott Williams and Wilkins Asia.

Ng TP, Broekman BF, Niti M, Gwee X, Kua EH. Determinants of successful aging using a multidimensional definition among Chinese elderly in Singapore. Am J Geriatr Psychiatry. 2009 May;17(5):407-16.

NICE guidelines for dementia http://guidance.nice.org.uk/CG42

NICE guidelines for self harm
http://guidance.nice.org.uk/CG16

Niti M, Yap KB, Kua EH, Ng TP. APOE-epsilon4, depressive symptoms, and cognitive decline in Chinese older adults: Singapore Longitudinal Aging Studies. J Gerontol A Biol Sci Med Sci. 2009 Feb;64(2):306-11. Epub 2009 Jan 30.

Roman GC, Tatemichi TK, Erkinjuntti T, et al. (1993) Vascular dementia: diagnostic criteria for research studies. Report of the NINDS-AIREN International Workshop. *Neurology*, **43**:250-60.

Sahadevan S, Saw SM, Gao W, Tan LC, Chin JJ, Hong CY, Venketasubramanian N. Ethnic differences in Singapore's dementia prevalence: the stroke, Parkinson's disease, epilepsy, and dementia in Singapore study. J Am Geriatr Soc. 2008 Nov;56(11):2061-8.

Schwingel A, Niti MM, Tang C, Ng TP. Continued work employment and volunteerism and mental well-being of older adults: Singapore longitudinal ageing studies. Age Ageing. 2009 Sep;38(5):531-7. Epub 2009 May 27.

Semple D, Smyth R, Burns J, Darjee R and McIntosh (2005) *Oxford handbook of Psychiatry*. Oxford: Oxford University Press.

Taylor D, Paton C, Kapur S (2009) *The Maudsley prescribing guidelines*. 10th edition London: Informa healthcare.

Tan LL, Wong HB, Allen H. The impact of neuropsychiatric symptoms of dementia on distress in family and professional caregivers in Singapore. Int Psychogeriatr. 2005 Jun;17(2):253-63.

Tew CW, Tan LF, Luo N, Ng WY, Yap P. Why family caregivers choose to institutionalize a loved one with dementia: a Singapore perspective. Dement Geriatr Cogn Disord. 2010;30(6):509-16. Epub 2011 Jan 20.

World Health Organisation (1992) ICD-10 : The ICD-10 Classification of Mental and Behavioural Disorders : Clinical Descriptions and *Diagnostic Guidelines*. Geneva: World Health Organisation.

Chapter 16 Child and adolescent psychiatry and intellectual disabilities

Autism	Gilles de la Tourette syndrome
Asperger's syndrome	Enuresis and encopresis
Attention deficit and hyperkinetic disorder	Common psychiatric disorders in childhood and adolescent
Conduct disorder and oppositional defiant disorder	Down syndrome
School refusal and truancy	Fragile X syndrome
Academic difficulties	Revision MEQ, MCQ & EMIS

Autism or autistic disorder

Onset

Onset is before the first 3 years of life. 70% of cases do not have normal development. 30% of cases have clear 'setback' in the second or third year of life.

Epidemiology

Western countries:

The prevalence is between 7-28/10,000. The M:F ratio is 4:1, accounting for 25% to 60% of all autistic disorders. There is no clear association with socio-economic status.

Singapore:

Bernard-Opitz et al (2011) studied 167 autistic children and found the following:

1. A predominance of boys over girls in autism and a low incidence of birth complications.
2. 60% of the children were diagnosed before the age of 3 years.
3. A high frequency of caregivers for autistic children were foreign maids.
4. The Academy of Medicine Singapore-Ministry of Health Clinical Practice Guidelines Workgroup on Autism Spectrum Disorders developed a guidelines and published in 2010. The recommendations from the MOH guidelines are incorporated in the following sections.

Aetiology

Autism has various causes. Aetiology is obscure in most instance.

1) **Genetic causes:** Heritability is over 90% with higher concordance in monozygotic twins. The recurrence rate in siblings is roughly 3% for narrowly defined autism, but is about 10-20% for milder variant of autism. Chromosome 2q and 7q are suspected locus.
2) **Environmental causes**: There is a moderately increase in rate of perinatal injury.

Mnemonic RFT+PKU: (Rubella, Fragile X syndrome, Tuberous sclerosis, PKU: Phenylketonuria), CMV, neurofibromatosis and infantile spasms are recognised causes.

Theoretical background:

One influential theory suggests that the primary deficit in autism lies in the theory of mind. The theory of mind refers to the capacity a person which attribute independent mental state to oneself and other people in order to predict and explain the actions of others.

Signs of autism

Figure 1 Clinical signs of autism

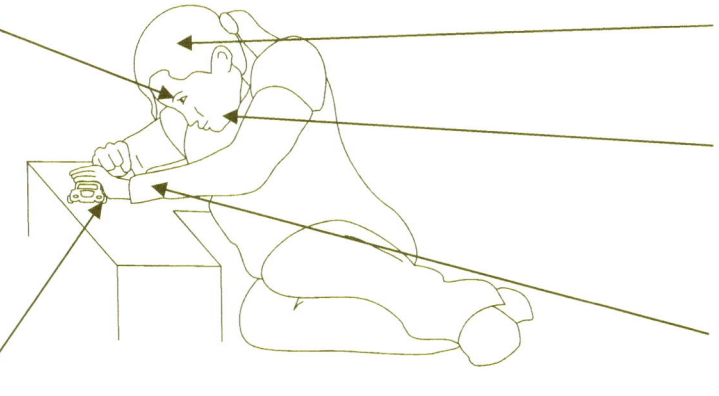

Impairment in making eye contact but able to do so when asked, lack of social reciprocity and having difficulty to identify mental state of other people.

Attachment to odd objects or non soft objects, rigid and resistant to change in routine and lack of imaginative play

Play is distorted in autism. Play provide an avenue to assess cognitive and affective development of a child. Normal children often have imaginary playmates and play monologue. Children suffering from autism are lack of imaginary play.

Lack of creativity and fantasy in thoughts.

Echolalia, palilalia (repeating the same phase at the end of a setence), pronominal (pronoun) reversal, lack of social usage of language.

Self injury (e.g. wrist biting) and stereotyped behaviour (e.g. hand-flapping, nodding, rocking).

All professionals involved in diagnosing autistic disorder in children should consider using either the ICD-10 or DSMIV-T systems of classification (The MOH guidelines 2010).

The ICD – 10 diagnostic criteria include:

1) **The presence of abnormal development is manifested before the age of 3 years:** including abnormal receptive or expressive language, abnormal selective or reciprocal social interaction, abnormal functional or symbolic play.
2) **Abnormal reciprocal social interactions** include failure in eye gaze and body language, failure in development of peer relationship, lack of socio-emotional reciprocity and lack of spontaneous sharing with other people.
3) **Abnormal communication** includes lack of development of spoken language, lack of social imitative play, failure to initiate or sustain conversational interchange, stereotyped and repetitive use of language.
4) **Restricted, stereotyped and repetitive behaviour** include preoccupation with stereotyped interest, compulsive adherence to rituals, motor mannerisms, preoccupation with part-objects or non functional elements of play materials.
5) **Other non-specific problems include**: phobias, sleeping and eating disturbances, temper tantrums and self-directed aggression.
6) **The absence** of other pervasive development disorder, socio-emotional problems, intellectual disability and schizophrenia-like symptoms.

Autism spectrum disorder
This is now considered a new DSM-5 terminology that now comprises of 4 previous DSM-IV-TR conditions (autistic disorder, asperger's disorder, childhood disintegrative disorder and pervasive developmental disorder not otherwise specified).

Autism Spectrum Disorder
The DSM-5 Criteria states that an individual diagnosed with Autism spectrum disorder would have marked difficulties in terms of communication and engagement with others across multiple social situations. These difficulties include:

(a) difficulties with demonstrating appropriate behaviours in social contexts, (b) difficulties associated with non verbal communications used for social interactions and (c) difficulties associated with failure to initiate or adapt to social interactions.

In addition, the DSM-5 also requires the individuals to have characteristic repetitive behavioural patterns, such as (a) Repeated stereotypical movements or (b) Highly ritualized behavioural patterns. DSM-5 criteria specified that these behaviours must have started since the early developmental period and has resulted in marked impairments in terms of functioning. The DSM-5 also advises clinicians to distinguish clearly between intellectual disability and autism spectrum behaviour. There are situations to which intellectual disability and autism might co-occur.

The MOH guidelines (2010) recommend that children with one or more of the following clinical features must be referred promptly for comprehensive developmental evaluation:
   A. No babble, pointing or other gestures by 12 months.
   B. No single words by 18 months.
   C. No spontaneous (non-echoed) 2-word phrases by 24 months.
   D. Any loss of language or social skills at any age.

Assessment:

The MOH guidelines (2010) recommend that diagnostic evaluation of a child suspected to have autism should be carried out by a multi-disciplinary team or professional who is trained and experienced with diagnosis of autism. Evaluation includes:
a) An autism-specific developmental history
b) Direct observations
c) Obtaining wider contextual and functional information.

Instruments

1) Autism Diagnostic Observation Schedule (ADOS) for patient and Autism Diagnostic Interview (ADI) for parents.
2) Child Behaviour Checklist (CBCL): In Singapore, Ooi et al (2011) constructed an Autism Spectrum Disorder scale based on 9 items from CBCL. High scores in the following categories: withdrawn/depressed, social problems, and thought problems syndromes significantly discriminated autism children from other disorders in Singapore.
3) IQ test: performance IQ is better than verbal IQ.

Investigations

The MOH guidelines recommend that:

1) Children with ASD with the following features should have a genetic evaluation:
   a) Microcephaly or macrocephaly
   b) A positive family history (of a genetic syndrome)
   c) Dysmorphic features.

2) Electro-encephalography (EEG) is not routinely recommended in children with ASD but should be considered if any of the following are present:
   a) Clinical seizures
   b) Symptoms suggestive of sub-clinical seizures
   such as staring spells
   c) A history of developmental regression.

Management

The MOH guidelines (2010) recommend that:

For the child:

1) Every pre-school child diagnosed with autism should have an individualised intervention plan that sets out

the goals, type(s), frequency and intensity of intervention, in order to address particular developmental and educational needs.
2) An individualised intervention plan should consist of a variety of quality programmes and activities. This includes attendance in comprehensive early intervention programmes, programmes targeting specific needs and also positive engagement with parents and/or caregivers.
3) Alternative-augmentative communication systems may be recommended for pre-school children with autism because the expanded (spoken or written) communication may stimulate speech acquisition in non-verbal children and enhance expression in verbal children.
4) Visual strategies are useful interventions for children with autism because they offer visual support to communication, increase spontaneous imitation and socially communicative behaviour.
5) Social skills programmes depend on the functioning level of the preschool child with autism and may include:
    - Assessment and teaching of social skills interaction in natural settings.
    - Provision of structure, visual cues and predictability.
    - Making abstract concepts more "concrete".
    - Activities that enable purposeful and appropriate interaction with typically developing peers.
    - Goals focusing on fostering self-appreciation and self esteem.
6) Pharmacological treatments are indicated:
    - Fluvoxamine may be considered for repetitive thought and maladaptive behaviour but should be used with caution in children with autism because of limited efficacy and poor tolerance.
    - Risperidone is recommended for the management of irritability, hyperactivity and stereotypic behaviour when used as short term treatment for children with autism.
    - Methylphenidate may be considered for treating hyperactivity in children with autism, although the magnitude of response is often less than that seen in typically developing children with attention deficit hyperactivity disorder.
    - Melatonin may be considered in the management of the disturbed sleep patterns of children with autism.

For the parents

1) Parents and caregivers should be encouraged to discuss the need for practical emotional support. This enables information to be provided, referrals made and support services made available
2) Parents and caregivers are recommended to consult appropriate professionals when considering educational placement for their child with autism, e.g. child and educational psychologists who are informed of the special educational provisions in Singapore.

Comorbidity:

- Generalised intellectual disability (50% have IQ< 50; 70% have IQ < 70; 100% have IQ < 100),
- Seizure (25%): peak age of seizure is 11-14-year-old.
- OCD (10%).
- Hyperactivity,
- Emotional problems,
- Temper tantrums.

Prognosis

1) **Western countries:** The most important predictor is childhood IQ and presence of speech by 5-years. 50% do not develop useful speech. Non-verbal IQ < 60 is associated with severe social impairment and lack of independent living. Only 10% are able to work independently.
2) **Singapore:** The MOH guidelines state that for autistic children diagnosed with a before three years old, parents should be advised that it is difficult to reliably predict prognosis, because individual outcomes are extremely variable and depend on many factors.

Asperger's syndrome

Epidemiology: 3-4 per 1000 children; M:F = 9: 1

ICD-10 diagnostic criteria:

1) Relatively normal early development and the child is noted to have lack of warmth and interest in social relationships around the third year of life.

2) Language development is <u>not</u> delayed and single word should have developed by age of 2 and communicate phrases by age of 3. Motor milestones are delayed.

Other salient clinical features:

1) Preoccupation with restricted, stereotyped and repetitive interests and associated activities. The extensive information is often acquired in a mechanical fashion. Classic interests include scientific fields (e.g. trains, weather, dinosaurs), but interests in other fields such as arts or music are seen as well.

2) They are good with logic, rules and routine. People with Asperger's syndrome tend to see the details (and thus argue over minute details with others), without seeing the 'big picture'.

3) They have problems in distinguishing people with different social roles, as well as recognizing social boundaries. They may talk to the principal and classmates in the same manner. They have troubles with normal social conventions, troubles making and keeping friendships, tend to be introverted and have less need for friendships.

Comorbidity: Anxiety, depression, OCD, possible development of schizophrenia (uncommon).

IQ test: verbal IQ is better than performance IQ (reversal of autism).

Treatment:
1) Psychoeducation should be offered to parents to enhance acceptance and maintain routines at home and school.
2) As the child gets older, he can be helped by supportive counselling.
3) The patient is encouraged to obtain employment in routine job (e.g. librarian). Shelter employment and residence are reserved for severe cases.

Prognosis: IQ>70 indicates for good prognosis. Most individuals will be able to obtain employment in a fairly routine job. Successful relationship with the opposite sex leading to marriage is uncommon.

Attention deficit and hyperkinetic disorder (ADHD)

Epidemiology

- Epidemiology of ADHD: Prevalence in the UK is 1-3 % (requires both hyperactivity and inattention) and prevalence in the US is higher: 5% (requires only either hyperactivity or inattention).
- Male to female ratio is 3:1.
- Peak age of onset is between 3-8 years.

Diagnostic criteria

- ADHD is an American term. In the UK, it is often known as "hyperkinetic disorder". The diagnostic criteria listed under the diagnostic criteria for ICD-10 and DSM-IV-TR are similar.
- The onset has to be younger than 7 years in more than two settings (e.g. school and home).
- Symptoms of inattention include *(Mnemonic: SOLID)*:

- Starts tasks or activities but not able to follow through and finish.
- Organization of tasks or activities is impaired.
- Loses things necessary for tasks and activities such as school assignments or stationary.
- Instructions are not followed.
- Distraction by external stimuli.
- Other features: careless mistakes, forgetfulness in daily activities.

- Symptoms of hyperactivity and impulsivity include (*Mnemonic: WORST FAIL*):
  - Waiting for in lines or await turns in game cause frustration. (impulsivity)
  - On the move most of the time such as running and climbing. (hyperactivity)
  - Restlessness and jitteriness. (hyperactivity)
  - Squirms on seat. (hyperactivity)
  - Talk excessively without appropriate response to social constraints. (impulsivity)

  - Fidgets with hands and feet. (hyperactivity)
  - Answers are blurted out before questions. (impulsivity)
  - Interruption of other people's conversations. (impulsivity)
  - Loud noise in playing. (hyperactivity)

Attention-Deficit/ Hyperactivity Disorder
The DSM-5 criteria states that an individual diagnosed with attention-deficit and hyperactivity disorder would have difficulties in terms of functioning, mainly due to (a) inattention and/or (b) hyperactivity and impulsivity.
The DSM-5 states that these inattentive and/or hyperactivity and impulsivity symptoms must be present before the age of 12 years old. In addition, these symptoms must have resulted in impaired functioning in at least 2 different social situations. The DSM-5 requires the individual to fulfill at least 6 of the following signs and symptoms of inattention: 1. Failing to pay close attention to details 2. Concentration difficulties 3. Difficulties with sustaining attention at tasks 4. Daydreaming and does not seem to be able to follow normal conversations 5. Difficulties with organization of tasks 6. Reluctance to participate in tasks that involve much attention 7. Frequently loses important objects 8. Easily distractible and 9. Forgetfulness about daily activities. The DSM-5 also specified that only 5 of the above signs and symptoms of inattention need to be fulfilled if individuals are 17 years of age and above.

The DSM-5 also requires the individual to fulfil at least 6 of the following signs and symptoms of hyperactivity and impulsivity: 1. Moving about and unable to sit still 2. Leaves seat even when required to remain seated 3. Climbs or runs about in inappropriate situations 4. Always having excessive energy and always on the move 5. Chats excessively 6. Impulsive and gives answers even before being asked to 7. Having difficulties waiting for his/her turn 8. Unable to carry out normal conversation due to frequent interruptions.
Clinicians would need to know the subtypes of ADHD, which include
a. Combined presentation - both clusters of signs and symptoms are met for the past 6 months
b. Predominantly inattentive presentation - Just inattentive clusters of signs and symptoms for the past 6 months
c. Predominantly hyperactive/impulsive presentation - Just the hyperactive clusters of signs and symptoms for the past 6 months
The DSM-5 has specified that only 5 of the above signs and symptoms of hyperactivity and impulsivity need to be fulfilled if individuals are above 17 years of age.

Other specified attention-deficit / hyperactivity disorder
The DSM-5 has this diagnostic category for individuals who have signs and symptoms suggestive of attention-deficit / hyperactivity disorder, but do not meet the full set of criteria.

Further investigations

- Gather information about behaviours from school and home environment.
- Administer Connor's Performance Test and Connor's Rating scales for teachers and parents.
- Arrange direct school observation by a member of the child psychiatry team in school whom the child has not met before.
- Psychometric testing such as IQ assessment or academic assessment if there is evidence of intellectual disability.
- Make arrangement for physical examination, baseline height and weight measurement for the child.

- Social investigations and engagement of medical social worker if there are underlying social issues.

Comorbidity

- Lim et al (2008) reported that 18.5% of ADHD children in Singapore experienced sleep disturbance related to medication, with 13.0% reporting daytime somnolence and 5.5% reporting insomnia.
- Conduct disorder / oppositional defiant disorder
- Depression and anxiety
- Substance abuse in adolescents.

Management

The NICE guidelines (UK) recommend that pharmacological treatment (e.g. stimulant) should be offered as the first-line treatment in school-age children and young people with severe ADHD. Parents should also be offered a group based parent-training or education programme.

Pharamcological treatment

There are two types of medication: stimulants and non-stimulants. Look for exercise syncope, breathlessness and other cardiovascular symptoms in the history. Physical examination includes measurement of heart rate and blood pressure (plot on a centile chart) and examination of the cardiovascular system. Electrocardiogram (ECG) is required if there is past history of cardiac disease.

1) **Methylphenidate (stimulant)**
    - Pharmacodynamics: dopamine reuptake inhibition and direct release of dopamine.
    - There are two types of methylphenidate in Singapore: Ritalin (price = $0.27 / 10 mg); Concerta (long-acting) (price = $2.88 / 20mg).
    - Methylphenidate is indicated for ADHD or ADHD with comorbid conduct disorder.
    - Beneficial effects of methylphenidate include improving on attention span and hyperactivity for a certain number of hours while in the school setting.
    - Monitor height, weight, cardiovascular status (including blood pressure) and seizure throughout treatment. For methylphenidate, the child psychiatrist needs to specifically look for tics, psychosis and anxiety symptoms.
    - The child psychiatrist can consider prescribing modified release preparations (long-acting methylphenidate or Concerta-XL) if the parents can afford. Long-acting methylphenidate allows single daily dosage and enhancing adherence while reducing stigma. (Starts with 18mg OM and titrate up to 54 mg/day).
    - Explain the common side effects associated with stimulant such as insomnia and headache. Other side effects include stomach pain, nausea, decrease in appetite and slow growth process.
    - Serious side effect such as liver impairment, leucopenia and death are rare.
    - There is a risk of misuse in adolescents with history of stimulant misuse.

2) **Atomoxetine (other non-stimulants)**
    - Pharmacodynamics: noradrenaline reuptake inhibition.
    - Indications: ADHD with tics, high risk of misuse of methylphenidate or poor response to methylphenidate.
    - Starting dose is 0.5mg/kg/day and increases the dose to 1.2mg/kg/day.
    - The child psychiatrist needs to monitor agitation, irritability, suicidal thinking and self-harming behaviour, and unusual changes in behaviour, particularly during the initial months of treatment, or after a dose change.
    - Height, weight, cardiac function, blood pressure and seizure should be monitored on regular basis.
    - Other non-stimulants include TCA such as imipramine which was widely used in the past but TCA leads to anticholinergic side effects and has weaker evidence in his efficacy against ADHD.

3) **Parent-training/ education programmes**
    - Refer parents to educational programmes to learn about ADHD, its management and coping strategy.
    - For parents, the programme should include individual or group-based parent-training/education programmes.

- For children and youth with ADHD, the programme should include CBT or social skill training.
- Offer training to the teachers on behavioural interventions in the classroom to help the child to cope with ADHD.

4) Behaviour therapy
- **Positive reinforcement** of positive behaviour including reward system and praises. The parents can consider using a star chart to promote positive behaviour at home.
- **Environmental modifications** aim at improving attention e.g. placing the child in the front row of class, minimising distractions etc.
- **Combination of behaviour therapy and medication is better than medication alone.**

Prognosis
- Many patients do not require medications when they get older.
- Although symptoms of hyperactivity and impulsivity often improve as the child grows older, inattentive symptoms are likely to persist (50% persist).
- 25% of ADHD children still have symptoms at the age of 30.
- It is appropriate to continue treatment in adults whose symptoms remain disabling.
- 20% of ADHD children ultimately develop antisocial personality disorder. 15% develop substance misuse.

Conduct disorder (CD) and opposition defiant disorder (ODD)

Epidemiology and aetiology

1. CD was diagnosed in 4% of children in the Isle of Wight study (UK). The prevalence is higher in socially deprived inner city areas and large families. M:F = 3:1. CD begins in middle childhood.
2. 5-10% of children suffer from problems with a mixture of ODD and CD symptoms.
3. Genetic factors: CD clusters in families as a result of shared environment or inheritance of antisocial trait from parents with criminal behaviours.
4. Biological factors: low plasma 5-HT level, testosterone excess (↑non-aggressive CD symptoms), low cholesterol and low skin tolerance are associated with CD.
5. Psychological factors: Fearlessness theory states that the children with CD exhibit a lack of anxiety and fear. Stimulation-seeking theory states that children with CD often have low arousal level and need to engage in antisocial behaviour to increase arousal levels. CD is associated with reading difficulty.
6. Social factors: Uncaring school, parental psychiatric disorder, parental criminality, hostility towards the child, avoidance of parental demands and getting more parental attention by antisocial behaviour are important causes. Family factors associated with CD include single parent (death of same gender parent), parental psychopathology and overcrowding (> 4 children) environment.
7. In the UK, the peak age of offending is between 14-17 years and the age of criminal responsibility is 10 years. Few juvenile offenders have an ICD-10 diagnosis. The most common type of offence is property offences and violent crimes accounting for 5% of the total offence. 5% of primary school children steal at least once. Comfort stealing occurs in socialised children who feel unloved by their parents but not associated with antisocial behaviour. In adolescence, the goal of stealing is to obtain money for personal pursuits.

Table 1 Compare and contrast the ICD -10 criteria for conduct and opposition defiant disorders

	CD	ODD
General criteria	There is a repetitive and persistent pattern of behaviour in which either the basic rights of the others or major age appropriate societal rules are violated. It lasts for at least 6 months.	The general criteria for CD must be met.
Individual symptoms	**The child:** displaying severe temper tantrums, being angry and spiteful, often telling lies and breaking promises. **To adults:** frequent argument, refusing adults' requests or defying rules and staying out after dark against parental prohibition (onset < 13 year old).	At least 4 symptoms from the left and must have been present for 6 months. Focus more on temper tantrums, being angry and spiteful, argument with adults, defying rules and blaming the others. Should not have more than 2 symptoms in

	To others: annoying them deliberately, blaming them for his mistakes, initiating fights with the others, using weapons to harm the others, exhibiting physical cruelty (also to animals), confronting victims during a crime, forcing another person into sexual activity and frequently bulling the others. **To objects or properties**: deliberately destroy them, set fire, steal objects of value within home or outside and breaks into someone's house. **Running away**: from school (truant < 13-year-old) and parental or parental surrogate home (at least twice). Underlined criteria only need to occur once for the diagnosis of CD to be made. ICD – 10 classifies CD into mild, moderate and severe. Substance abuse is **not** a diagnostic criteria for CD.	physical assault, damage of property, running away from school and home.
Types of CD	F90.0 CD confined to the family context. **F91.1 Unsocialized CD:** (poor relationships with the individual's peer group, as evidenced by isolation, rejection, unpopularity and lack of lasting reciprocal relationship). **F91.2 Socialized CD:** normal peer relationship. **F92.0: Depressive CD:** both criteria of CD and mood disorders are met. **F92.8: Other mixed disorders of conduct and emotions**. Criteria of CD and one of the neurotic, stress related, somatoform, disorders or childhood emotional disorders are met. **F93: Mixed disorders and emotions, unspecified.**	No sub-category. Although ICD – 10 does not specific the age, ODD patients are usually less than 10-year-old with onset at the age 3-8 years and a duration of 6 months. Defiant behaviours usually occur at home with familiar people.

Conduct Disorder – under the section on Disruptive, Impulse-control and Conduct disorders in DSM V
The DSM-5 diagnostic criteria states that there must be repeated and persistent pattern of behaviour in which the basic rights of others or societal norms are violated.
This could be manifested by the presence of at least 3 of the following, over a period of 12 months, with at least one criteria being fulfilled in the past 6 months.
a. Aggression to People and Animals as manifested by: Bullying, intimidating others, or initiating physical fights, or having used a weapon to cause serious physical harm to others, or have been physically cruel to people or animals, or has stolen while confronting a victim, or has forced a victim into sexual activity.
b. Destruction of Property as manifested by: Engaging in fire setting with the intention of causing serious damage, or has intentionally destroyed other's property (by other means apart from setting fire)
c. Deceitfulness or theft as manifested by: Having broken into someone's home, or have had lie to obtain goods or favors to avoid obligations or have had stolen items of value without confronting a victim
d. Serious violations of rules as manifested by: Often staying out at night despite parental prohibitions, or has had ran away from home overnight at least twice or has had played truant from school.
Clinicians needs to be aware that for individuals who are 18 years and older, the criteria must not be met for antisocial personality disorder.

Opposition Defiant Disorder– under the section on Disruptive, Impulse-control and Conduct disorders in DSM V
The DSM-5 diagnostic criteria specified that there must be a pattern of irritable mood, argumentative behaviour and vindictiveness for a duration of at least 6 months, accompanied by at least 4 of the following symptoms:
- Irritable Mood: Loses temper easily, or being annoyed easily, or always angry and resentful
- Argumentative behaviour: Often gets into trouble with authority figures, or often defiant towards requests made by authority figures, or often does things to annoy others, or often blames others for his or her mistakes or misbehaviour.
- Vindictiveness: Been spiteful or vindictive at least twice within the past 6 months.

OSCE station

Name: Charles	Age: 14-year-old

Charles presents with a 5-year history of increasingly significant infringements of the law, truancy, repeated fighting and expulsion from two boarding schools. He is the younger of the two children with a 17-year-old sister, who is a high achiever. Both parents have a university education.

Task: Take a history from his father to establish diagnosis of conduct disorder.

OSCE grid

A) Gathering more information on the referral	A1) The source of referral: e.g. Via the GP, legal authorities and school authorities etc.	A2) Timing of the referral: Why being referred now given Charles has a 5-year history of behavioural disturbance?	A3) Onset of the problem Whether the behaviour had begun suddenly 4 years ago (as a result of particular precipitant, e.g. parental, marital disharmony, changes in the family, history of abuse or rejection)	A4) Identify long pre-existing history of behaviours Adverse family environment and interaction style such as vague instructions, harsh, physical punishment with inconsistent discipline.
B) Assessment of behavioural disturbances	B1) Form of disturbed behaviour Students should follow Table 1 to inquire about the symptoms of CD. Other disturbed behaviours such as fire-setting, bed-wetting, school refusal and their time of onset in relationship to CD.	B2) The severity of behaviour and types of CD: Students should gather enough information to classify his CD as mild, moderate or severe and classify his CD as socialised or unsocialised; child or adolescent onset.	B3) The effects of his behaviour on his family and the others Explore Charles's relationship with his elder sister. Look for sibling rivalry.	B4) Reaction from parents How parents have previously dealt with Charles's behaviours and what methods had been effective or ineffective? (Common reactions include nagging, more harsh punishment, coercive interactions or lack of insight of Charles's CD) Look for maintaining factors for Charles' problems.
C) Assessment of psychiatric symptoms	C1) Other child psychiatric problems: e.g. ADHD, past history of school refusal etc.	C2) Emotional state whether Charles is depressed, with biological and cognitive symptoms of depression. Evidence for self harm (Has the father noticed any laceration wound?) and suicidal ideation.	C3) Psychotic features: Whether Charles currently or previously experiencing auditory or visual hallucination. Explore any form of delusion.	C4) History of substance abuse e.g. cannabis, solvent and alcohol misuse. Explore smoking habit.
D) Problems	D1) Why he had	D2) Academic	D3) Accommodation	D4) Legal problems

with school, accommodation and the legal system	been expelled from schools? (Explore both public and private schools) Explore school environment (poor organisation or being unfriendly to Charles?) Was he referred to see a counsellor or educational psychologist? Is he in any school at this moment?	performance Does Charles have any reading, spelling or mathematic problems? Ask the father to comment on his results in primary and secondary school? Are there any differences?	Whether he is now living at home or elsewhere (e.g. with uncle because his father cannot discipline him).	Ask the parent to take you through his current and past involvement with the legal system.
E) Background history	E1) Developmental history: eliciting a history of perinatal difficulties, developmental delay, low IQ, hearing impairment, early behavioural problems, difficult temperament, history of poor co-ordination and motor skills.	E2) Family history of psychiatric illness Particularly anti-social personality traits and forensic record in the parents, a history of parental alcoholism, parental psychiatric illness or parents who themselves suffered from conduct disorders in the past.	E3) Parenting, marriage and expectation: Identify occupation of parents. Are the parents working for long hours with little time for Charles? Are there any marital problems and parental discord or any disagreement over parenting? Explore parental expectation on Charles and in particular to the level of parental education and the fact that his older daughter is a high achiever.	E4) Past medical history: Has any doctor commented on abnormality in his nervous system or any atypical facial appearance (such as 'neurological soft signs' or syndromes associated with mental retardation) Does Charles suffer from any chronic medical illness? (e.g. epilepsy)

Closure of the interview: Thanks the father and invite him to ask further questions. Emphasise the importance of assessing Charles alone and Charles together with the family including his elder sister. You will need to obtain further information from the school or other sources, e.g. the GP.

Management for CD and ODD

1. **Setting of the management**: e.g. inpatient (sometimes for CD) or outpatient (CD and ODD). Identify the indications for hospitalisation (e.g. detoxification if the adolescent is dependent on substances or present with risk to harm the others).
2. **Safety issues:** To evaluate potential risk to the child or others (e.g. parents, siblings)
3. **Parent management training** is the most evidence-based treatment by getting parents to pay attention to desired behaviour rather than being caught up in lengthy confrontation with the child. Positive aspects of parent-child relationships are promoted and parents are also taught effective techniques for handling undesired behaviour (e.g. by video demonstrations). Parents are advised to join local support group.
4. **Management strategies for the family**: couple therapy can address marital discord. Family therapy can target at specific pattern of family dysfunction, to encourage consistency in parenting style and to develop appropriate disciplinary strategies with limit setting.
5. **Liaison with school and educational authorities**: to address specific academic difficulty such as reading and encourage the child to return to school early if he or she is a truant.

Prognosis
- 33% of children suffering CD develop antisocial symptoms which persist into adulthood and become antisocial personality disorder.

School refusal and truancy

Table 2 Compare and contrast school refusal and truancy

	School refusal	Truancy
Family history of psychiatric disorders	Anxiety disorders and failure of parents to separate from own families of origin.	Antisocial personality disorder or forensic history
Family size	Small family and being the youngest child	Large family size
Parenting style	Over-protective parenting or unassertive parents (ineffective father, over anxious mother)	Inconsistent discipline
Age of child and aetiology	**3 peaks**: age 5 (manifestation of separation anxiety at school entry), age 11 (triggering by transfer to secondary school or avoidance character) and age 14-16 (manifestation of depression or phobia such as agoraphobia or social phobia). For children under 11-years, it is often triggered by illness or death of family member. For children over 11-years, the child may have avoidance character with long-standing tendency to withdraw in challenging environment with poor self esteem. They are unassertive outside family with poor give-and-take in peer relationships.	More common in adolescents than younger children.
Symptoms	Overt anxiety at the time of going to school with "somatic disguise" such as abdominal pain	Not associated with psychiatric symptoms but wilful intention to skip classes.
Location when absent from school	Usually at home with parental permission. Parent are aware of their whereabouts.	Usually outside home, engage in alternative activities without parental permission and awareness.
Academic performance	Satisfactory academic performance	Poor academic performance

Academic difficulties

The following abilities are 2 standard deviations (SD) below the lower limit of child's age and 1 SD below the non-verbal IQ based on the ICD-10 criteria:

- Specific speech articulation disorder
- Expressive language disorder
- Receptive language disorder
- Specific reading disorder
- Specific spelling disorder
- Specific disorder of arithmetical skills
- Specific developmental disorder of motor function.

Specific Learning Disorder
The DSM-5 defined individuals as having specific learning disorder if they have marked difficulties with learning and utilizing skills learnt, even with remediation and interventions and have had persisted for a total duration of at least 6

months.

Specific learning disorder might include difficulties with regards to a. spelling, b. written expression c. comprehending meaning of what is read d. word reading e. mastering numerics and calculation and mathematical reasoning.

Clinicians must be aware that this diagnosis cannot be formulated if the individual has intellectual disabilities. In addition, the learning and utilizing skills must be significant below the average expected based on an individual's chronological age

Elective mutism

1. 0.1%, peak age: 6-10 years, 50% have minor speech problems.
2. Language expression and comprehension is normal but there is consistent failure to speak in social situations when the child is expected to speak.
3. Longer than 4 weeks in duration.
4. Aetiology: overprotective mother, distant father or trauma. Not associated with social adversity.
5. Treatment: psychotherapy.
6. Poor prognosis if duration is longer than 12 months. 50% improve in 5 years.

Gilles de la Tourette's syndrome (GTS)

Age of onset: 5-8 years; M:F = 3:1; 10 -25% of children manifest simple tics but Tourette's Disorder is rare (0.05%).

Aetiology:
1. Family loading with OCD
2. Neurotransmitter dysregulation (dopamine excess, ↑D_2 receptor sensitivity and noradrenaline may plays a role)
3. Neuroanatomical area involved: basal ganglia
4. PANDAS (Paediatric Autoimmune Neuropsychiatric Disorders Associated with Streptococcal infections)
5. Provoked by stimulants e.g. methylphenidate.

Clinical features:

1. Multiple motor tics and at least one vocal tic in the past.
2. Tics are sudden, rapid and involuntary movements of circumscribed muscles without any purpose.
3. Vocal tics include grunting, snarling, coprolalia (involuntary utterance of obscene words, vulgarities) and echolalia (repeat speech of examiner).
4. Absence of neurological disorder and absence during sleep.
5. Fluctuating course.
6. Two-thirds of patients have EEG abnormalities.

ICD-10 criteria: Tics occur many times per day, most days for over 12 months without remission > 2 months. Onset is before 18 years of age.

Tic Disorder
The DSM-5 has sub-classified Tic Disorders into 3 main subtypes, which includes:
a. Tourette's disorder
b. Persistent (chronic) motor or tic disorder
c. Provisional Tic disorder

Tourette's disorder
The definition of a tic is that of a sudden and repetitive motor movement or vocalization.
For Tourette's disorder, the individual must have experienced tics that have lasted for at least 1 year since the initial onset. Individuals usually have either multiple motor or vocal tics. DSM-5 has specified that the disorder should have its onset prior to the age of 18, and must not be due to an underlying substance usage or medical condition.

Persistent (chronic) Motor or Vocal tic disorder
Similarly, the onset is before the age of 18 years old, but the difference is that individuals usually present with either single or multiple motor or vocal tics that have lasted for at least 1 year in duration.

Provisional Tic disorder
DSM-5 specified that for provisional tic disorder, the individual has experienced either single or multiple motor and/or vocal tics, but the duration of the symptoms has not met the 1 year criteria.

Associated features: obsessions, compulsions, learning difficulties, impulsivity, problems with attention and emotional disturbance.

DDX:

1. **Transient tic disorder** (10-20% of children, onset < 18 years, involving blinks, frowns, grimaces, head flicks, grunts, throat clearing, sniffs from 4 weeks to 1 year in duration with good prognosis).
2. **Chronic motor or vocal tic disorder** (rare, onset < 18 years, duration> 1 year with remission < 2 months), associated with neurological disorders and pervasive development disorders.

Management

General management strategies: Assessment of the biopsychosocial aspects of GTS with consideration of the developmental stage of the patient and impact on family.

Pharmacological treatment:
Risperidone and sulpiride are effective and better tolerated than haloperidol.

Psychological treatment: massed practice involves repeating tics as much as possible as an attempt to reduce them. Habit reversal involves performing simultaneous incompatible movements to reduce unwanted movements.

Social treatment: Remedial academic help, behavioural modification, family therapy, self-esteem building and social skill training.

Nonorganic enuresis

The DSM-5 criteria states that there must be:
1. Repetitive involuntary or intentional voiding of urine into bed or clothes
2. A frequency of at least 2 times a week for the past 3 months
This behaviour has resulted in impairment in terms of functioning.
The DSM-5 states that the diagnosis can be made for children of at least 5 years of age.

Prevalence: 10% at age 5, 5% at age 10, 1% at age 18. M:F = 2:1; 25% with psychiatric disturbance.

Primary enuresis (lifelong bedwetting): Family history, small bladder, large family size, social adversity, low IQ and institutional upbringing.

Secondary enuresis: free of bedwetting for 6 months but having incontinence due to stress, urinary tract infection, diabetes and chronic renal failure. Also classified as diurnal or nocturnal enuresis.

Age and frequency: Onset > 5 years. 5-7 year old (bedwetting at least > 2 times / month), > 7 year old (at least 1 time / month)

Duration: at least 3 months. ICD 10 criteria exclude organic causes.

Investigations: PR examination, intravenous urogram (IVU), age of onset >15 years: requiring urodynamic study

Management:
1. Fluid restriction at night
2. **Star chart**: effective in one-third of cases
3. **Alarm**: Child must wake up and urinate. It takes 8 weeks to produce dryness, effective in 70-90% of cases. One-third will relapse and need second alarm trial.
4. **Medication**: Desmopressin, ADH analogue given nasally and TCA (anticholinergic effect cause urinary retention).

Nonorganic encopresis
The DSM-5 criteria states that there must be:
1. Repetitive passage of faeces into inappropriate places, whether involuntary or intentional
2. A frequency of at least once each month for the past 3 months
The DSM-5 states that the diagnosis can be made for children of at least 4 years of age.

Epidemiology: 1% in school children; more common in boys; stronger association with psychiatric disorders than enuresis.

Aetiology: Retentive encopresis as a result anxiety (fear of soiling) or anger (protest against parents), constipation, Hirschsprung's disease, poor toilet training (continuous encopresis) and pervasive development disorder.

Onset of onset: at least > 4 years. (bowel control is acquired before bladder control); frequency is at least 1 episode per month for 6 months.

Investigations: barium enema, thyroid function test to look for hypothyroidism.

Management: Explain the soling process to the parents and the role of rectal loading, regular defaecation by oral laxatives and star charts.

Other childhood psychiatric disorders

Pica

- Persistent eating of non-nutritive substances for at least twice per week for a total duration of 1 month.
- Chronological and mental age > 2 years.
- Associated with mental retardation, psychosis and social deprivation.

Sleep disorders in children and adolescents

1. 20-30% of children and adolescents in the general population have sleep problems that are of concern to their parents or GPs. 50% of children with either sleep walking or night terror have family history.
2. About 1 in 6 pre-school children have difficulty to sleep at night.
3. Typical day and night cycle occurs at 1-year-old. Full term new born sleeps 16 hours per day and drops to 13 hours in 6 months. By 6 months, 80% of babies sleep right through the night and 10% of children at age 1 are still wakening every night. Transitional objects may help the child to settle at night. A 2-year-old child sleeps 13 hours per day. A 5-year-old child sleeps 11 hours per day. A 9-year-old sleeps 10 hours per day.

Figure 2 Summary of sleep disorders in children and adolescents

```
                              Sleep disorders
    ┌────────────┬─────────────┬─────────┬──────────────┬──────────────┐
```

Excessive daytime sleepiness	Dyssomnias:	Night terror:	Somnambulism (Sleep walking):	Nightmares:
Causes: asthma, sleep apnoea (2%) narcolepsy (1 in 10000) and Kleine – Levin syndrome	Associated with maternal depression Treatment is sleep hygiene (avoiding daytime naps, reducing stimulation prior to sleep, consistent bed time) and consider melatonin treatment.	In 6% of children, peak age: 1.5 to 6 years. Occurs in stage 4 of sleep (NREM sleep or at the early part of the night), the child wakes up terrified with a 'bloodcurdling' scream which typically terrifies parents and family. Does not have any recollection on awakening. Main intervention is reassurance to the parents. Help the child deal with any life stresses, and ensure regular sleep routines. Waking up a child during a night terror usually only worsens agitation. Pre-emptively waking the child has been described but there is little evidence to support this.	10-15% of children, peak age: 4-8 years. The child gets out of bed, walks while asleep with no recollection upon awakening. Treatment is to ensure safety and lock doors and windows. Usually resolves on its own.	25-50% of children, peak age: 3-6 years, in REM sleep or second half of the night. The child can recall the nightmare.

Separation anxiety disorders of childhood

Onset is before 6-year-old and duration is at least 4 weeks.

Table 3 Summary of the ICD-10 criteria for separation anxiety disorder

Unrealistic persistent worry	Symptoms in the day	Symptoms at night
Possible harm befalling major attachment figures or about loss of such figures. The child will have anticipatory anxiety of separation (e.g. tantrums, persistent reluctance to leave home, excessive need to talk with parents and desire to return home when going out)	Persistent reluctance or refusal to go to school because of the fear over separation from a major attachment figure in order to stay at home. Repeated occurrence of physical symptoms (e.g. nausea, stomach-ache, headache and vomiting).	Difficulty in separating at night as manifested by persistent reluctance or refusal to go to sleep without being near to the attachment figure. The child also has repeated nightmares on the theme of separation.

Prognosis: there seems to be a link between separation anxiety in childhood and panic disorder in adult.

Sibling rivalry disorder

The child has abnormally intense negative feelings towards an immediately younger sibling. Emotional disturbance is shown by regression, tantrums, dysphoria, sleep difficulties, oppositional behaviour, attention seeking behaviour with one or both parents. Onset is within 6 months of the birth of an immediately younger sibling and duration of disorder is 4 weeks.

Bullying

Bullying is common in schools. With the advance in information technology, cyber-bullying is common nowadays. The main treatment is counselling which allows the victim to work toward a solution within the school environment, to be supported by friends and parents and to stay away from difficult situations.

Schizophrenia with adolescent onset

Epidemiology: 5% of adult experiencing their first episode of psychosis before the age of 15.

Very early onset schizophrenia = 13-year-old. (0.9/100,000). Early onset schizophrenia = 18-year-old (17.6/100,000). Onset is commonest after 15 years and prognosis is worse with younger onset.

Causes: Genetic predisposition, neurodevelopmental hypothesis with reduction in cortical volumes and increase in ventricular size, drug induced psychosis and organic causes (e.g. CNS infection in adolescents)

Salient features: Visual and auditory hallucinations are common and 80% show exaggerated anxiety when admitted to the psychiatric ward. Look for mood disturbance, persecutory ideation, abnormal perceptual experience, cognitive and social impairment.

Compared to the adults: passivity, well formed delusions, thought disorders, first rank symptoms are less common in young people.

DDX: Conduct disorder, severe emotional disorder, encephalitis

Treatment: Treatment of choice is second generation antipsychotics (e.g. risperidone). Close liaison with paediatrician is necessary if the psychosis is resulted from an organic cause.

Adolescent depression

Epidemiology: Similar to adult. Prevalence is 8%, M:F = 1:4, mean duration of illness is 7-9 months.

Aetiology: 60-70% are caused adverse life events, arguments with parents and multiple family disadvantages. Positive family history of depression is common

Clinical symptoms: Usually similar to adult but promiscuity may be the presenting feature.

Compared to the adults: young people are less likely to have psychomotor retardation.

Management (NICE guidelines, UK):

Mild depression: Watchful waiting for 2 weeks. Offer one of the following for 2-3 months: supportive psychotherapy, CBT, guided self help. No antidepressant is required.

Moderate to severe depression (including psychotic depression): individual CBT and short term family therapy are the first-line treatment. If the adolescent does not respond, consider alternative psychotherapy or add fluoxetine after multidisciplinary review. If still unresponsive after combination of treatment, arrange another MDT review and consider either systemic family therapy (at least 15 fortnightly sessions) or individual child psychotherapy (30 weekly sessions)

Use of antidepressants: Fluoxetine 10mg/day is the first line treatment for young people who need antidepressant.

Escitalopram and sertraline can be used if there is clear evidence of failure with fluoxetine and psychotherapy. Antidepressants should be discontinued slowly over 6-12 weeks to reduce discontinuation symptoms.

Prognosis: High remission rate: 90% by 2years and high recurrence rate: 40% by 2 years.
Development of hypomania after taking antidepressant is an important predictor for bipolar disorder in the future.

Suicide and deliberate self harm

Epidemiology:

UK: 20,000 young people in England and Wales are referred to hospital for assessment of DSH each year. M:F in DSH = 1:6. Suicide is common between 14-16 year old. M:F in suicide = 4:1. Suicide is the third commonest cause of death after accident and homicide.

Singapore: Loh et al (2012) found that the suicide rate for adolescent was 5.7 per 100,000, with gender ratio of 1:1 and higher rates among ethnic Indians.

Aetiology and risk factors for suicide: Psychiatric disorder (e.g. depression, psychosis, substance abuse, CD), isolation, low self-esteem and physical illness). Family issues like loss of parent in childhood, family dysfunction, abuse and neglect.

Singapore: Loh et al (2012) reported that mental health service use was associated with unemployment, previous suicide attempts, family history of suicide, more use of lethal methods, lack of identifiable stressor, and less likely to leave suicide notes in adolescents who committed suicide.

Increase in adolescence suicide is a result of:

1. Factors influencing reporting ("copy-cat" suicides resulting from media coverage, the fostering of illusions and ideals through internet suicide groups and pop culture)

2. Factors influencing incidence of psychiatric problems (Problems with identify formation, depression, substance abuse and teenage pregnancy)

3. Social factors Bullying, the impact of unemployment for older adolescents, poverty, loosening of family structures, living away from home, migration, parental separation and divorce.

Methods used

UK:

DSH: Cutting and scratching are common impulsive gestures. Cutting often has a dysphoric reducing effect.
Suicide: self-poisoning.

Singapore:
Loh et al (2012) reported that psychosocial stressors and suicide by jumping from height were common in suicide victims.

Management of DSH (NICE guidelines, UK)

e.g. self laceration: Offer physical treatment with adequate anaesthesia. Do not delay psychosocial assessment and explain care process. For those who repeatedly self poison, do not offer minimisation advice on self poisoning as there is no safe limit for overdose. For those who self – injure repeatedly, teach self management strategies on superficial injuries, harm minimisation techniques and alternative coping strategies.

Management of suicidal adolescent: Consider inpatient treatment after balancing the benefits against loss of personal freedom. The doctor should involve the young person in the admission process. ECT may be used in adolescents with very severe depression and suicidal behaviour which is not responding to other treatments.

Prognosis: DSH: 10% will repeat in 1 year (higher risk of repetition in older males). 4% of girls and 11% of boys will kill themselves in 5 years after first episode of failed suicide attempt.

Anxiety disorders in childhood

Epidemiology: The prevalence for separation anxiety disorder is 3.6%. At least 50% of adult cases of anxiety symptoms have their onset in childhood. 2% of children have phobia.

Aetiology: In infancy, fear and anxiety are provoked by sensory stimuli. In early childhood, fear is evoked by stranger and separation anxiety. In late childhood, anxiety is caused by fear of dark, animals (more common in girls) and imaginary creatures.

Phobic anxiety disorder of childhood
The individual manifests a persistent or recurrent fear that is developmentally appropriate but causes significant social impairment. Duration of symptoms is at least 4 weeks based on the ICD-10 criteria.

Separation anxiety disorder

The symptoms of separation anxiety disorder include:

- Significant distress and worry at the time of separation from the caregiver (e.g. the time when the child goes to school or the caregiver needs to leave home for work).
- Clinging to the caregiver, crying, displaying temper tantrums, and refusing to participate in activities (e.g. refuse to enter the kindergarten and join classroom activities).
- Fear of harm to the caregiver when caregiver goes to work.
- Difficulties in sleeping at night without the presence of caregiver.
- Development of somatic symptoms such as gastric discomfort, nausea, or headaches.

Separation anxiety disorder is associated with school refusal. Although this disorder is more common in children, it may occur in adolescents.

Social anxiety disorder of childhood
Persistent anxiety in social situations where the child is exposed to unfamiliar people including peers. The child exhibits self-consciousness, embarrassment and over-concern about the appropriateness of his or her behaviour. The child has satisfying social relationship with familiar figure but there is significant interference with peer relationships. Onset of the disorder coincides with the developmental phase and duration of symptoms is at least 4 weeks based on the ICD-10 criteria.

Childhood emotional disorder
Prevalence: 2.5%; more common in girls; the presence of anxiety and somatic complaints indicate good prognosis.

Salient features for young children with anxiety disorders
In general, sleeping difficulties and headache are common but panic attacks are less common. Behaviour therapy is the mainstay of treatment.

Childhood onset OCD

Epidemiology: boys have earlier onset than girls and it is a rare condition.

Aetiology: Genetic cause for OCD (5% of parents have OCD), Paediatric Autoimmune Neuropsychiatric Disorders Associated with Streptococcal infections (PANDAS) is caused by haemolytic streptococcal infection.

Salient features: insidious onset, fear of contamination, checking rituals, worry about harm to self. Doctors need to bear in mind that mild obsessions and rituals (e.g. magical number of repetition to get good exam result) is part of the normal development.

Comorbidity: Tourette's syndrome, ADHD, anxiety, depression, bedwetting, sleep disturbance, motor changes and joint pain.

Investigations for PANDAS: Streptococcus titre can be used. The anti-streptolysin O (ASO) titre, which rises 3-6 weeks after a streptococcus Infection, and the anti streptococcal DNAase B (AntiDNAse-B) titre, which rises 6-8 weeks after a streptococcus Infection.

Treatment: CBT is the first-line treatment. Sertraline has the licence to treat OCD from the age of 6 years and fluvoxamine from the age of 8 years. Fluoxetine is indicated for OCD with significant comorbid depression. Avoid TCA (except clomipramine), SNRIs, MAOIs and prescribing only antipsychotics without antidepressants. There is a delayed onset of action up to 4 weeks and full therapeutic effect up to 8-12 weeks. Duration of treatment is at least for 6 months. In childhood body dysmorphic disorder, fluoxetine is the drug of choice.

PTSD and adolescents

Trauma: Occur in young people exposing to traumatic event e.g. physical or sexual abuse, fatal accident happens to friends or relatives.

Salient clinical features:

Younger patients: compulsive repetitive play representing part of the trauma, failing to relieve anxiety and loss of acquired developmental skills in language and toilet training and emergency of new separation anxiety.

Adolescents: depression is common in older children and adolescents and new aggression may emerge.

Other adult symptoms such as nightmares, social withdrawal, numbing are common in young people.

Treatment: CBT focusing on psychoeducation and anxiety management is the main treatment. SSRIs may be indicated for severe anxiety and depression.

Somatisation in childhood

Epidemiology: Recurrent abdominal pain occurs in 10-25% of children between the age of 3 to 9 years. Hysteria in childhood is rare with equal gender ratio.

Aetiology: Stress (bullying) and anxiety can initiate and amplify somatic symptoms. Somatic symptoms are commonly associated with school refusal. Risk factors in the child include obsessional, sensitive, insecure or anxious personality. Risk factors in the family include over-involved parent, parental disharmony and overprotection, rigid rules and communication problems.

Salient features: Recurrent localised abdominal pain lasting for few hours is commonly associated with emotional disorder. For hysteria, the child may present with disorders of gait or loss of limb function. Secondary gain is often implicated.

Treatment: Reassurance, psychoeducation and behavioural therapy.

Childhood eating disorders

Epidemiology: One-third of British children have mild to moderate eating difficulties by the age of 5 years.

Risk factors: Male gender, low birth weight, developmental delay, early onset of the feeding problem, history of vomiting for long duration and higher social class.

Classification of childhood onset eating disorders:
- **Childhood-onset anorexia nervosa** : weight loss, abnormal cognitions on weight and shape

- **Childhood-onset bulimia nervosa**: recurrent binges, lack of control, abnormal cognition
- **Food avoidance emotional disorder**: weight loss, mood disturbance and no anorexia nervosa features
- **Selective eating:** narrow range of food and unwillingly to try new food
- **Restrictive eating**: smaller amount than expected, diet normal in terms of nutritional content
- **Food refusal:** episodic, intermittent or situational
- **Functional dysphagia**: Fear of swallowing, choking and vomiting
- **Pervasive refusal syndrome**: refusal to eat, drink, walk, talk or self care

Treatment: Behavioural therapy can enhance the motivation to eat, reduce behaviours which expel food and time out for negative behaviours. Social skill training is indicated for the child. Family therapy is useful for younger patients with shorter duration of illness.

Prognosis for childhood onset anorexia nervosa: two-thirds of children make good recovery.

Substance abuse in adolescents

3 types of abuse: Experimental (initial use due to curiosity), recreational (continuous) and dependent (strong compulsion to take the drugs)

Aetiology:

1. Environmental factors: widespread drug availability, high crime rate, poverty, cultural acceptance of drugs
2. School factors: peer rejection and school failure
3. Family factors: parental substance abuse and family conflict
4. Individual risk factors: low self esteem, high sensation seeking and self destruction.

Smoking: two-thirds of 15-16-year-old adolescents smoke.

Alcohol abuse in adolescence (based on UK findings):
Equal sex incidence; 50% of 16 to 19-year-olds are regular drinkers; 50% of adolescents taste the first alcohol at home. Alcohol dependence is associated with adolescent suicide.

Volatile substances: (solvent or glue) 21% have tried and this is a common method used by Singaporean adolescents. The effects include euphoria, disinhibition, impulsiveness, giddiness, nausea, vomiting, slurred speech, visual hallucination and paranoid delusion. Secondary school students like this method as they can be intoxicated in school. Chronic use leads to tolerance and withdrawal symptoms.

Cannabis: Most adolescents in the UK, US and Australia have tried cannabis and most will stop using cannabis in their 20s. In Singapore, there were university students who took cannabis during their electives overseas and developed psychosis after returning to Singapore. About 1 in 10 students use it on a daily basis. It may precipitate acute schizophrenia.

General treatment involves referring to the substance misuse service. It should be a user friendly service to promote self referral and enhance collaboration with school and other agencies. Strategies include harm reduction, motivational enhancement and family therapy.

Abuse

Around 4% of children up to the age of 12 are brought to the attention of professionals because of suspected abuse. The following are common contexts where the assessment of abuse takes place.

1. Suspicion of abuse arising in the course of assessment or reported directly by the child pending further confirmation (e.g. Abuse was suspected because the child had said something, changes occur in the child's behaviour and the presence of unexplained physical signs)
2. If abuse has already been confirmed, the doctor needs to assess the child to determine the effects of abuse, safety issues and recommend further treatment.
3. Allegations of abuse may arise for the first time during ongoing therapy of a child or an adolescent.

Physical abuse may present with physical injuries or showing psychological consequences (like unhappiness, wariness, watchful excessively inhibited, or aggressive, and provocative and poor academic achievement.

Emotional abuse may involve persistent negative attitudes, use of guilt and fear as disciplinary practices, ignorance or exploitation of the child's immature developmental status.

Sexual abuse: most cases arise within the family. Girls are more frequently abused than boys. The overall rate is 10% for both genders combined and this figure includes marginal sexual abuse. The prevalence of cases involving physical / genital contact is around 1%. 20-40% of abused children also have physical evidence of abuse. Any disclosure of childhood sexual abuse should be taken seriously. Obligation to protect child overrides medical confidentiality and doctor can initiate police / social service referral. In terms of prognosis, poor self esteem is common. The victim may have difficulty with future sexual relationships. The person may become a perpetrator for future sexual abuse.

Singapore perspectives: Cai and Fung (2003) studied 38 cases of childhood sexual abuse in Singapore and found that most of the children were young (74% below age 9) and female (78.9%) with perpetrators who are males and usually known to the victims. Back et al (2003) compared American and Singaporean college female students who were victims of abuse and reported the following:

1. American women were more likely to report a history of child sexual abuse, and to report experiencing more severe forms of sexual abuse.
2. Women in Singapore were more likely than women in the US to report a history of child physical abuse, to report experiencing injury as a result of the abuse, and to disclose the abuse. Singaporean women with a history of child sexual abuse reported elevated psychological symptom levels relative to their non-abused peers and to US women with a history of child sexual abuse, even after controlling for exposure to other types of traumatic events.
3. No significant differences in symptomatology with regard to child physical abuse were observed between two ethnic groups.

Ethical issue in child and adolescent psychiatry

Gillick competence: A child below 16-year-old can give consent to treatment without parental agreement (e.g. contraception) provided that child has achieved sufficient maturity to understand fully the treatment proposed. The child has no right to refuse treatment that is in his or her best interests.

Intellectual Disability (Intellectual developmental disorder)
The DSM-5 Criteria states that an individual must have deficits in the following main areas, which are:
1. Skills required for intellectual functioning. Impairments in development of intellectual functioning must be confirmed by means of standardized psychometric testing
2. Skills required for independent social living, such as having the ability to handle activities of daily living independently.
The DSM-5 Criteria specified that the onset of these deficits must be within the developmental period of any individual. Intellectual disability has been further sub-classified within the DSM-5 as Profound, Severe, Moderate and Mild types. The current terminology of Intellectual disability corresponds to the ICD-10 terminology of intellectual developmental disorders.
It is crucial to take note that the classification of subtypes is based on the levels of support an individual requires with regards to independent social living, and not in accordance to the psychometric scores achieved on psychometric testing.

History:
1. The reproductive and obstetric history of mother
2. Family history of intellectual disability
3. Aetiology of intellectual disability (a syndrome, head injury, birth complications, severe neonatal infection)
4. Family's attitude and coping toward the condition
5. Previous highest level of functioning
6. Recent life events
7. Circumstances just before presenting complaint if any (e.g. temper tantrums, aggression, behaviour problems).

Techniques to interviewing people with intellectual disabilities:

1. Talk to the child first
2. Use simple and short sentences
3. Less verbal language
4. Attend to non-verbal cues
5. Use open and closed questions
6. Avoid leading questions and beware of suggestibility
7. Make comments rather than questions
8. Allow time for the patient to respond
9. Try to interview in a familiar setting if possible

Management
1. Treatment involves comprehensive multi-disciplinary assessment
2. Consent issues must be properly addressed.
3. Target symptoms/signs identified and behaviour therapy is the first-line treatment.
4. For challenging behaviours, a psychiatrist should be consulted for the possibility of initiating psychotropic medications (e.g. antipsychotics or mood stabilisers).

Down syndrome

Epidemiology

Western countries: The most common cytogenic cause of intellectual disability. Down syndrome accounts for 30% of all children with mental retardation and occurs in 1 in 800 live births (1/2500 if mothers < 30 years old and 1/80 if > 40 years old).

Singapore: Lai et al (2002) reported that the live birth prevalence of Down Syndrome in Singapore had fallen from 1.17/1000 live births in 1993 to 0.89/1000 live births in 1998 as a result of antenatal diagnosis and selective termination.

Genetic mechanism: 94% of Down's syndrome is due to meiotic non-dysjunction or trisomy 21. 5% is due to translocation of chromosome 21 and 1% is due to mosaicism. Robertsonian translocations are due to fusion of chromosome 14 and 21. The extra chromosome is usually of maternal origin (90%) and most maternal error occurs during the first meiotic division. The risk of recurrence of translocation is 10%.

IQ (Children & young adults): 40 – 45; Social skills are more advanced than intellectual skills.

Comorbidity: Alzheimer's disease (50-59-year-old = 36-40%; 60-69-year-old = 55%), Over the age of 40, there is high incidence of neurofibrillary tangles and plaques. Immune system is impaired and high risk for diabetes (due to autoimmune reasons). Hypothyroidism is common. The most common cause of death is chest infection. Common psychiatry comorbidities include OCD (especially presenting with a need for excessive order/tidiness), autism, bipolar disorder, psychosis and sleep apnoea.

Fragile X syndrome

Epidemiology: Most common inherited cause of intellectual disability. Affects 1 in 2000 females and 1 in 700 females is a carrier. Affects between 1 in 4000 and 1 in 6000 live male births.

Mode of inheritance: X-linked dominant with low penetrance; affecting both males & females; tinucleotide repeats (TNR) "CGG" on long arm of X chromosome which leads to methylation and turning off of the FMR-1 gene. Tan et al (2005) proposes that there is a need to incorporate fragile X testing in routine screening of patients with developmental delay in Singapore.

IQ: variable mild LD in affected females, moderate to severe LD in males (In 80% males, IQ<70). The length of TNRs is inversely related to IQ as shorter length (<200 triplets) may not cause methylation and hence, better IQ scores.

Clinical phenotypes: Men are more likely than women to exhibit the typical physical features (only 1 in 5 males affected by mutation at fragile site are phenotypically and intellectually unaffected).

Other behaviour: hyperactivity, self-injury and social anxiety disorder.

Revision MEQ

You are a resident in paediatrics. A 10-year-old boy is referred to you as he has been very active and disruptive in school. The school is considering suspending him because of his behaviour. Developmental milestones are known to be normal. His mother is very concerned and bought his son to the children emergency for assessment.

Q.1 State three differential diagnosis.

1. Attention deficit and hyperkinetic disorder (ADHD).
2. Conduct disorder or oppositional defiant disorder.
3. Intellectual disability and behaviour problems.

Q.2 If you suspect that he suffers from ADHD. Name 5 inattentive symptoms you will look for based on DSM-IV-TR diagnostic criteria.

1. Starts tasks or activities but not able to follow through and finish.
2. Organization of tasks or activities is impaired.
3. Loses things necessary for tasks and activities such as school assignments or stationary.
4. Instructions are not followed.
5. Distraction by external stimuli.
6. Other features include careless mistakes, forgetfulness in daily activities.

Q.3 How would you gather the feedbacks of his behaviour in school.
1. Advise the parents to pass Connor's Performance scale to the teacher to assess.
2. Send an educational psychologist to observe the child's behaviour in class with permission of school and parents.

Q.4 You are concerned about the safety of the childhood. What are the potential risk associated with ADHD in this boy?
1. Risk of accident (e.g. road traffic accident).
2. Risk of fall or head injury.

Q.5 The mother read an article which states that artificial colouring and additives cause ADHD in children. What is your advice to the mother?

1. You should emphasize the value of a balanced diet, good nutrition and regular exercise for the child.
2. There is no need to eliminate certain food from diet.
3. The parents can keep a diary if there are foods or drinks which appear to affect the child's behaviour. They can consult a dietician if necessary.

Revision MCQ

Q.1 You are the GP in an HDB estate. A mother brought her 2-year-old boy because she worries that he may suffer from autism. Which of the following the characteristics is not a criteria to refer this child for specialist assessment based on MOH guidelines?

A. He has no babble, pointing or other gestures by 12 months.
B. He cannot say a single word by 18 months.
C. He has rapid loss of language or social skills at the age of 20 months.
D. He does not have spontaneous 2-word phrases by 24 months.
E. He cannot sing in the playgroup with other children at the age of 26 months

The answer is E. Option A, B, D and E are MOH referral criteria.

2. The mother of a 14-year-old boy with ADHD consults you about the side effect of methylphenidate. Which of the following statement is false?

A. Methylphenidate almost always delays physical growth.
B. Drug holiday is required to facilitate growth.
C. Methylphenidate suppresses appetite.
D. Methylphenidate causes insomnia
E. Tolerance is very common in patients who continue methylphenidate for long term.

The answer is E. Option E is false because most patients do not develop tolerance (i.e. needs to take a higher dose of the medication to achieve the effect) of methylphenidate if the medication is used for long-term.

3 A 13-year-old boy has dropped out from secondary school. He is odd and not accepted by his classmates. Your consultant recommends to rule out Asperger's syndrome. Which of the following features does not support this diagnosis?

A. Language delay.
B. Marked clumsiness
C. Restricted and repetitive behaviours
D. Socially withdrawn
E. Worries about the welfare of his classmates

The answer is A. Asperger's syndrome is different from autism as the former condition is not associated with language delay.

EMIS:
A – Elective mutism
B – Akinetic mutism
C – Hysterical mutism

A 6 year old girl speaks very clearly and fluently with her friends at school but becomes mute when at home. This is known as – Elective mutism.

References

American Psychiatric Association (1994) *DSM-IV-TR: Diagnostic and Statistical Manual of Mental Disorders (Diagnostic & Statistical Manual of Mental Disorders)* (4th edition). Washington: American Psychiatric Association Publishing Inc.

American Psychiatric Association (2013) *DSM-5: Diagnostic and Statistical Manual of Mental Disorders (Diagnostic & Statistical Manual of Mental Disorders)* (5th edition). Washington: American Psychiatric Association Publishing Inc.

Academy of Medicine Singapore-Ministry of Health Clinical Practice Guidelines Workgroup on Autism Spectrum Disorders. Academy of Medicine Singapore-Ministry of Health clinical practice guidelines: Autism Spectrum Disorders in pre-school children. Singapore Med J. 2010 Mar;51(3):255-63.

Back SE, Jackson JL, Fitzgerald M, Shaffer A, Salstrom S, Osman MM. Child sexual and physical abuse among college students in Singapore and the United States. Child Abuse Negl. 2003 Nov;27(11):1259-75.

Baker P. (2004) *Basic Child Psychiatry 7th edition*. London: Blackwell.

Bernard-Opitz V, Kwook KW, Sapuan S. Epidemiology of autism in Singapore: findings of the first autism survey. Int J Rehabil Res. 2001 Mar;24(1):1-6.

Black D & Cottrell D (1993) *Seminars in Child and Adolescent Psychiatry*. Gaskell: London.

Cai, Y, Fung D. Child sexual abuse in Singapore with special reference to medico-legal implications: a review of 38 cases. Med Sci Law. 2003 Jul;43(3):260-6.

Chiswick, D and Cope R. (1995) *Seminars in forensic psychiatry*. London: Gaskell

Fox C & Carol J (2001) *Childhood-onset eating problems*. London: Gaskell.

Forrest R (2009) Are child and adolescent psychiatrists right or wrong to call badly behaved children and young people conduct disordered? *Child and Adolescent Faculty and Executive Newsletter of Royal College of Psychiatrists*, **36**: 8 – 18.

Fraser W and Kerr M (2003) *Seminars in the Psychiatry of Learning Disabilities* (2nd edition). Gaskell: London.

Goldbloom DS, Davine J (2010) Psychiatry in Primary Care. Centre for Addiction and Mental Health: Canada.

Golubchik P, Mozes T, Vered Y, Weizman A (2009) Platelet poor plasma serotonin level in delinquent adolescents diagnosed with conduct disorder. *Prog Neuropsychopharmacol Biol Psychiatry*, **33**:1223-5.

Goodman R and Scott S (2005) *Child psychiatry (2nd edition)*. Oxford: Blackwell Publishing.

Lai FM, Woo BH, Tan KH, Huang J, Lee ST, Yan TB, Tan BH, Chew SK, Yeo GS. Birth prevalence of Down syndrome in Singapore from 1993 to 1998. Singapore Med J. 2002 Feb;43(2):070-6.

Lim CG, Ooi YP, Fung DS, Mahendran R, Kaur A. Sleep disturbances in Singaporean children with attention deficit hyperactivity disorder. Ann Acad Med Singapore. 2008 Aug;37(8):655-61.

Loh C, Tai BC, Ng WY, Chia A, Chia BH. 2012 Suicide in young singaporeans aged 10-24 years between 2000 to 2004. Arch Suicide Res. Apr;16(2):174-82.

Molly Douglas, Helen Casey, Harriet Walker (2012) Psychiatry on the Move London: Hodder education.

McGuffin P, Owen MJ, O'Donovan MC, Thapar A, Gottesman II (1994) *Seminars in Psychiatric Genetics*. London: Gaskell.

NICE guidelines for attention deficit hyperactivity disorder http://guidance.nice.org.uk/CG72.

NICE guidelines for depression in children and young people http://guidance.nice.org.uk/CG28

NICE guidelines for self harm http://guidance.nice.org.uk/CG16.

Ooi YP, Rescorla L, Ang RP, Woo B, Fung DS. Identification of autism spectrum disorders using the Child Behavior Checklist in Singapore. J Autism Dev Disord. 2011 Sep;41(9):1147-56.

Puri BK (2000) *Psychiatry 2nd edition:* London: Saunders.

Puri BK, Hall A, Ho RC (2013) Revision Notes in Psychiatry 3rd ed. London: Hodder Arnold.

Sadock BJ, Sadock VA. (2003) *Kaplan and Sadock's Comprehensive Textbook of Psychiatry*. (9th ed.) Philadelphia (PA): Lippincott Williams and Wilkins.

Tan BS, Law HY, Zhao Y, Yoon CS, Ng IS. DNA testing for fragile X syndrome in 255 males from special schools in Singapore. Ann Acad Med Singapore. 2000 Mar;29(2):207-12.

Taylor D, Paton C, Kapur S (2009) *The Maudsley prescribing guidelines.*10th edition London: Informa healthcare.

World Health Organisation (1992) *ICD-10 : The ICD-10 Classification of Mental and Behavioural Disorders : Clinical Descriptions and Diagnostic Guidelines.* Geneva: World Health Organisation.

Chapter 17 Forensic psychiatry and local mental health legislation

Psychiatric aspects of other offences: shoplifting and fire-setting	The Mental Capacity Act
Relationship between psychiatric disorders and criminal behaviour	The Advanced Medical Directive Act
Sexual offences	Revision MEQ and MCQ
The Mental Disorder and Treatment Act	

Psychiatric aspects criminal acts: shoplifting and fire-setting

Shoplifting

Most shoplifting is a conscious and goal-directed activity without psychiatric disorder. Possible contributing factors include gain, organised criminal activity, poverty, low self-esteem, frustration, boredom and thrill seeking.

60% of shoplifting is committed by women but convictions are more numerous for men based on the UK figures. Only 2% are referred for psychiatric assessment. There are two peaks in terms of age groups: 50-60 years and teenagers.

Amongst those with psychiatric disorders, depression, acute stress reaction and adjustment disorder are common. This act may represent a "cry for help" or a wish to be punished in depressed people who have excessive guilt feelings. Shoplifters with a depressive illness make little effort to conceal their actions. Absentmindedness as a result of poor concentration in depressive illness is an acceptable defence.

Other psychiatric disorders include personality disorder (17%), acting on delusions (15%), intellectual disability (11%), dementia and delirium (5%), mania, influence from drugs (3%) and dissociative states. Organic state such as epilepsy is a recognised cause.

Distressed children often steal for self-comfort. Conduct disorder or antisocial personality traits may be associated with criminal stealing and the peak age is 15.

Kleptomania

Kleptomania, an impulse control disorder, is characterised by repeated failure to resist the impulse to steal in which tension is relieved by stealing. The gender ratio, F:M is 4:1.

Classically, the compulsion is characterised by a feeling of tension associated with a particular urge to steal, excitement during theft and relief after committing the act. At the same time, the urge is recognised as senseless and wrong and the act is followed by guilt. The stealing is not an expression of anger or a part of antisocial personality trait.

Pure kleptomania is extremely rare. Such compulsions are associated with depression, anxiety, bulimia nervosa, sexual dysfunction and fetishistic stealing (e.g. women's underwear).

Objects are not acquired for personal use (same set of T-shirts) or monetary gain and may be discarded or given away after stealing or hoarded.

Arson (fire-setting)

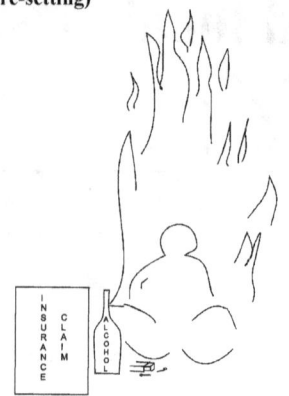

The most common motives for arson are revenge and fraud insurance claims. The other causes include anger and the need to relieve tension by fire setting.

There is a higher representation of men with intellectual disability (IQ: 70-79) because they display passive aggression and a sense of power or excitement. 20-30% of arsonists have psychiatric disorders (e.g. alcohol misuse - more common, schizophrenia - uncommon). Pyromania is a rare condition where the arsonist derives sexual satisfaction through fire-setting.

Common psychiatric disorders and crime

Schizophrenia patients show a similar rate of offending to the rest of the population but they have a minor increase in risk to commit violent crime. They are more likely to be arrested than other offenders. Schizophrenia patients usually assault a known person when they are violent. The delusional ideas often motivate violent behaviours and the patients usually admit experiencing command hallucinations during the act. Those with negative symptoms commit violent offences inadvertently and neglectfully.

If a schizophrenia patient commits murder under the influence of psychotic experiences (e.g. command hallucination) at the time of the act, he or she will be exempted from capital sentence as a result of diminished responsibility of McNaughton's rule.

Table 1 Summary of the relationship between psychiatric disorders and crimes

Psychiatric disorders	Criminal behaviour
Affective disorder	Shoplifting in middle-aged offenders may be associated with depression. Violent offending in depressive disorder is rare. Offending is more common in mania and hypomania than in depression.
Personality disorder	The term psychopathic disorder should only be used as a legal category. Personality disorders in forensic psychiatry are usually mixed in types. A wide range of personality traits such immaturity, inadequacy, hostility and aggression may contribute to offending behaviour.
Intellectual disability	Offending is more likely in mild and moderate mental retardation than in severe mental retardation. Property offences are often committed with a lack of forethought and are opportunistic. Offences are broadly similar to offenders without mental retardation although there have been reported for increases rates of sex offending and fire-setting in patients with mental retardation.
Organic state	Personality change is a frequent early feature of dementia such as frontal lobe dementia. In general, offending by dementia patients is uncommon. The most common offence is theft. Antisocial behaviour may appear before any sign of neurological or psychiatric disturbance in Huntington's disease. For patients with epilepsy, the rate and type of offending is similar to those of offenders in general. There is no excessive of violent crimes in epileptic prisoners. The increase of prevalence of epilepsy in prisoners (2 times of the general population) is a result of common social and biological adversity leading to both epilepsy and crime.
Substance abuse	Alcohol misuse is commonly seen in the more than 50% of perpetrators and victims of violence and rape. Substance misuse is a commonly associated finding in offenders with antisocial personality disorder.

Sexual offences

Epidemiology

1. Exhibitionism is one of the most common sexual offences committed by men and it is a summary offence (minor offence). It is a form.
2. 50% of exhibitionists appearing in court have no previous conviction.
3. Common age group: 25-35-year-old men
4. Exhibitionism characteristically takes place at a distance and is rarely associated with learning disability or psychiatric disorders.

Figure 1 Factors which suggest high risk of re-offending in indecent exposure

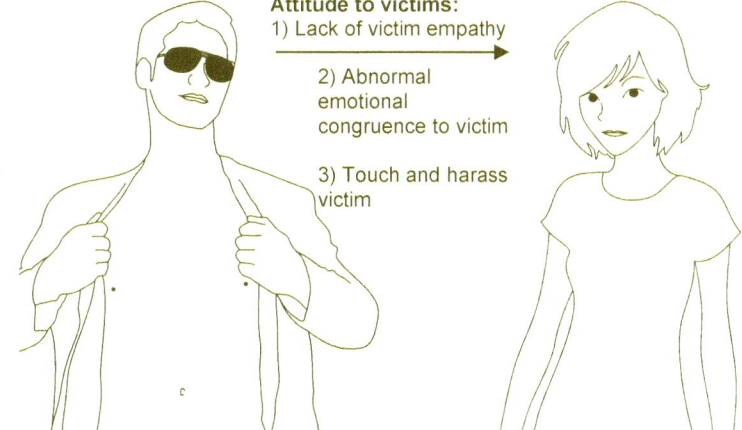

Management of exhibitionists

1. Counselling or CBT to think of negative consequences of exhibitionism.
2. Behavioural techniques on self-monitoring
3. Anti-libidinal medication e.g. cyproterone acetate. Possible side-effects of cyproterone acetate include asthenia, lassitude, gynaecomastia, depression, inhibition of spermatogenesis and erection.

Epidemiology of adult sex offenders:

1. Recent study identified 35% of sex offenders have a mental illness in the UK.
2. Sex offenders with mental illnesses are less likely to target children and victims are also less likely to be women.
3. Unsocialised rapists have high incidence of conduct disorder.
4. Approximately 90% of rapists do not commit a second rape. It is believed that the great majority of perpetrators of sexual crimes begin their offending behaviour in adolescence.
5. 20% of victims develop chronic anxiety and depressive symptoms.

Facts about sex offenders:

1. There is often a history of cold and affectionless upbringing by unloving parents (e.g. a violent father and an over-involved mother).
2. The majority of rapists fail to ejaculate and many rapists suffer relative impotence during the act.

Figure 2   Overview of management of sex offenders

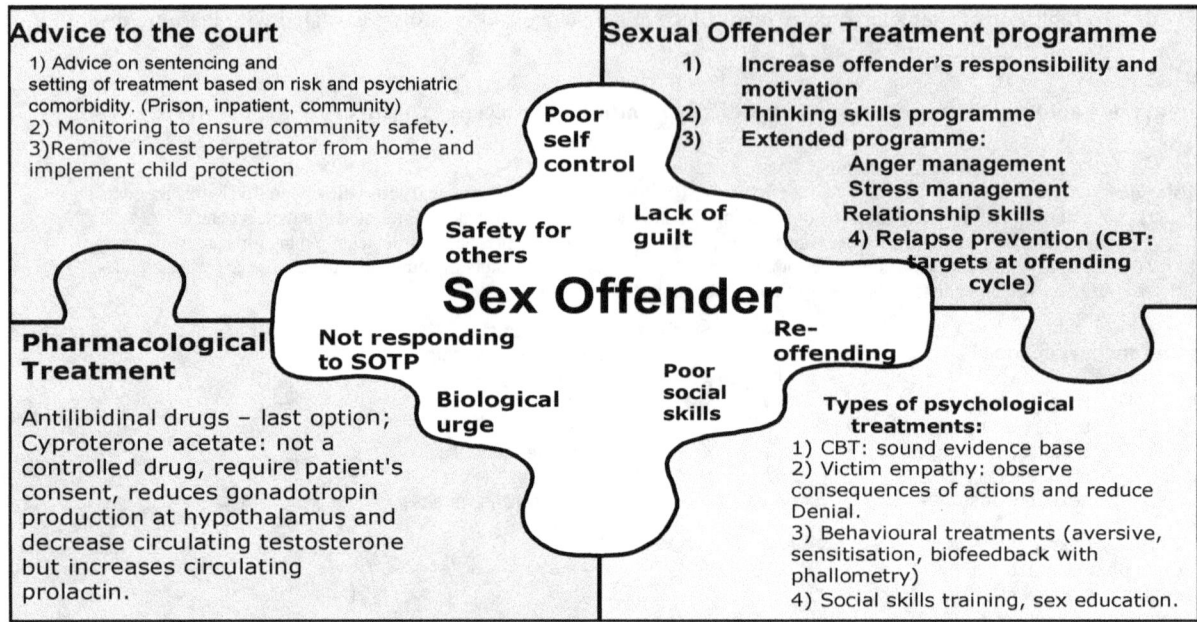

SOTP = Sex offender treatment programme which involves a CBT approach.

Local mental health legislations

Mental disorder and treatment act (MDTA)

The MDTA can only be applied at the Institute of Mental Health (IMH) in Singapore because this is the only gazetted mental hospital at the time of writing.

Table 2 Summary of Mental disorder and treatment act (MDTA) in Singapore

Admission for treatment	1) A person may be admitted to a psychiatric institution and there detained for treatment in accordance with the provisions of this Act for the period allowed by the provisions of this Act. (2) Nothing in this Act shall be construed as preventing a person who requires treatment for any mental disorder — (a) from being admitted to a psychiatric institution without any order or directive rendering him liable to be detained at a psychiatric institution; or (b) from remaining in a psychiatric institution after he has ceased to be so liable to be detained.
Apprehension of mentally disordered person	It shall be the duty of every police officer to apprehend any person who is reported to be mentally disordered and is believed to be dangerous to himself or other persons by reason of mental disorder and take the person together with a report of the facts of the case without delay to — (a) any medical practitioner for an examination and the medical practitioner may thereafter act in accordance with section 9; or (b) any designated medical practitioner at a psychiatric institution and the designated medical practitioner may thereafter act in accordance with section 10.
Mentally disordered person	Where a medical practitioner has under his care a person believed to be mentally disordered or to require

may be referred to psychiatric institution	psychiatric treatment, he may send the person to a designated medical practitioner at a psychiatric institution for treatment and that designated medical practitioner may thereafter act in accordance with section 10.
General provisions as to admission and detention for treatment	1) A designated medical practitioner at a psychiatric institution who has examined any person who is suffering from a mental disorder and is of the opinion that he should be treated, or continue to be treated, as an inpatient at the psychiatric institution may at any time sign an order in accordance with **Form 1** in the First Schedule — (a) for the admission of the person into the psychiatric institution for treatment; or (b) in the case of an inpatient, for the detention and further treatment of the person, and the person may be detained for a period of 72 hours commencing from the time the designated medical practitioner signed the order. (2) A patient who has been admitted for treatment or detained for further treatment under an order made under subsection (1) may be detained for a further period of one month commencing from the expiration of the period of 72 hours referred to in that subsection if — (a) before the expiration of the period of **72 hours**, the patient has been examined by another designated medical practitioner at the psychiatric institution and that designated medical practitioner is of the opinion that the patient requires further treatment at the psychiatric institution; and (b) that designated medical practitioner signs an order in accordance with **Form 2 in** the First Schedule. (3) A patient who has been detained for further treatment under an order made under subsection (2) shall not be detained for any further period at the psychiatric institution for treatment unless before the expiration of the period of **one month** referred to in that subsection, the patient has been brought before 2 designated medical practitioners working at the psychiatric institution, one of whom shall be a psychiatrist, who have examined the patient separately and who are both satisfied that he requires further treatment at the psychiatric institution. (4) Each of the designated medical practitioners referred to in subsection (3) shall sign an order in accordance with **Form 3** in the First Schedule. (5) Two orders signed in accordance with subsection (4) shall be sufficient authority for the detention of the patient to whom they refer for a period not exceeding **6 months** commencing from the date of the order. (6) A person shall not be detained at a psychiatric institution for treatment unless — (a) he is suffering from a mental disorder which warrants the detention of the person in a psychiatric institution for treatment; and (b) it is necessary in the interests of the health or safety of the person or for the protection of other persons that the person should be so detained.

The Mental Capacity Act (MCA)

Principles of the MCA
The key principles of the MCA are derived from the fundamental right of every patient to autonomy. This means that every patient should be assumed to have the capacity to make their own decisions (e.g. treatment, investigations) unless proved otherwise.

Assessment of mental capacity

The assessment consists of two stages:
1. Does the person have an impairment or a disturbance in the functioning of their brain? For example, this may include conditions associated with mental illness (e.g. dementia), head injury, or symptoms of drug or alcohol abuse.
2. Does this impairment or disturbance affect the person's ability to make a specific decision? Before applying this stage of the test, education and information must be first given to the patient to help them make the decision.

In making a decision, a person must be able to understand the information that is relevant to the decision, retain it and apply it as part of the decision-making process and then communicate their decision (verbal, writing, or nodding). A failure in any of the above processes indicates that the person lacks capacity.

Lasting power of attorney (LPA)

A person can appoint a LPA to make the decision on his or her behalf when the person loses his or her mental capacity. The LPA is expected to make decision based on the person's best interest. If a person did not appoint a LPA before losing the mental capacity, a committee of person can be appointed by the court to make the decision based on the person's best interest.

The Advanced Medical Directive Act

Singaporeans aged 21 or over have been able to make advanced refusals of medical treatment, should a patient become terminally ill or lose mental capacity to make a decision. The Advance Medical Directive Act states that doctors must be explicit to patients with regards to what they are agreeing to (e.g. DNR – do not resuscitate) – both sides must be clear about the treatment (e.g. resuscitation) to be refused and the circumstances (e.g. terminal cancer or vegetative state) in which it will apply.

Revision MEQ

You are the resident receiving emergency medicine training at a general hospital. One night, a 30-year-old man is brought in by his relatives because he wanted to jump down from his HDB flat after his girl friend left him. He still wants to jump down if he has a chance. He has history of severe depressive disorder and he has stopped the antidepressant for 6 months. Your hospital does not have a psychiatric ward and the psychiatrist-on-call is not available.

Q.1 Do you think this man needs to be admitted to a psychiatric ward and what is your rationale?

Yes, he has to be admitted because he is at risk of harming himself (i.e. suicide) as a result of mental illness (i.e. depression).

Q.2 If this man actively refuses admission, can you admit him to a psychiatric ward involuntarily in Singapore?

Yes.

Q.3 Which is the only hospital where he can admitted involuntarily in Singapore?

Institute of Mental Health (IMH).

Q.4 Which local mental health legislation states that this man can be admitted for psychiatric treatment against his will?

The Mental Disorder and Treatment Act

Q.5 Can you sign the Form 1 in a general hospital?

No, the Form 1 can only be signed at IMH.

Q.6 What is the procedure to transfer this man to IMH Emergency Department for further assessment?

1. Perform necessary investigations and ensure that this man is medically fit for transfer.
2. Call 6389 2000 (IMH) and speak to psychiatric registrar on call and discuss with him about the transferral.
3. Prepare a memo for transferral.
4. Transfer the patient by ambulance (not by taxi or family car). The patient should be escorted by a nurse. If the patient is at medical risk, he or she has to be escorted by a medical doctor and nurse.
5. Apply physical or chemical restraint (e.g. oral or intramuscular lorazepam) if necessary.

Revision MCQ

Q.1 The man mentioned in the MEQ scenario arrived at IMH. The attending psychiatrist finds that he is of high risk and signs the Form 1 for admission. Based on the Mental Disorder and Treatment Act in Singapore, what is the maximum duration that this patient will be kept for observation?
A. 24 hours
B. 48 hours
C. 72 hours
D. 96 hours
E. 120 hours.
The answer is C.

Q.2 The man mentioned in the MEQ scenario has stayed in IMH for 2 weeks. The attending psychiatrist finds that he suffers from severe depression with poor insight. He signs the Form 2 to prolong his stay. Based on the Mental Disorder and Treatment Act in Singapore, what is the maximum duration that this patient will be kept for treatment?
A. 1 month
B. 2 months
C. 3 months
D. 4 months
E. 5 months.
The answer is A.

Q.3 A 50-year-old with chronic renal failure actively refuses dialysis. He is assessed to have mental capacity to make decision. The renal consultant wants to apply a local health legislature to force him to have dialysis involuntarily. Which of the following legislation apply in this case to carry out the consultant's recommendation?
A. Advanced Medical Directive Act
B. Community Treatment Act
C. Mental Capacity Act
D. Mental Disorder and Treatment Act
E. No legislation is applicable in this case.
The answer is E. Mental Disorder and Treatment Act does not apply on medical treatment.

Q.4 A woman is referred by her lawyer to establish the diagnosis of kleptomania after arrested for shoplifting. Which of the following does not suggest this diagnosis?
A. The stolen items are useless for her.
B. This is the first episode of shoplifting in her whole life.
C. She committed shoplifting for many times but she was not caught.
D. She consulted psychiatrist for compulsive stealing.
E. She threw away the stolen items immediately.
The answer is B.

References

Books

Perry G P, Orchard J. (1992) Assessment and treatment of adolescent sex offenders. Sarasota: Professional Resource Press.

Stone JH, Roberts M, O'Grady J, Taylor AV, O'Shea K.(2000) *Faulk's Black Forensic Psychiatry* (3rd edition). Oxford: Blackwell Science.

Gelder M, Mayou R and Cowen P (2001). *Shorter Oxford Textbook of Psychiatry*. Oxford: Oxford University Press.

[Cooper JE (2001) *ICD – 10 Classification of Mental and Behavioural Disorders with Glossary and Diagnostic Criteria for Research*. London: Churchill Livingstone.

Olumoroti OJ, Kassim AA (2005) *Patient management problems in psychiatry*. London: Churchill Livingstone.

Puri BK, Hall A, Ho RC (2013) Revision Notes in Psychiatry. London: Hodder Arnold.

Puri BK, Brown RA, McKee HJ. Treasaden IH (2005) *Mental Health Law. A practical guide*. London: Hodder Arnold.

Sims A (2002) *Symptoms in the mind: an introduction to descriptive psychopathology.* 3rd edition. London: Saunders.

Royal College of Psychiatrists' Council Report CR 53. (1996) *Assessment and clinical management of risk of harm to other people*. RCPsych: London.

Chiswick, D and Cope R. (1995) *Seminars in forensic psychiatry*. Gaskell: London.

Johnstone EC, Cunningham ODG, Lawrie SM, Sharpe M, Freeman CPL. (2004) *Companion to Psychiatric studies*. (7th edition). London: Churchill Livingstone.

Royal College of Psychiatrists (2006). *Good psychiatric practice: confidentiality and information sharing.* Council Report CR133.

Bluglass R, Bowden P. (1990) *Principles and practice of forensic psychiatry*. Edinburgh: Churchill Livingstone.

Gunn J, Taylor PJ. (1993) *Forensic psychiatry. Clinical, legal and ethical issues*. Oxford: Butterworth-Heinemann.

Casey P (2007) Clinical features of the personality and impulse control disorder In Stein G & Wilkinson G. *Seminars in general adult psychiatry*. London: Gaskell.

Coid J. (1986) Alcohol, rape and sexual assault: (b) socioculture factors in alcohol related aggression. In Brain P.F. *Alcohol and Aggression*. London: Croom Helm.

Gareth Gillespie (2010) Assessing mental capacity. MPS Casebook Vol. 18 no. 3. London: Medical Protection Society.

Journals

Coxe R, Holmes W (2009) A comparative study of two groups of sex offenders identified as high and low risk on the static-99. *J Child Sex Abus.* **18**:137-53.

Murray G, Briggs D, Davis C (1992) Psychopathic disordered, mentally ill and mentally handicapped sex offenders: a comparative study. *Medicine, Science and Law*, 32, 331-336.

Shaw K, Hunt IM, Flynn S, Meehan J, Robinson J, Bickley H, Parsons R, McCann K, Burns J, Amos T, Kapur N, Appleby L (2006) Rates of mental disorder in people convicted of homicide: National clinical survey. *The British Journal of Psychiatry.* **188**: 143 - 147.

Swinson N (2007) National Confidential Inquiry into Suicide and Homicide by People with Mental Illness: new directions. *Psychiatric Bulletin.* **31**: 161-163.

D'Orban PT. (1979) Women who kill their children. *British Journal of Psychiatry.* **134**: 560-571.

McKerracher DW, Street DR, Segal LJ. (1966) A comparison of the behaviour problems presented by male and female subnormal offenders. *Br J Psychiatry.* **490**:891-7.

Black DW, Gunter T, Allen J, Blum N, Arndt S, Wenman G, Sieleni B. (2007) Borderline personality disorder in male and female offenders newly committed to prison. <u>Comprehensive Psychiatry.</u> **48(5)**:400-5.

[26] Rasmussen, K., Almvik, R., & Levander, S. (2001). Attention deficit hyperactivity disorder, reading disability, and personality disorders in a prison population. *Journal of the American Academy of Psychiatry and the Law*, **29**, 186-193.

Davies S, Clarke M, Hollin C, Duggan C. (2007) Long-term outcomes after discharge from medium secure care: a cause for concern. *Br J Psychiatry.* **191**:70-4.

Kingham M and Gordon H. (2004) Aspects of morbid jealousy. *Advances in Psychiatric Treatment*, **10**, 207 - 215

Mullen PE, Pathé M, and Purcell R (2001) The management of stalkers. *Advances in Psychiatric Treatment*, **7**, 335 - 342.

Taylor PJ & Gunn J. (1984) Violence and psychosis. I: Risk of violence among psychotic men. *British Medical Journal.* **288**, 1945-1949.

Whitman. S, Coleman. TE, Patmon, C, et al. (1984) Epilepsy in prisons: elevated prevalence and no relationship to violence. *Neurology*, **34**, 775-782.

Website
Mental Disorder and Treatment Act:
http://statutes.agc.gov.sg/non_version/cgi-bin/cgi_legdisp.pl?actno=2008-ACT-21-N&doctitle=MENTAL%20HEALTH%20(CARE%20AND%20TREATMENT)%20ACT%202008%0A&date=latest&method

=part&sl=1

Chapter 18   Psychiatric ethics

The four ethical principles	Capacity assessment and involuntary treatment
Confidentiality and psychiatry	Boundary violation
The Torasoff's rules	Revision MEQ and MCQ

The four ethical principles

Table 1 Summary of the four ethical principles

Ethical principles	Definition of moral principles	Examples
1) Autonomy	Autonomy refers to the obligation of a doctor to respect a patient's rights to make his or her own choices in accordance with his or her beliefs and responsibilities.	To obtain an informed consent from a patient before ECT.
2) Non-maleficence	Non-maleficence refers to the obligation of a doctor to avoid harm.	When there was no psychotropic drug, psychiatrist in the past had to use a dangerous treatment known as insulin coma therapy. This practice is considered to be unethical in modern psychiatry because this treatment may cause harm in patients.
3) Beneficence	Beneficence refers to the fundamental commitment of a doctor to provide benefits to patients and to balance benefits against risks when making decisions.	A doctor decides to transfer a highly suicidal patient who refuses treatment to the Institute of Mental Health for compulsory treatment because he thinks that the patient will benefit from psychiatric treatment. At this moment, the patient refuses treatment because he is lack of insight or understanding of his own mental illness. Once the patient recovers, he will be able to see the benefits of receiving compulsory treatment which can save his life from suicide.
4) Justice	Justice refers to fair distribution of psychiatric service or resources.	The psychiatric ward needs to reserve a few empty beds for patients to be admitted from the emergency department although some families want to book a bed for stable patients because the caregivers want to take a break.

Fiduciary duty refers to the duty that a doctor must act in the patient's best interests. A doctor (i.e. the fiduciary) is in a legal contract with a patient (i.e. the beneficiary). A fiduciary duty exists when a patient places confidence in the doctor and relies upon that the doctor to exercise his or her expertise in acting for the patient.

Confidentiality and psychiatry

Figure 1 Ethical issues surrounding confidentiality

Breaching confidentiality in an emergency

Common scenarios include patient is about to attempt suicide, extremely drowsy after intoxication of substance and in critical condition after hanging.

Psychiatrists should act in the best interests of the patients to prevent potential serious harms to the health and safety of the patient.

Psychiatrists may need to obtain history from informants when patients cannot provide a history in the emergency situation.

Incompetent patients / patients who are lack of capacity

Psychiatrists should document efforts to obtain patient's consent and seek permission for actions related to release of clinical information in emergency situation. The psychiatrists need to document the reasons for failure to seek a consent from incompetent patients.

Confidentiality issues in minors

1) Provide adequate information to parents/guardians (e.g. in order to obtain consent for treatment)

2) Explain to parents at outset that child's psychotherapy is confidential and the disclosure of information conveyed by child will depend on clinical judgement.

3) Psychiatrist needs to work with custodial parents in case divorce takes place and obtain legal permit to disclose information to non-custodial parents if necessary.

4) For adolescents, maturity and independence of needs to be considered when deciding on how much information is to be released to the parents.

The concept of privileges

Patient has the privilege to control his or her clinical information.

Confidentiality refers to the process when a patient entrusts his or her own doctor in keeping information private and the patient can consent to or not to release of confidential information to the third party. It is a right owned by the patient. It is an obligation on the psychiatrist to safeguard confidentiality.

Confidentiality can be viewed as a subset of privacy.

Confidentiality and precautions in various communication devices

Do not release patient's information via telephone and facsimile.

Do not save patients' information in ancillary storage devices (e.g. storing discharge summaries on a thumb drive as it can be easily lost or stolen)

To release minimal information to other professionals after seeking patient's permission.

Reveal minimum necessary data in writing.

The need for de-identification in maintaining confidence in case write-up.

Mandatory reporting

Confidence limited by legislation requiring mandatory reporting when protection of community outweighing the duty to an individual. (e.g. child protection, firearms possession, registration boards, fitness to drive, certain infectious diseases such as HIV)

Breaching confidentiality and duty to inform/ protect patient and others

1) Duties to community safety override confidentiality. Includes passing information to government officials if public safety is at risk (e.g. homicide, passing communicable diseases such as HIV to others, dangerous driving).
2) Psychiatrists should alert patient that information is to be released and discuss the basis for decision.
3) Psychiatrists should try to re-develop therapeutic relationship afterwards.

Confidentiality and requests from the court

Balance the duty to inform the court with the duty of care to the patient. If the testimony is damaging to the therapeutic relationship, the patient should be evaluated by an independent psychiatrist.

Inform patient prior to assessment of purpose of assessment and to whom information will be given.

The Torasoff's rule and breach of confidentiality

The case of Tarasoff is an influential case in psychiatry to guide psychiatrists and psychologists worldwide on their duty to protect the other people who are at risk by breach of confidentiality. This case occurred in US during the 1960s. Prosenjit Poddar was a university student and he fell in love with a female student called Tatiana Tarasoff. Poddar told the university psychologist that he wanted to kill Tarasoff in a psychotherapy session. Without any precedent, the psychologist decided to maintain the confidentiality of Poddar's homicidal plan. Poddar eventually murdered Tarasoff. The Tarasoff family sued the psychologist from the University of California for not informing Tatiana Tarasoff and the police about Poddar's homicidal plan. In 1976, the California Supreme Court concluded that the mental health professionals should have a duty to protect someone who may be harmed by patients. The Tarasoff's rules has two components: Tarasoff I: duty to warn and Tarasoff II: duty to protect.

Capacity assessment

It is important to offer an informed consent before a major procedure (e.g. ECT, surgery or dialysis). It is the duty of the doctor to provide accurate information about an illness, the proposed treatment, advantages and disadvantages of the proposed treatment and alternatives.

Patients suffering from major psychiatric illnesses such as dementia (any form: Alzheimer's disease, vascular dementia), delirium or acute confusional state, schizophrenia, delusional disorder, bipolar disorder with psychotic features and severe depressive disorder with psychotic features are at higher risk of having reduction in capacity to give consent. The doctor must assess each patient carefully. Having the above illness does not mean that the patient automatically being disqualified from giving an informed consent for his or her own treatment. The doctor needs to ask the following questions to determine the patient's capacity to give consent.

1) What is the nature of your medical condition?
2) What is the purpose of the proposed treatment?
3) Can you tell me the benefits of the proposed treatment?
4) Can you tell me the risks/side effects of the proposed treatment?
5) What happens if you do not get the proposed treatment?
6) Are there alternative to the proposed treatment?
7) If the patient is continuing a chronic treatment (e.g. dialysis), why do you want to discontinue the treatment?

A patient may refuse the proposed treatment and has the capacity to do so. It is not the negative answer which determines one's capacity. It is the rationale behind such decision to determine one's capacity. A schizophrenia patient believes that she has chronic renal failure and the risk of renal function deterioration if dialysis is discontinued. She still refuses dialysis because of financial concern and she cannot afford the treatment. This patient has the capacity to make decision. On the other hand, a schizophrenia patient does not believe that she has renal failure and believes that the renal physician has a plot to kill her through dialysis. This patient does not have the capacity to give consent.

For testamentary capacity (i.e. the capacity to write a will), the doctor can modify the above questions and change the focus to the following: the purpose of a will, the procedure to write a will, the advantages of having a will, the disadvantages of not having a will, personal knowledge of his or her own assets and the details of beneficiary.

Involuntary treatment

Figure 2 Involuntary treatment and ethical issues

Clinical characteristics which indicate involuntary treatment is appropriate:	Ethical issues behind involuntary admission
1. Severe psychiatric illness (schizophrenia, severe depression, bipolar disorder) and florid symptoms 2. Dangerousness to self or others	1. Autonomy versus paternalism or beneficence 2. The duty of care to the family and to the wider society 3. Right of public to peace and freedom from harm

3. Failure of community treatment 4. Capacity to give consent 5. Lack of insight and non – compliance 6. Previous side effects associated with psychotropic medication 7. No less restrictive alternative 8. Admission was shown to be previously helpful 9. Barriers to successful community treatment e.g. homelessness, no social support.	4. Non-maleficence 5. Competence 6. Capacity to give consent 7. Balance between rights of patients and others 8. Rights of psychiatric patients to be treated despite inability to consent 9. The need for natural justice and independent review

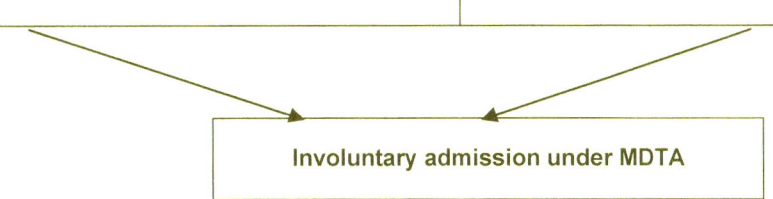

Involuntary admission under MDTA

Boundary violation

Based on medical ethics, a doctor cannot develop any sexual relationship with his or her patients. In psychiatry, the boundary between doctor and patient is well defined and extend to non-sexual boundary. A doctor should not make friend with his or her psychiatric patients. A doctor cannot meet a psychiatric patient outside the clinical setting for non-clinical purpose (e.g. having a coffee together in a coffee shop). Home visit with the community psychiatric team is an exception because it has a clinical purpose and the psychiatrist does not go there alone. A doctor should not have financial transaction with his or her psychiatric patients (e.g. lending money to patients or renting a flat from the patient). The reason why a doctor needs to maintain clear boundary with his or her psychiatric patients is because of the fact that psychiatric patients represent a vulnerable group and the psychiatric treatment (e.g. psychotherapy) is an intensive programme which triggers emotion in both patient and the psychotherapist or psychiatrist. The therapeutic relationship will become complicated if other non-therapeutic factors are involved.

Revision MEQ

You are a resident receiving family medicine training in a polyclinic. A 25-year-old female patient has been seeing you for management of depression. She has been cutting her wrist regularly. She seems to suffer from borderline personality disorder. Recently, she has broken up with her boy friend. She feels very lonely and empty. She feels abandoned. She does not like to take antidepressant. She needs a professional who can listen to her and keeps her occupied.

Q.1 You have been seeing her for a few times. During one consultation, she invites you to go for a coffee in Holland village. She says that she will be very upset if you reject her offer and she will hate you forever. Will you go with her and what is your rationale?

No, I will not go for a coffee with this patient because it is considered to be boundary violation.

Q.2 She is also asking for your handphone number. She needs your personal handphone number because she wants to call you if she needs urgent help. She promises that she only calls you if she is in an emergency and she will not call you for no reason. Will you give her your number and what is your rationale?

No, I will not give her my number. Patient is testing whether you concern her in an inappropriate way. She will call you in the middle of the night and expect that you go and save her. It is beyond a doctor's capacity to offer such service and it leads to uncertainty (e.g. the patient may say she took an overdose at home and you need to go to her flat. The situation can be uncontrollable once you enter her flat).

Q.3 You find that this patient is testing the boundary between doctor and patient relationship. You are very stressed. Name 3 actions to prevent further boundary violation.

1. Establish a treatment contract with the patient which states clearly what is allowed and not allowed between a doctor and his/her patient.
2. Seek supervision from a senior family doctor or psychiatrist.
3. Refer the patient to see a psychologist for psychotherapy and manage this patient in a team with other professionals such as social worker.
4. Refer the patient to see a psychiatrist.

Revision MCQ

Q.1 An individual with bipolar affective disorder and long-term alcohol dependence develops haematemesis. He is advised by the gastroenterologist to go for an oesophagogastroduodenoscopy (OGD) to identify the bleeding site. On assessment, he appears to have the capacity to make that decision and he is not manic. The psychiatrist advises acceding to his wish not to have the OGD. This case illustrates which of the following?
A. The principle of respect for a person's autonomy
B. The principle of beneficence
C. The principle of non-maleficence
D. Paternalism approach
E. Utilitarian approach.

The answer is A. Although this man has history of bipolar disorder, he has the capacity to make a decision. Hence, his autonomy is respected.

Q.2 Which of the following statements about capacity is false based on scenario in Q.1?
A. Capacity implies that the patient understands the relevant information about OGD given by the gastroenterologist
B. Capacity is a clinical opinion given by a clinician
C. Capacity is a legal term
D. Capacity requires the mental ability from the patient to make and communicate a decision
E. The gastroenterologist is expected to provide all relevant information about OGD

The answer is C. Capacity is not a legal term but competence is. A person is deemed to be competent if he or she has the capacity to understand and act reasonably. Competence is a legal term and determined by the legal system.

Q.3 The case Tarasoff v. Regents of the University of California is related to which of the following ethical principles:
A. Autonomy
B. Capacity
C. Confidentiality
D. Consent
E. Equality.

The answer is C.

Q.4 A patient has his or her own right to decide what happens to his or her own personal information. This concept is best known as:
A. Capacity
B. Confidentiality
C. Fiduciary
D. Privacy
E. Privilege.

The answer is E. Privilege refers to a patient's right to decide what happens to their personal information. The personal information or clinical information belongs to the patient but not the hospital or doctors.

References

Bloch S, Chodoff P, Green SA (1998) *Psychiatric Ethics* (3rd edition). Oxford: Oxford University Press.

Joseph DI & Onet J (1998) Confidentiality in Psychiatry in Bloch S, Chodoff P, Green SA (1998) *Psychiatric Ethics* (3rd edition). Oxford: Oxford University Press.

Beauchamp TL (1998) The philosophical basis of psychiatric ethics in Bloch S, Chodoff P, Green SA (1998) *Psychiatric Ethics* (3rd edition). Oxford: Oxford University Press.

Musto DF (1998) A historical perspective in Bloch S, Chodoff P, Green SA (1998) *Psychiatric Ethics* (3rd edition). Oxford: Oxford University Press.

Green SA (1998) The ethics of managed mental health care in Bloch S, Chodoff P, Green SA (1998) *Psychiatric Ethics* (3rd edition). Oxford: Oxford University Press.

Gelder M, Mayou R and Cowen P (2001) *Shorter Oxford Textbook of Psychiatry*. Oxford: Oxford University Press.

Brouillette MJ, Roy DJ Legal and ethical issues in Citron K, Brouillette MJ, Beckett A (2005) *HIV and Psychiatry – A training and resource manual*. Cambridge: Cambridge University Press.

American Psychiatric Association. AIDS policy: Policy Guidelines for Inpatient Psychiatric Units. Americarl Journal of Psychiatry, 150(5) (1993): 853. as quoted in Brouillette MJ, Roy DJ Legal and ethical issues in Citron K, Brouillette MJ, Beckett A (2005) HIV *and Psychiatry – A training and resource manual*. Cambridge: Cambridge University Press.

Appelbaum PS, Gutheil TG (2007) *Clinical handbook of Psychiatry and the Law* (3rd edition) Philadelphia: Lippincott Williams and Wilkins:

Scott AIF (2004) *The ECT handbook* 2nd Edition. London: Royal College of Psychiatrists.

American Psychiatric Association (1978) *Electroconvulsive therapy. Task Force Report No. 14.* Washington: American Psychiatric Association Press.

Johnstone EC, Cunningham ODG, Lawrie SM, Sharpe M, Freeman CPL. (2004) *Companion to Psychiatric studies*. (7th edition). London: Churchill Livingstone.

Chapter 19 Community psychiatry and local mental health resources

The Community mental health team	Psychiatric rehabilitation
The local mental health resources in the community	Stigma and psychiatry

The Community Mental Health Team (CMHT)

The CMHT is multidisciplinary in nature, comprising of mental health workers such as psychologists, social workers, psychiatrists, case managers, rehabilitation therapists, occupational therapists and nurses. Each team is unique in that there is no set framework for which of these professionals and how many can band together to form a community mental health team. The few general principles governing the formation of these teams is that each of the team players should have a broad understanding of mental health issues and treatment and they should be able to employ their professional skills within the dynamics of a team, which may sometimes result in blurring of the distinction between roles and sharing of their respective skills, i.e. nurses and occupational therapists may be able to do basic supportive psychotherapy. Not all of these teams have a trained psychiatrist at its helm, due to the lack of resources, so initial assessment may be undertaken by a trained member of the team. However, the general consensus is as much as possible for each patient to have an assessment done by a trained psychiatrist. In addition, the principle of collective responsibility applies in CMHT as the responsibilities are shared among team members.

The CMHT can be formed based on the characteristic of the service users, e.g. old people with dementia, young people with substance misuse, and people with severe mental illnesses. The main roles of the CMHT are to provide a thorough initial assessment, home based care for people who fail to attend outpatient appointments, with systematic reviews, outreach to people with mental illnesses and their families in the community and continued outpatient assessments to review their progress. In order for community care to be successful, an adequate in-patient facilities in the trust is necessary for treatment of acute illnesses and crisis interventions.

The local mental health resources in the community

For children and adolescents

Community Health Assessment Team (CHAT)

The CHAT works closely with the post-secondary educational institutions i.e. polytechnics, universities and vocational institutions, social agencies and voluntary welfare groups to promote awareness of mental wellness among youths, help them to be aware of mental health services available to them, and encourage them to seek help early if they have mental or emotional health issues. It provides a one-stop resource hub is set in a non-traditional setting to help youths in distress as well as destigmatise mental illness.

Source of information and website: http://www.youthinmind.sg

Response Early Intervention and Assessment in Community Mental Health (REACH)

REACH is a mobile multidisciplinary mental health team that works closely with school clusters to:

1. Improve mental health of children and adolescents in schools
2. Provide early interventions, support and training to school counsellors on mental health disorders
3. Develop a mental health network for children and adolescents in the community involving:
    - General practitioners (family doctors and community paediatricians)
    - Full-time School Counsellors (FTFC), and/or
    - Voluntary Welfare Organisations and Family Services Centre

Source of information and website: http://reachforstudents.com

For Adults

Adult Community Mental Health Team Service

The Adult Community Mental Health Team (CMHT) is funded by Ministry of Health under the National Mental Health Blueprint. The primary aim of the CMHT service is to maintain adult persons with mental illness (18 to 65 years old) in the community for as long as possible and reduce hospital readmissions and length of stay. The CMHT is a multidisciplinary team (consists of psychiatrists, psychologists, occupational therapists, medical social workers, community nurses and counsellors) and provides the following services:

1. *Assertive Care Management*
(psycho-social rehabilitation of patients who are high users of inpatient services)

This team provides community-based treatment to IMH patients with severe and persistent psychiatric illnesses (such as schizophrenia, delusional disorder and manic-depressive psychosis), so that they may continue to live in the community while working towards recovery.

2. *Mobile Crisis Team*

IMH patients and their caregivers in crisis situations can call a Mobile Crisis hotline for help. The hotline puts them through to a qualified counsellor for immediate assistance and advice. In critical situations, they will be put in touch with the Mobile Crisis Team who will accompany the patient for admission to IMH if necessary.

3. *Community Psychiatric Nursing Service*

A team of Community Psychiatric Nurses helps to provide continuity of care for discharged patients in their homes and counsel patients on compliance with medication. The team also provides psychological support to caregivers. During home visits, the Community Psychiatric Nurse assesses the mental state of the patient and observes the therapeutic effects and any side effects of medications. Feedback gathered from the patient is shared with caregivers to help them with their care management. This service caters to patients of all age groups.

Source of information and website: http://www.imh.com.sg/clinical/page.aspx?id=272

Community Rehabilitation Support and Service (CRSS)

The Community Rehabilitation Support & Service (CRSS) programme for individuals with psychiatric disabilities started in January 2006. A community project of the Singapore Anglican Community Services (SACS) and supported by the Ministry of Health (MOH) and the National Council of Social Service (NCSS), the programme involves a mobile-team of professionals providing essential services to clients at their place of residence in the community. The objective of the CRSS programme is to enable people with psychiatric disabilities to live safely in the community, and that they are meaningfully engaged in work, studies or other meaningful activities of their choice.

The services provided by the CRSS mobile-rehabilitative team includes:
1) Mental Wellness
2) Services Coordination
3) ADL Training
4) Caregivers' Training
5) Group Work
6) Treatment Therapy
7) Community Integration.

Source of information and website: http://www.sacs.org.sg/crss.htm

Residential mental health centres

Some psychiatric patients discharged from psychiatric ward need to stay in a care centre for 1 to 6 months to prepare integration to society.

Simei Care Centre

Simei Care Centre (SCC) is a community based, purpose-built rehabilitation centre operated by the Singapore Anglican Community Services, a voluntary welfare organization. It caters to the various needs of persons with psychiatric disabilities in Singapore.

Completed in December 2004, it has a capacity of 156 residential members and another 90 day care members.

SCC hopes to help each of the members rediscover skills and resources needed to successfully live, learn and work in the community with the least amount of professional assistance.

Source of information and website: http://www.sacs.org.sg/scc.htm

Hougang Care Centre

Hougang Care Centre (HCC) is based on the Clubhouse model of psycho-social rehabilitation. This model focuses on the strengths of persons with mental illness. It holds that everyone can be productive in his or her own way, and that participating in meaningful work has a regenerative effect. It offers opportunities for members to engage in meaningful activities that will help to develop life, work, and social skills through their involvement in the running of the Clubhouse. Clients are called 'members' to emphasise a sense of inclusion and mutual acceptance, and are expected to work side-by-side with staff in their own recovery process.

Source of information and website: http://www.sacs.org.sg/hcc.htm

Singapore Association of Mental Health

The Singapore Association for Mental Health is a voluntary welfare organisation (non-government and non-profit organisation), which seeks to promote the social and mental well-being of the people of Singapore. More specifically, the organisation aims to promote mental health, prevent mental illness, improve the care and rehabilitation of the mentally ill and the emotionally disturbed, and to reduce the misconception and social stigma that surround mental illness. It provides counselling service, a daycare centre and a support group for eating disorder.

Source of information and website: http://www.samhealth.org.sg/index.php

For old people

Aged Psychiatry Community Assessment and Treatment Service (APCATS) - IMH

Aged Psychiatry Community Assessment and Treatment Service (APCATS) is a community-oriented psycho-geriatric outreach service. It has two programmes: Clinical Service (CS) and Regional Eldercare Agencies Partnership (REAP).

APCATS Clinical Service (CS) provides assessment and treatment for homebound elderly patients. It also provides education and support for patients and caregivers.

APCATS Regional Eldercare Agencies Partnership (REAP) is a newly launched initiative in 2008 where APCATS partner community eldercare agencies and family physicians for training, consultation and support. The team also assists in the coordination of services to improve the continuity of care for the elderly with mental illnesses.

Source of information and website: http://www.imh.com.sg/clinical/page.aspx?id=273

G-Race Community Programme (GCP) - NUH

GCP is a community-oriented programme developed to provide support for older persons with mental health challenges so that they can continue living in their own homes.

1. Provides mental health services for the treatment of psychogeriatric disorders in elderly patients from NUH by helping them transit back to their homes seamlessly.

2. Improves the quality of life of elderly patients and alleviates the stress of caregivers through support and education.

Source of information and website: http://www.nuh.com.sg/umc/about-us/about-us/department-of-psychological-medicine/clinical-services/programmes/grace-community-programme-gcp.html

Psychiatric rehabilitation

Psychiatric disorders cause impairment (interference with the function of a system), disability (interfere the function of the person as a whole) and handicap (social disadvantages resulted from impairment and disability). Psychiatric rehabilitation aims to restore and improve function and maintain the function at an optimal level. Hence, psychiatric rehabilitation aims to reduce impairment, disability and handicap. Common rehabilitation strategies include cognitive rehabilitation, training in independent living, use of community facilities, enhancing social interaction and psychoeducation. There are many examples to prove that the severity of psychiatric symptoms may not always correlate to the success of the rehabilitation. For example, a person suffering from schizophrenia with severe paranoid delusions may still be able to hold a job and maintain independence in his activities of daily living. Conversely, a person with schizophrenia whose symptoms are stable on medication, may decompensate in terms of social functioning in light of psychosocial neglect or stress.

Employment for psychiatric patients

Temasek Cares Employment Support Services is a programme of Singapore Anglican Community Services. An evidence based support employment programme, it is established to help persons with mental illness to find competitive jobs.

Source of information and website: http://www.sacs.org.sg/employment.htm

Stigma in psychiatry

Stigma is a prejudice (negative attitude) based on stereotypes (the linkage of differences to undesirable characteristics) usually leading to discrimination. Discrimination is a form of behaviour ranges from simple avoidance of people with mental illness to social discrimination (e.g. people with mental illness are rejected when they apply for insurance). Hence, stigma is caused with lack of knowledge (ignorance), problems with attitudes (prejudice) and problems with behaviour (discrimination).

Historically, stigma in psychiatry has been divided into two types by Corrigan and Watson: public and self. The former refers to the attitude of the general public towards the mentally ill. The latter is the prejudice that the mentally ill internalise due the reaction of the public towards them and the disability they suffer from their mental illness. Stigma also makes people with mental illness feel angry, emotionally hurt, depressed and disappointed. Factors contributing to stigma include the abnormal behaviour associated schizophrenia or mania, iatrogenic stigma such as side effects of psychotropic medications (e.g. tardive dyskinesia), the association by the media of some forms of psychiatric illness with criminal behaviour, the low financial priority given to the mental health services and the social consequences of disclosing mental illness. It is important to note knowing someone with a mental illness does not necessarily always reduce the prejudice or the stigma amongst the general population. However people who have had contact with patients who have recovered or improved with treatment have shown to have less discriminatory attitudes towards the mentally ill.

Certain classes of mental disorders are associated with higher rates of blame and prejudice such as substance misuse as other often think that the problems are self-inflicted. In 1996-2003, the Royal College of Psychiatrists launched an anti-stigma campaign which focused on the following disorders: anxiety disorders, depressive disorder, dementia, schizophrenia, eating disorders and substance misuse. The audience groups of this programme are doctors, young people, employers, media and the general public in the UK. The programme led to significant changes in the public opinion regarding dangerousness associated with mental illnesses and people became keener to talk about mental illnesses. The stigma in mental illnesses can be seen in many aspects of the patient's lives, from employment, to housing and even amongst their own family members.

References

Burns T (2004) *Community Mental Health Teams: a guide to current practices.* Oxford: Oxford University Press.

Byrne P (2009) Stigma - in 100 words *The British Journal of Psychiatry*; 195: 80.

Humayun A (1998) Community care In Williams C & Trigewell P (1998) *Pass the MRCPsych Essay*. London: Saunders.

Puri BK, Hall A, Ho RC (2013) Revision Notes in Psychiatry. London: Hodder Arnold.

Thornicroft G (2006) *Shunned: discrimination against people with mental illness.* Oxford: Oxford University Press.

Mastering Psychiatry: A Core Textbook for Undergraduates
Clinical OSCE Interview Grids

Psychopathology

OSCE station

A 20-year-old university student was brought in by the counsellor to the emergency department. He was found sitting in the lift of the residential hall the whole day and refused to attend classes. He claims that he hears voices talking to him.

Task: assess his hallucinations.

OSCE grid: approach to hallucinations:
Candidates are advised to ask about hallucinations in other modalities as you will discover more information. This person may have used recreational drugs causing visual or tactile hallucinations. The person can be paranoid and suspicious of doctors. Candidates have to spend time establishing rapport at the beginning.

A) Auditory (second person auditory hallucination in depression and mania)	A1) Introduction When people are under stress, they may have unusual experiences. I understand from your counselor that you have been hearing voices.	A2) Open questions about hallucinations Can you tell me more about the voices? Can you give me an example?	A3) Nature of the voices How do they address you? Do they speak directly to you?	A4) Command hallucinations Do they give you orders? Do you obey them?	A5) Congruence with mood Does your mood influence the content of the voices? For example, when you are sad, you hear the voices say sad things.
B) Auditory (third person auditory hallucination in schizophrenia)	B1) Number of voices B1) Do you hear more than one voice talking about you?	B2) Content of the hallucination What do they say?	B3) Running commentary Do they comment on what you are doing or thinking?	B4) Audible thoughts Are the voices saying out your thoughts aloud?	B5) Echo de la pense Do those voices echo your thoughts after a few seconds?
C) Confirmation of the nature of hallucination (rule out pseudohallucination)	C1) Where do these voices come from? Do you come from external space?	C2) Do the voices come from inside or outside your head?	C3) Do you feel that the voices are real? Are the voices as clear as my voice?	C4) Can you stop them? Can you distract yourself from the voices?	C5) When do these voices occur? Were you falling asleep or waking up? (to rule out hypnagogic and hypnopompic hallucinations)
D) Hallucinations in other modalities and assess insight	D1) Visual hallucinations Have you seen things that other people can't see? What do you see? Can you give me an example?	D2) Olfactory hallucinations Is there anything wrong about the way we smell? Can you tell me more about it? If yes, who sent the gas to you?	D3) Gustatory hallucinations Have you noticed that food or drink seems to have a different taste recently?	D4) Tactile / haptic hallucinations Have you had any strange feelings in your body? How about people touching you? How about insects crawl	D5) Assess insight Do you have any explanation of above experience? Do you need help (e.g. taking medication to reduce the voices?)

				ing?	
E) Course/ Comorbidity/ risk assessment	E1) Course of hallucinations How long have you experienced those voices?	E2) Assess impact How does it affect your life How is your mood? Has your mood been affected by those voices?	E3) Assess other first rank symptoms Do you worry that those voices are part of a plot? (e.g. harming you or controlling you) Do you feel that your thoughts are being interfered? (e.g. thoughts are being inserted, withdrawn or broadcasted)	E4) Assess substance misuse Do you use recreational drugs or alcohol?	E5) Risk assessment I can imagine that you are stressed by the voices. Some people may want to give up. Do you have thought of ending your life?

Schizophrenia & Psychosis

OSCE: Interview a patient to establish a diagnosis of schizophrenia

A 22 – year – old university student is brought by his counsellor to the Accident and Emergency Department. He was seen by the psychiatrist at the University clinic and diagnosed to suffer from schizophrenia. You are the resident working in the Accident and Emergency Department.

Task: Take a history to elicit first rank symptoms and other related symptoms to establish the diagnosis of schizophrenia.

A. Introduction and assess hallucinations.	A1. Introduction 'I would like to ask you some questions. Some of them *may appear a bit strange.* Is that all right with you? I gather that you had been through a lot of stress recently. When under stress sometimes people have *certain unusual experiences.* Have you had such experiences?'	A2. Assess auditory hallucinations By unusual experience, I mean, for example, some people hearing voices when no one around. If yes,……… 'Do the voices speak directly to you?' (2nd person) 'Do they speak among themselves?' (3rd person) 'What sort of things do the voices say?' **Mnemonics: EAR** **E**choes of own thoughts: 'Do the voices repeat your thoughts?' **A**rguing voices involve at least 2 voices. **R**unning commentary on patient: 'Do these voices describe or comment upon what you are doing or thinking?' Do the voices tell you to do things? (command	A3. Other hallucinations Visual hallucinations: 'Have you ever had experiences during which you saw things or shadows that others could not see?' Tactile hallucinations: 'Do you feel that there are strange sensations within you, as if something is crawling within your body?' Olfactory / gustatory hallucinations: 'Have you ever had experiences during which you smell or experience strange tastes that others do not experience?'

		hallucinations)	
B. Assess thought interferences.	**B1. Thought insertion** 'Is there any interference with your thoughts?' 'Do others put or force their thoughts into your mind?'	**B2. Thought withdrawal** 'Could someone take your thoughts out of your head?'	**B3. Thought broadcasting** 'Do other people know what you think in your mind?' 'Do you feel that your thoughts are broadcasted to other people?'
C. Delusions insight, mood and substance misuse.	**C1. Delusion of control or passivity experience.** 'Has there been any difficulty with feelings, actions, or bodily sensation? Is there someone or something trying to control you in the following areas:' **Mnemonic – WEA**: W- Will (impulses) E – Emotions (Feelings or affect) A – Actions (volitional)	**C2 Other delusions.** 'When under stress some people find that someone is playing tricks on them. Have you had any such experiences?' Delusion of Persecution: 'Are there some people who try to harm you or make your life miserable?' Delusion of reference: 'Do you think that someone is watching, following or spying on you?' Delusion of grandeur: 'Do you have any special powers or abilities that others don't have?' Delusion of guilt: 'Do you feel like you deserve punishment for mistakes you made in the past? Can you tell me the nature of the mistakes and punishment you deserve?'	**C3 Assess insight and mood.** 'What do you think is the cause for these experiences?' 'Could you suffer from an illness in your mind?' 'Do you think treatment would help to reduce those experiences?' 'How do you describe your mood? Do you feel sad?' 'When you feel sad, do you have thought of harming yourself?' 'I encounter some students use recreational drugs when they go for overseas exchange or party. Have you used recreational drugs recently?'
D. Assess negative symptoms and academic disturbances (Note: alogia/lack of speech and flatten affect is observed during interview)	**D1. Apathy (lack of motivation)** 'Do you encounter any difficulty in looking after yourself? How often do you tend to take a shower or a bath?' 'Has anyone complained about the state the flat is in? Is it difficult to stay tidy or to keep the flat the way you	**D2. Anhedonia (lack of interests)** 'Have you spent any time with friends lately?' 'Do you find it difficult to feel close emotionally to others?' 'Do you have any activity you enjoy to do nowadays?' 'What were your main interests or hobbies in the past?'	**D3. Assess social and academic deterioration.** 'How do you find your academic performance recently?' 'Can you concentrate in your study?' 'It seems that your academic performance is not as good as in the past. How long has it been?'

		would like it?'		

OSCE grid: Assess depression.

You are the resident working in the AED. A 30 – year - old teacher is referred by polyclinic for management of depression. He cannot cope with the workload and he also has interpersonal problems with the school principal.

Task: To a history to establish the diagnosis of depressive disorder.

Please note that forgetting to have a brief assessment of suicidal risk in a depressed patient may result in a failure.

A.	Assess core symptoms of depression.	A1) Assess mood. 'During the past month, how often have you been bothered by feeling down or depressed? 'Can you rate your current mood from a scale of 1 to 10? 1 means very depressed and 10 means very happy.' 'Which part of the day is the worst?' (Elicit diurnal variation of mood)	A2) Assess energy level. 'Have your energy levels been recently?' 'Do you feel tired most of the time?'	A3) Assess interest. 'Can you tell me more about your interests and hobbies before the current depressive episode?' 'During the past month, how often have you been bothered by having little interest or pleasure in doing things?'
B.	Assess biological symptoms of depression.	B1. Assess sleep. 'How has your sleep been lately?' 'Can you fall asleep? If not, how long does it take?' 'How many times do you wake up in the middle of the night? (exclude urination)' 'What time do you wake up in the morning? (look for early morning wakening). If you wake up, can you fall asleep again?'	B2. Assess appetite and weight. 'Has your appetite changed recently? If yes, do you tend to eat less or more?' 'Has your weight changed recently? If so, have you lost weight or gained weight. If yes, how many kilograms were involved?'	B3. Assess sexual functions. I hope you would not mind if I ask you some sensitive questions such as sexual problems as depression may affect sexual function. Is it OK with you? 'Have there been any changes in your sexual function recently? If yes, can you tell me more about the nature of sexual dysfunction?' 'When did the sexual dysfunction start? (Does it coincide with the onset of depression?)'
C.	Assess cognitive symptoms.	C1. Assess cognitive impairment. 'What has your	C2. Assess feelings towards self and future. 'How do you see	C3. Explore common cognitive bias. 'Can you tell me more about your

		concentration been like recently? Can you concentrate when you teach?' 'How has your memory been?'	yourself?' 'Do you see yourself a failure?' 'How do you see your future? Do you feel hopeless?'	negative thoughts?' Look for selective abstraction, overgeneralization or catastrophes thinking. Depending on the patient's response, you may gently challenge patient's belief or provide an alternative explanation to seek his or her view?
D.	**Assess risk, psychotic features, insight.**	**D1. Assess suicide risk.** 'Have you felt that life is not worth living?' 'Would you do anything to harm yourself or hurt yourself?' 'Have you done anything of that sort?' Have you made any plans? Have your told anybody about it?'	**D2. Assess psychotic features.** 'When people are under stress, they complain of hearing voices or believing that other people are doing something to harm him. Do you have such experiences?'	**D3. Assess insight** 'What is your view of the current problem? Do you think that you may suffer from a depressive illness?'
E.	**Explore causes and background.**	**E1. Explore family history of depression.** 'Do you have any biologically related relative suffer from depression?' 'Do you have any biologically related relative attempt or commit suicide in the past?'	**E2. Explore past psychiatric history and relevant medical illnesses.** 'Did you seek help from a psychiatrist or GP in the past for your low mood?' 'Did you receive any treatment from a psychiatrist? If yes, can you tell me more about the medication and side effects?' 'How anxious do you feel in yourself?' (explore comorbidity) 'Do you drink alcohol on a daily basis to cope with stress or help you to sleep?' 'Do you suffer from any chronic medical illness?'	**E3. Assess support system.** 'Can you tell me the person who is providing emotional support to you at this moment?' 'Is there a person in the school whom you can talk to?' 'What is your career plan at this moment?' 'Have you sought help from Ministry of Education?'

OSCE video – Explain ECT (Refer to Clinical OSCE Videos)
You are the resident and you have admitted an elderly woman suffering from severe depressive episode with delusion of guilt. She does not respond to the antidepressant and antipsychotic drug. Your consultant has recommended ECT and her daughter is very concerned and wants to

> speak to you.
>
> Task: Talk to her daughter and address her concerns.

- **Approach**: Express empathy. (e.g. I can imagine the idea of ECT sounds very scary for you, and it's clear you want the best care for your mother. I would like to discuss what ECT involves, because it is very different than what is portrayed in the media. This way, you can make an informed decision)
- **Core information about ECT:**
    - ECT involves inducing a fit, while the patient is under general anaesthesia.
    - ECT is the most effective treatment for depression, particularly for those who have high risk of suicide, very poor appetite and not responding to oral medication; sometimes in pregnant women because it has no side effects to the foetus.
      It is very safe and has been with us for the past 50 years.
- **Will my mother be awake during ECT?** No, your mother will be given **anaesthesia** to put her into sleep and a medication **that paralyze muscles**, so the risk of breaking bones is rare. The patient is given **oxygen** before the procedure. The patient's blood pressure, heart rhythm, and medical status is monitored throughout the procedure and when she comes out of the anaesthesia.
- **How often will my mother get ECT and for how long?** 3 times per week, Mon, Wed, Fri and for 6 sessions (2 weeks); some patients may need 9 to 12 sessions..
- **How do you know the ECT is successful or not?** We will monitor the duration of her fit. It has to be at least 25 second in duration. We will monitor her muscle movement through electrical recordings (i.e. EEG). If response is poor, we will increase the energy level 5% each time.
- **How do you decide on the dose of ECT?** By age-based dosing: Energy level = patient's age divided by 2.
- **What tests do you include in your pre-ECT work-up?** Physical exam, FBC, RFT, ECG, CXR. Assess patient's dentition, especially for elderly or those who have inadequate dental care.
- **What is the preparation for the night before the ECT?** Fasting is required after 12:00 midnight and she should avoid sleeping pills if possible.
- **What is the risk involved?** ECT itself is safe. Risk is associated with anaesthesia.
- **How does ECT affect memory?**
    - Anterograde and retrograde amnesia can occur, though in the majority of patients this does not last more than a few months following the last ECT treatment.
    - Amnesia of events immediately preceding and following ECT treatments may be permanent (reassure the relative those memory is not important).
    - Anterograde amnesia is always transient. In a very small number of patients, the symptoms of retrograde amnesia may be permanent.
- **What are other common side effects?** Memory problems, confusion, nausea, muscle aches and headache are the most common in the morning after the ECT.
- **What are the risk factors associated with confusion after ECT?** Old age; prior cognitive impairment; lithium; anticholinergic and bilateral placement.
- **How would you reduce confusion after ECT?** Unilateral treatment on right – side of the brain, lower electrical energy, increasing the time between ECT treatments and holding off lithium or sleeping pills.

What is the mortality rate associated with ECT? The mortality rate is very low, and is the same as that for general anaesthesia, which is 1 in every 20 000 people

> **OSCE grid: Assess bipolar disorder**
>
> You have been asked to see a 28-year-old unemployed man who has not slept for 5 day and claims to have full energy. He claims to be the President of Singapore and his plan is to unite all the world leaders to fight for poverty in developing countries.
>
> Task: Take a history to establish the diagnosis of bipolar disorder..

A. Assess mood symptoms.	A1. Assess mood. How's your mood today? If I ask you to rate your mood from 1 to 10, 1 means very depressed and 10 means very happy, how would you rate your mood today? How long have you been feeling high? Do you have mood swings? How about feeling low? If so, roughly how many low or high episodes you would experience in a year?	A2. Assess irritability. How do you get on with people recently? Do you feel that they annoy you? Do you lose your temper easily? What would you do if these people irritate you?	A3. Assess grandiosity How would you compare yourself with other people? Are you special? If yes, please tell me more. Could your special ability be a misunderstanding? Can you provide more evidence about it? Do you feel that you are at the top of the world (i.e. above all the other people)?
B. Assess biological symptoms.	B1. Assess sleep and energy. How has your sleep been lately? What is your energy level like? Do you feel that you need much less sleep but full of energy?	B2. Assess appetite and weight. How has your appetite been lately? Have you lost weight recently?	B3. Assess sexual function and contraceptive method. I am going to ask you some sensitive questions. How has your interest in sex been lately? Do you have sex with new partners? Do you take any precaution to protect yourself (e.g. condom)? If your patient is a woman, you need to ask LMP and chance of pregnancy.
C. Assess cognitive and psychotic symptoms.	C1. Assess interests and plans. Could you tell me about your interests? Have you developed any new interests lately? Do you have any new plan or commitment at this moment? (for example, starting a new business or investment)	C2. Assess thought and speech. Has there been any change in your thinking lately? Have you noticed that your thoughts speed up? Do you find your thoughts racing in your mind? Do your family members say that the topics in your speech change so fast and they cannot follow.	C3. Assess psychotic features. When people are under stress, they have unusual experiences such as hearing a voice talking to them but cannot see the person. Do you encounter such experiences? If so, what did the voices say? How many voices spoke at one time? Do you believe that you have special power or status which other people do not have? If yes, can you tell me about your special power or status? Are you very certain that you have

			such ability or status?
D. Assess risk and insight	**D1. Assess risk.** Have you been buying a lot of things? Have you incurred a lot of debts (e.g. credit card debts?) Do you drive? Have you been involved in speeding or traffic offences? Have you been in trouble with the police lately? (e.g. due to violence). When you feel sad, have you thought of harming yourself?	**D2. Explore comorbidity.** Do you take recreational drugs on a regular basis to get the high feelings? How about alcohol? Do you drink on a regular basis?	**D3. Assess insight.** Is there any reason why you encounter those experiences? Do you think there is a illness in your mind? For example, this illness affects your mood? If so, do you think you need treatment?

OSCE video – Explain lithium and side effects

A patient was admitted to the psychiatric ward after a manic episode. The consultant psychiatrist has advised him to consider taking lithium as a maintenance treatment. The patient is very concerned about bipolar disorder and lithium after reading the information from internet.

Task: address his concerns about lithium treatment.

1. **Why do you want to prescribe lithium?** Lithium is used to stabilise your mood. After my assessment, your mood seems to be elevated and you suffer from a condition called mania in the context of bipolar disorder.
2. **What is mania?** Feeling high, irritable, full of energy, very good appetite, no need for sleep, high sexual drive, racing thoughts, grandiose ideas, overspending, poor judgement, dangerous behaviour and unusual experiences such as hearing voices.
3. **Why do I sometimes feel depressed?** Periods of depression occur in bipolar disorder. Your mood will go up and down.
4. **What exactly is lithium?** It is a type of salt and can be found naturally.
5. **How long have psychiatrists been using lithium?** 50 years already.
6. **What is the usual dose of lithium?** Starting dose 400mg a day, increase slowly to 800mg to 1200mg per day.
7. **How do you decide the right dose for me?** Based on serum levels 0.4 – 0.8 mmol/L; clinical response.
8. **What time of the day should I take lithium?** Usually at night. The modern lithium has long release version and can last for whole day.
9. **What should I do if I miss a dose?** If you forget a dose, take it ASAP as you remember.
10. **Can I take lithium now?** No, we need to do some blood tests for you.
11. **What do you need those blood tests?** To check it is safe for you to take lithium. Your kidney and thyroid have to be in good condition.
12. **Do I only need to have those blood tests once?** The lithium may affect the function of kidney and thyroid. We have to check every 6 months.
13. **Lithium sounds scary. How do you know it is safe for me to take?** It is usually safe if your kidney and thyroid are in good condition. Extra care if you take pain killer, medication containing sodium.
14. **How do I know lithium works for me?** Your highs and lows become less extreme. It will reduce thoughts of harming oneself. It may take weeks or months to appreciate the beneficial effects of lithium..
15. **Can I mix alcohol with lithium?** No, it will lead to drowsiness if lithium combines with alcohol, ↑ fall risk & accidents. Avoid alcohol in 1st & 2nd months; if you need to drink

socially, try a small amount & see how you feel. Don't drink and take lithium when you drive.
16. **When I feel better, can I stop taking lithium?** You should not stop suddenly. Need to consult your doctor. Lithium is usually a long-term treatment.
17. **Is lithium addictive?** No, it is because you do not need to take more and more lithium to achieve the same effect.
18. **Do I need to know anything else as I stay in Singapore?** Drink enough water in hot weather. Lack of water in body may cause more side effects.
19. **My younger brother likes to steal my medicine. What would happen to him if he swallows a large amount of lithium?** Lithium is toxic if a person takes an overdose. A person will first present with loose stool/vomiting, then very shaky hands, unsteady walking, confusion and may die. You need to send the person to the Emergency Department immediately.
20. **What are the other alternatives besides lithium?** There are other medications which can stabilise patient's mood which are anti-fit / epilepsy medication.

CASC grid – Assess suicide risk

A 24-year-old woman took an overdose of 20 tablets of paracetamol. She is brought in by her partner to the Accident and Emergency Department and you are the resident on duty at the Accident and Emergency Department.

Task: Assess her suicide risk.

A. Assess her suicide plan and intent	A1. Introduction. I am Dr. XXX. I can imagine that you have gone through some difficult experiences. Can you tell me more about it? Can you tell me why you took the 20 tablets of paracetamol tonight? Was there any life event leading to this suicide attempt?	A2. Assess her plan. Was the overdose planned? If yes, how long have you thought about it? How did you collect the paracetamol tablets? What did you think would happen when you took the paracetamol?	A3. Assess intent. Have you thought about taking your own life by the overdose?
B. Assess circumstances of suicide attempt.	B1. Assess location of the suicide attempt. Where did you take the medication? Was anyone else there/were you likely to be found? Did you lock the door or take precaution to avoid discovery?	B2. Assess severity of overdose and other self-harm. Besides the paracetamol, did you take other tablets? Did you mixed the paracetamol with alcohol? Did you harm yourself by other means? (e.g. cutting yourself)	B3. Suicide note or good-bye message. Was a suicidal note left? Did you send a SMS or email to say 'good-bye' to your partner or family members?
C. Assess events after suicide	C1. The discovery How did you	C2. Assess physical complications.	C3. Assess current suicide risk.

attempts.	come to be in A&E? Were you discovered by other people? If yes, how did they discover you?	Did the overdose lead to any discomfort? E.g. severe vomiting. Did you have a period of black out?	How do you feel about it now? Are you regretful of your suicide attempt? Would you do it again?
D. Assess other risk factors or protective factors.	D1. Past history of suicide. Have you attempted suicide previously? If yes, how many times? What are the usual causes of suicide attempts? Did you try other methods like hanging, stabbing yourself, jumping from heights or drowning?	D2. Past psychiatric / medical history. Do you have a history of mental illness? (e.g. depression) and take a brief mood history and past treatment if depression is present. Are you suffering from any other illnesses? (e.g. chronic pain)	D3. Assess protective factors. We have discussed quite a lot on the overdose and some of the unhappy events. Are there things in life you are looking forward to? Who are the people supporting you at this moment? How about religion?

OSCE video: Assess anxiety, panic attacks and phobia

You are the resident working at the Accident and Emergency Department. A 26-year-old married man is referred by her GP because of his fear that he is going to lose control with hyperventilation in his office. He seems to be very stressed.

Task: Assess anxiety, panic attack and phobia.

OSCE grid: assess anxiety, panic attack and phobia.

A. Introduce and assess generalised anxiety	A1. Introduction I am Dr. XXX, a resident of the Accident and Emergency Department. I understand that your GP has referred you because you are afraid that you are losing control. I can imagine that it is a terrible experience. In the next 7 minutes, I want to find out more about your experiences. Is it ok with you? Can you tell me more about your stress?	A2. Assess generalised anxiety Do you tend to worry a lot? If yes, how many days in the last month? Do you worry about anything in particular?	A3) Assess physical symptoms What sort of symptoms do you get when you feel worried? • Do you feel shaky? • Do you sweat a lot? • Do you have difficulties with breathing? • Do you feel that your heart is beating very fast? • Do you have loose stools? • Do you feel dizzy or light-headed?
B. Assess panic attacks and agoraphobia	B1. Assess panic attacks Have you ever had the experience that you felt as if you might have a heart	B2. Assess triggers Is there anything that trigger the attacks?	B3. Assess agoraphobia. Do you have tend to feel anxious in crowded places or public transport?

	attack or that you might even lose control? If yes, can you describe the symptoms to me. How frequent have these attacks been for you? Do you always anticipate about another attack? (anticipatory anxiety)	Tell me how you felt when you knew the attack was coming along? Are you very concerned and worried about these attacks?	Do you have fear when away from home? Can you tell me what happens when you have this fear? Do you avoid those places?
C. Assess social phobia, specific phobia, comorbidity, past history	**C1. Assess social phobia.** Do you worry about social situations where you are being the focus of attention? Do you feel that other people are observing you and you feel very uncomfortable? Can you tell more about your concern?	**C2. Assess specific phobia.** Are you scared of other situations or object? Can you tell more about the situation or object?	**C3. Assess comorbidity and past psychiatric history** I am sorry to hear that you are affected by the above signs and symptoms. How does this condition affect your life? How do you cope? Did you seek help from your GP or psychiatrist? Did they offer you any treatment? If yes, what is the effect on your condition? How is your mood? How is your sleep and appetite? Do you need to drink alcohol or take sleeping pills to overcome those symptoms? Can you tell me more about your medical history? So you suffer from thyroid or heart disorders?

OSCE grid: Assess obsessive compulsive disorder

A GP has referred a 26-year-old woman to you who has severely chapped hands due to repeated hand washing. She is very concerned about contamination.

Task: Take a history to establish the diagnosis of OCD.

A) Assess obsessions	A1) Introduction and assess the reasons for excessive hand washing. I am Dr. XXX. The GP has referred your case to me due to excessive hand washing. Can you tell me why you need to wash your hands so many times a	A2) Assess the nature of obsessions and resistance. Do you feel that your thoughts are excessive? Are those ideas reasonable? Do you feel unpleasant	A3) Assess obsessional doubts. Do you ask yourself the same question over and over again? For example, you cannot be certain whether you have closed the door even though you	A4) Assess obsessional impulses. Do you have impulses which you cannot control? (e.g. impulse to do inappropriate thing) Do you have recurrent thoughts of	A5) Explore other obsessions: Do you like things to be in a special order? Do you feel upset if someone changes it?

	day? Can you tell me more about your concerns? Do you come up with this thought? (Assume the patient tells you that she is concerned about contamination).	about those thoughts? Do you want to stop those thoughts?	have checked a few times.	harming yourself or others?	
B) Assess compulsions	**B1. Assess compulsive washing.** Can you tell me how many times you need to wash your hands per day? Why do you need to wash your hands so many times a day? How long does it take for you to take a bath? Why does it take so long? What do you do inside the bathroom?	**B2. Assess compulsive checking.** Do you need to check things over and over again? What kinds of items do you check? (e.g. windows, doors). How long does it take for you to finish checking all items before leaving your house?	**B3. Assess compulsive counting.** Do you count things over and over again? If yes, why? Is there a number you like or do not like?	**B4) Assess other rituals** Do you perform a regular ritual or ceremony to prevent something bad from happening?	**B5) Assess nature of compulsions** How do you find the repetitive behaviours? Are they excessive? Are they pleasurable? What would happen if you do not clean your hands? How long have you been washing your hands excessively?
C). Assess impact, comorbidity, risk and insight.	**C1) Assess psychosocial impact.** Since you wash your hands very frequent, does it affect your work? Does it affect your relationship with other people? Are you slow at work? Can you tell me your water bill? Is it very high?	**C2)_Assess comorbidity.** Do you feel stressed or nervous? How's your mood? How's your sleep and appetite? How you thought of ending your life? Can you tell me more about your character? Are you a perfectionistic person?	**C3) Assess biological complications** Since you wash your hands many times a day, do you have any complication on your skin? Do you consult a dermatologist?	**C4) Assess insight.** What is your view of the current problem? Do you think you have an illness in your mind? Do you think you need check to reduce the hand washing behaviour?	**Assess her insight and expectations** Does she think that she has a psychiatric illness? If not, what are her views and explanations? Has she read any information on OCD? What are her expectations on treatment? Are they realistic or achievable? What type of treatment does she prefer? (Medication,

		Do you have abnormal twitching movement in your face? (Assess tics)			psychotherapy or both).

OSCE grid: Assess post traumatic stress disorder

The GP has referred a 35-year-old driver to you for assessment. He was almost killed in a road traffic accident 6 months ago and, he is suing the other party for compensation.

Task: take a history to establish the diagnosis of PTSD.

A) Exploration on his trauma	A1) Explore the accident. I am Dr. XXX. I am sorry to hear that you were involved in a road traffic accident. In the next 7 minutes,, I would like to find more about the recent event. Is it ok with you? Can you tell me what happens on that night? Were you driving the car alone or with someone? Can you describe the severity of the accident? Was it life threatening?	A2) Explore immediate outcome of the accident. How long did you wait for the rescue to come? Do you remember what happened next? Were you brought to the accident and emergency department? Were you admitted to the hospital? What kind of treatment did they offer? Did you undergo operation?	A3) Assess the extent of injury and suffering. Can you tell me some of the complications after the accident? Do you lose any ability or function? For example, memory, mobility or sensation. Are you in pain at this moment? If yes, for how long?	A4) Assess outcome of other people involved in the accident (if any) Were the other passengers injured? If so how many? What happened to them? What is your relationship with them? Do you feel sorry towards them?
B) Assess PTSD symptoms	B1) Re-experiencing Identity the latency period between the incident and the onset of PTSD symptoms. How does the memory relives itself? How vivid is it? Does the memory come in the form of repetitive distressing images? How often do those mental images come in a day? Do you have nightmares at night? Can you tell me more about it?	B2) Avoidance Do you try to avoid driving a car? How about sitting in a car? Do you try to avoid the place where the accident occurred?	B3) Hyperarousal "Are you always on the edge?" How about excessive sweating, fast heart beats and difficulty in breathing? How do you find your concentration?	B4) Emotional detachment: Are you able to describe your emotion? Do you feel blunted?

C) Assess comorbidity, vulnerability, compensation issues.	C1) Assess comorbidity. How is your mood? How do you see your future? Do you have thought of harming yourself? How do you cope? Do you turn to alcohol or recreational drugs?	C2) Assess vulnerability. Did you encounter any traumatic event when you were young? (e.g. abuse, past accident) Did you stay with your family when you were young? (explore social isolation) Can you tell me your education level? (low education is associated with PTSD).	C3) Explore compensation and legal procedure. What is the status of the legal procedure? Is your case due to be heard in court soon?	C4) Assess current support system and past treatment. I am very sorry to hear the road traffic accident and the complications you have gone through. Do you get any support from your partner or family members? Did you see a doctor for the anxiety symptoms? If yes, did the doctor offer treatment to you? How do you find the treatment?

Assessment Questionnaire for Alcohol addiction: Alcohol Use Disorders Identification Tool (AUDIT)

AUDIT questionnaire:

Introduction: Now I am going to ask you some questions about your use of alcoholic beverages during this past year. Please explain what is meant by "alcoholic beverages" by using local examples of beer, wine, vodka.

Hazardous alcohol misuse:
1. Frequency of drinking: "How often do you have a drink containing alcohol?"
2. Typical quantity: "How many drinks containing alcohol do you have on a typical day when you are drinking?"
3. Frequency of heavy drinking: "How often do you have six or more drinks on one occasion?"

Dependence symptoms:
4. Impaired control over drinking: "How often during the last year have you found that you were not able to stop drinking once you had started?"
5. Increased salience of drinking: "How often during the last year have you failed to do what was normally expected from you because of drinking?"
6. Morning drinking: "How often during the last year have you needed a first drink in the morning to get yourself going after a heavy drinking session?"

Harmful alcohol use
7. Guilt after drinking: "How often during the last year have you had a feeling of guilt or remorse after drinking?"
8. Blackouts: "How often during the last year have you been unable to remember what happened the night before because you had been drinking?"
9. Alcohol-related injuries: "Have you or someone else been injured as a result of your drinking?"
10. Others concerned about drinking: "Has a relative or friend or a doctor or another health worker been concerned about your drinking or suggested you cut down?"

AUDIT scores and management: 0-7: Alcohol education; 8-15: Simple advice; 16-19: Simple advice, brief counseling and continued monitoring; 20-40: Referral to specialist for diagnostic evaluation and treatment. A high AUDIT score is strongly associated with suicidality.

*OSCE: Alcohol dependence

Clinical Vignette: A 40 year-old man was admitted to the medical ward with minor head injury after he was drunk. Routine blood tests showed increased GGT and MCV. The physicians have sent a referral to you because the patient also accuses his wife of having an affair with another man which is not true.

Introduction	Introduction and establishment of therapeutic relationship with the patient. - Hi, I'm Dr. Michael. I	Establishment of drinking habits - Could you tell me when you first tasted alcohol?	Establishment of average alcohol consumption - Can you tell me on average how much	Establishment of social factors influencing drinking habits - Do you usually drink alone or with

	understand that you have been referred from your physician as you have had some blood abnormalities. There is also some concern about your relationship with your wife. - Could we spend some time to explore that in further details?	- Could you also tell me when you started drinking occasionally and regularly at weekends, evenings, lunchtimes and in the mornings?	do you drink everyday? - What do you usually drink? - What else do you drink?	your friends? - Do you have a tendency to indulge in more alcohol when you are drinking with your friends? - Do you always drink in the same hawker centre? - Do you always drink with the same company?
Tolerance and Withdrawal	**Establishment of tolerance** - Nowadays, do you need more alcohol to get drunk than what you needed in the past? - What is the maximum you have drunk in a day? - How much can you drink without feeling drunk?	**Establishment of withdrawal effects** - What happens if you miss your drink? - What would happen if you go without a drink for a day or two?	**Establishment of physical effects of alcohol withdrawal** - Do you feel shaky or sweat a lot?	**Elicit hallucinations** - Were there times when you were seeing or hearing things when you could not have your usual amount of alcohol?
Motivation to stop drinking (CAGE Questionnaire)	**C – Cut down drinking, requested by family members.** - Has anyone in your family advised you that you need to cut down on drinking?	**A – Annoyed by family members** - Do you feel that your family members are displeased with regards to your current drinking problem? - Have they told you that they are irritated by your drinking issues?	**G – Guilty of drinking** - Do you feel any guilt with regards to your current drinking issue?	**E – Eye opening in morning and then start to drink** - Do you find yourself having to resort to alcohol as an eye opener, or to kick start a day?
Complications of drinking	**Family and social issues** - Have you had issues with your family because of your drinking habits?	**Work and financial issues** - Has your drinking habit got you into issues with your work? - Do you have any problems currently financing your drinking habit?	**Forensic issues** - Have you got yourself into trouble because of your drinking? Do you have issues with drunk driving, drunk and disorderly behavior in the public or ended up in fights when you were drunk?	**Relationship issues** - Have there been any problems with your existing marital relationship?
Treatment & Motivational interviewing	**Treatment** - Have you undergone any specific treatment previously for your alcohol issues?	**Relapse after Treatment** - How long have you been successful without relying on alcohol? - Could you tell me more as to why you started drinking again? - When you restarted, how long did it take you before you were back at your normal level of consumption/	**Assessment for suitability currently to quit drinking** - Do you feel you have a problem with alcohol? - Have you ever thought of giving it up completely? - What do you think	

| | | | will happen if you give up completely? | |

OSCE (Assess AN)

Name: Sally **Age:** 15-year-old

Sally, a secondary school student with a two- year history of anorexia nervosa, is admitted to the hospital following a seizure after prolonged fasting. On admission, her BMI is 10 and her heart rate is 35 beats per minutes. You are approached by her parents who beg you to save Sally.

Task: To take a history from Sally to establish the aetiology and course of anorexia nervosa.

A) Severity of AN symptoms	A1) Dietary history	A2) Longitudinal weight history	A3) Methods to lose weight and binge eating	A4) Body image distortion	A5) Serious medical complications in the past:
	"Hello Sally, I am Dr. Tan. Can you take me through your diet habit on a typical day?" Look for the number of meal times, the content of food. "How long have you been eating in this way?" "Where do you learn this diet habit from?"	Take a history on Sally's weight. E.g. the lowest, highest and average weight in the past 2 years. "What is your ideal weight?"	Explore the methods used by Sally (e.g. avoidance of 'fattening foods', self-induced vomiting, purging and excessive exercise) Although Sally presents with AN, it is important to ask about binge eating	Assess how fixated Sally is on her overvalued idea (e.g. dread of fatness) and find out her self-imposed weight threshold. "How do you feel when you look into the mirror?" "Your BMI is only 10. How do you feel about it?" If patient still thinks she is too fat, gently challenge her belief and check her rationales.	Explore common neuropsychiatric complications (e.g. slowing of mental speed, fit), gastrointestinal (GI bleeding), and endocrine systems (no menstruation). Severe weight loss, very low heart rate and metabolic complications such as very low potassium or anaemia. Explore relevant past medical history, e.g. childhood obesity.
B) Aetiology of Sally's illness	B1) Identify predisposing factors Family dysfunction including marital disharmony and sibling	B2) Identify precipitating factors e.g. Sally may use her illness to get more attention from her	B3) Identify maintaining factors Identify the role of family in reinforcing and maintaining	B4) Development in adolescence Explore her cognitive and psychosexual development. Focus on	B5) School and peers Explore her interests and hobbies (e.g. ballet dancing, athletes) and academic performance.

	rivalry. Enmeshment, child abuse and rigidity in parenting may be present. Explore the family's views on food and weight.	parents and prevent them from arguing. This will positively reinforce her illness.	her abnormal eating behaviour.	common issues such as individuation.	Explore her peer and romantic relationships (previous bullying or rejection due to body image)
C) Course of Sally's illness, comorbidity and risk	C1) Previous treatment Explore both outpatient and inpatient treatment being offered to Sally. Explore previous use of medication (e.g. antidepressant, antipsychotics) and adherence to psychotherapy sessions.	C2) Outcomes of previous treatments Focus on the weight restoration and identify reasons resulting in failure (e.g. engagement difficulties with Sally)	C3) Sally's insight and feeling towards her illness Sally may have impaired insight and denies any illness. She would be aggrieved by repeated attempts by her family to get her to seek help.	C4) Explore comorbidity e.g. depression, anxiety, OCD, substance abuse and perfectionistic personality Explore how the comorbidity influences the response to treatment.	C5) Risk assessment: History of suicide and deliberate self harm.

Clinical OSCE on suicide assessment

Name: Miss C   **Age:** 22 – year –old   **University student**

Miss C is referred to you by her general practitioner after she took 30 tablets of paracetamol. She states that she lacks motivation in life. Life appears to be meaningless. Her existence is only postponing the inevitability of death. She has a history of repeated self-injury and she had two previous psychiatric admissions with her discharging herself. She claims that she has feels this way throughout her life

Task: 1) Perform a suicide risk assessment. 2) Explore the underlying cause for her suicidal ideation.

| A) Risk assessment | A1) Empathy statement: "I can imagine that you have gone through a difficult period. I am here to help you and listen to you." | A2) Current suicidal intention 1) Do you wish that you were dead? 2) Do you still have thoughts of ending your life? (If so, are they intermittent or | A3) Detailed assessment of suicide plan 1) Intent: 'Did you intend to end your life by taking an overdose?' 2) Detailed | A4) Negative aspects of life 1) 'Have you ever felt despaired about things?' 2) 'Have you ever felt life is a burden?' 3) 'Have you | A5) Positive aspects of life 1) 'Do you hope that things will turn out well?' 2) 'Do you get pleasure out of life?' 3) 'Can you |

| | | more persistent?)

3) How often do you act on these ideas?

4) How strongly are you able to resist those thoughts? | plans made: 'Did you plan for this suicide attempt? If yes, how long did you plan for it?'

3) Method considered and available: 'Besides overdose, did you harm yourself in other ways?'

4) 'Did you act alone or in front of the others?'

5) Did you inform anyone prior to suicide attempt?

6) Post-suicide attempt: 'Did you try to avoid discovery? Did you seek help?' | ever felt entrapped, defeated or hopeless?' | tell me more about your support system?'

4) 'Do you have any spiritual support? E.g. religion?' |
|---|---|---|---|---|---|
| **B) Underlying causes** | **B1) Current life stressors** e.g. adjustment to university life, study load and relationship problems. | **B2) Assessing problems from a developmental perspectives** e.g. childhood physical abuse, separation from parents, marital discord of parents, witnessing domestic violence or witnessing someone committing suicide in the | **B3) Existential crisis** 'Do you feel isolated?' 'How do you see the world? Do you feel the world is hostile and meaningless? If so, is suicide your final destiny?' | **B4) Past suicide attempts and self harm** History of self harm and suicide attempts. Explore common precipitating factors of previous suicide attempts. | **B5) Psychosocial problems** e.g. unplanned pregnancy, financial problems and poor coping. |

| C) Psychiatric comorbidity | C1) Depression and previous therapeutic relationships

Explore depressive symptoms in details (e.g. low mood, guilt, insomania, loss of interest). | family and risk of developing post traumatic stress disorder | | | |
|---|---|---|---|---|---|
| | | C2) Substance abuse

1) Do you take recreational drugs to cope with life?

2) How about alcohol or smoking? | C3) Eating disorder
1) How do you see your body image?

2) Have you put yourself on diet?

3) How about binge eating? | C4) Explore personality disorder

1) Borderline personality traits: chronic feeling of emptiness, unstable emotion, impulsiveness. | C5) Early psychosis or schizophrenia

1) Command hallucination (e.g. have you ever heard voices to ask you to harm yourself?) |

Assessment of sleep disturbances:

Assessment of sleep disorders requires eliciting a detailed medical and psychiatric history from the patient. Apart from having a detailed medical and psychiatric history, it is crucial to perform a more in-depth assessment of the sleep disturbances.

The following questions should be asked during assessment:

Daytime:
1. Do you feel sleepy during the day?
2. Do you take routine naps during the day?
3. Do you find yourself having difficulties with concentration during the day?

Night-time:
4. Could you describe to me a typical night of sleep (in terms of the number of hours you get, the quality of sleep etc.)?
5. Do you find yourself having difficulties with falling asleep?
6. Do you sleep well? Do you find yourself awake during the night? If so, what is the reason? Is it because of poor sleep or you need to go to toilet? (going to toilet twice per night is normal and not considered to be sleep disturbance)
7. Do you find yourself waking up much earlier in the morning?

Cause and course of sleeping problems:
8. Could you tell me how long you have had such sleeping difficulties?
9. What do you think might have precipitated such difficulties?
10. Do you wake up and sleep at the same time during weekends as weekdays?
11. Does your job currently require you to work on shifts or travel frequently?
12. Do you drink caffeinated beverages close to your desired sleeping time?
13. Do you have any other long standing medical problems apart from the difficulties you are experiencing currently with your sleep?

Treatment
14. Did you seek help for your sleep problems? (e.g. GP, psychiatrist, acupuncturist, traditional medicine practitioner)?
15. Are you on any chronic long-term medications to help yourself to sleep (e.g. sleeping pills)? Where do you get those medications?

In some cases, it is also crucial to obtain a sleep history from a sleep partner.

The following questions should be asked:
1. Have you noticed the change in sleep habits of your partner?
2. Have you noticed that your partner has been snoring during his sleep?
3. Does your partner exhibit any abnormal movements (e.g. kicking) during sleep? Were you injured?

OSCE station

Name: Mr. A   Age: 40 years old   Occupation: unemployed

You are the medical resident receiving training in hepatology. Mr. A, a hepatitis C carrier and an intravenous drug abuser, complains of severe weight and appetite losses, progressive lethargy, yellowing of the eyes and skin and abdominal distension. He consults a hepatologist who finds that he has deep jaundice and gross ascites. A CT scan of the abdomen reveals multiple liver masses and peritoneal deposits. Ascites fluid analysis shows malignant cells, accompanied by a very high serum α-fetoprotein protein level. The diagnosis of advanced hepatocellular carcinoma is made and Mr. A is informed that he has a very limited life expectancy.

Tasks: To address the end of life issues and his concerns

Table 7   Approach to counsel Mr. A

Possible questions raised by a cancer patient	Suggested approach
Why am I so unlucky to get this cancer?	Establish rapport and express empathically that you are sorry to hear what has happened.
Was it because I used drugs? Am I a bad person?	For issues of guilt, encourage the patient to avoid blaming himself. You can encourage him by saying that those who do not use drugs can also develop liver cancer e.g. people who get hepatitis B from birth.
I hate looking myself in the mirror. I look thin and my skin looks yellowish. What's wrong with me?	Need to explore his perception of his body image and look for possible jaundice.
The gastroenterologist says there is no cure for hepatitis C. Wouldn't it be better to give up?	Ask patient to think about the positive aspects of his life to look forward to and encourage him to fight the illness. Explore his view on his own death. Does he have any fear?
I am very worried that I will die soon. Will I die in severe pain? Can you just ask my doctor in charge to give me an injection and kill me? I don't want to suffer.	Address his suffering: Is he willing to ask for more pain control? Address the diversity of experiences of pain. Explain that euthanasia is illegal in Singapore and not an option. Explore his reasons for asking about this and discuss alternatives such as enhancing his pain management.
My mother doesn't know I have cancer. How should I tell her? She will be very sad. I am worried my family do not want me as I am a burden to them.	Explore his relationship with his family and his concerns around informing his mother of his diagnosis. Also inform him of the risks of hiding his illness from other family members. Address his concern as a burden and his concern of being abandoned.
Is there God? I have committed crimes and used drugs. Will I be forgiven?	Explore spiritual issues and his religious faith. Does he feel guilty about his past. Explore the need to be forgiven and who should forgive him e.g. God, family or friends.
Doctor, you are the psychiatrist. I	Explain boundary and your schedule in an empathetic way: you will

| need your emotional support. Will you stay with me until the day I die? | visit him regularly. Get other friends or caregivers involved to reach a conjoint effort. |

Assessment of dementia in OSCE exam:

A 70-year-old woman is brought by her son because she has become more forgetful.

Task: take a history to establish the diagnosis of dementia.

In clinical practice, dementia patients are often brought by concerned family members rather than complaining memory loss themselves. Dementia patients may not have insight or in denial of memory loss.

1. The onset of memory loss: gradual or sudden.
2. The extent of memory impairment: recent memory (more likely to be impaired) or long term memory (e.g. childhood history may not be affected).
3. Reactions to memory loss: confabulation (covering the memory loss by making up an answer), denial or catastrophic reaction (anger when being challenged of memory problems).
4. The extent of cognitive impairment: judgement, decision making, problem solving.
5. Explore aetiology: e.g. family history of AD, history of stroke, history of Parkinson's disease (e.g. resting tremor, shuffling gait, masked face) and history of multiple head injury.
6. Explore possible causes of reversible dementia: e.g. normal pressure hydrocephalus (gait abnormalities, urinary incontinence), dietary habits (vitamin B12 deficiency) and thyroid disorder.
7. Assess mood status: history of depression and possibility of pseudodementia as a result of depression (patient tends to give don't know answer). Assess sleep pattern and appetite.
8. Assess common psychotic features e.g. delusion of theft (accusing the domestic helper stealing an item because the patient cannot find), auditory or visual hallucination.
9. Assess behaviour problems: e.g. violent (e.g. attacking domestic helpers), disinhibition, wandering behaviour.
10. Assess risk: e.g. risk of having a fire or flooding at home as patient may forget to switch off stove or water tap, risk of fall, risk of financial exploitation, risk of self harm or suicide, risk of violence.
11. Assess activities of daily living (ADL): there are two types of ADL, basic ADL include bathing, feeding and toileting by oneself. Instrumental ADL include withdrawing money from the bank, shopping and using public transport.
12. Assess coping by patient (e.g. memory aids, reminders).
13. Assess coping by caregiver and strain on caregiver.
14. Explore past medical history and chronic medical treatment.
15. Assess education level and past occupation.

OSCE station

Name: Charles   Age: 14-year-old

Charles presents with a 5-year history of increasingly significant infringements of the law, truancy, repeated fighting and expulsion from two boarding schools. He is the younger of the two children with a 17-year-old sister, who is a high achiever. Both parents have a university education.

Task: Take a history from his father to establish diagnosis of conduct disorder.

OSCE grid

| A) Gathering more | A1) The source of referral: | A2) Timing of the referral: | A3) Onset of the problem | A4) Identify long pre-existing |

information on the referral	e.g. Via the GP, legal authorities and school authorities etc.	Why being referred now given Charles has a 5- year history of behavioural disturbance?	Whether the behaviour had begun suddenly 4 years ago (as a result of particular precipitant, e.g. parental, marital disharmony, changes in the family, history of abuse or rejection)	history of behaviours Adverse family environment and interaction style such as vague instructions, harsh, physical punishment with inconsistent discipline.
B) Assessment of behavioural disturbances	B1) Form of disturbed behaviour

Students should follow Table 1 to inquire about the symptoms of CD.

Other disturbed behaviours such as fire-setting, bed-wetting, school refusal and their time of onset in relationship to CD. | B2) The severity of behaviour and types of CD:

Students should gather enough information to classify his CD as mild, moderate or severe and classify his CD as socialised or unsocialised; child or adolescent onset. | B3) The effects of his behaviour on his family and the others

Explore Charles's relationship with his elder sister. Look for sibling rivalry. | B4) Reaction from parents

How parents have previously dealt with Charles's behaviours and what methods had been effective or ineffective? (Common reactions include nagging, more harsh punishment, coercive interactions or lack of insight of Charles's CD)

Look for maintaining factors for Charles' problems. |
| C) Assessment of psychiatric symptoms | C1) Other child psychiatric problems:

e.g. ADHD, past history of school refusal etc. | C2) Emotional state whether Charles is depressed, with biological and cognitive symptoms of depression.

Evidence for self harm (Has the father noticed any laceration wound?) and suicidal ideation. | C3) Psychotic features: Whether Charles currently or previously experiencing auditory or visual hallucination. Explore any form of delusion. | C4) History of substance abuse

e.g. cannabis, solvent and alcohol misuse. Explore smoking habit. |
| D) Problems with school, accommodation and the legal system | D1) Why he had been expelled from schools? (Explore both public and private schools)

Explore school environment (poor organisation or being unfriendly to Charles?)

Was he referred to | D2) Academic performance

Does Charles have any reading, spelling or mathematic problems?

Ask the father to comment on his results in primary and secondary school? Are there | D3) Accommodation

Whether he is now living at home or elsewhere (e.g. with uncle because his father cannot discipline him). | D4) Legal problems

Ask the parent to take you through his current and past involvement with the legal system. |

	see a counsellor or educational psychologist? Is he in any school at this moment?	any differences?		
E) Background history	**E1) Developmental history:** eliciting a history of perinatal difficulties, developmental delay, low IQ, hearing impairment, early behavioural problems, difficult temperament, history of poor co-ordination and motor skills.	**E2) Family history of psychiatric illness** Particularly anti-social personality traits and forensic record in the parents, a history of parental alcoholism, parental psychiatric illness or parents who themselves suffered from conduct disorders in the past.	**E3) Parenting, marriage and expectation:** Identify occupation of parents. Are the parents working for long hours with little time for Charles? Are there any marital problems and parental discord or any disagreement over parenting? Explore parental expectation on Charles and in particular to the level of parental education and the fact that his older daughter is a high achiever.	**E4) Past medical history:** Has any doctor commented on abnormality in his nervous system or any atypical facial appearance (such as 'neurological soft signs' or syndromes associated with mental retardation) Does Charles suffer from any chronic medical illness? (e.g. epilepsy)

Mastering Psychiatry: A Core Textbook for Undergraduates
Clinical Pharmacology Guide

The 'most' and the 'least' of the first generation and second generation of antipsychotics.

Antipsychotics	The most likely / High risk ↑	The least likely / Low risk ↓
1st generation		
First generation antipsychotics	1) In acute dystonia, young men at most risk. 2) In tardive dyskinesia, elderly women are at most risk.	Nil
Haloperidol	1) High risk for EPSE 2) High risk for galactorrhoea	1) Haloperidol - low risk of weight gain, postural hypotension and sedation.
Trifluoperazine	Nil	1) Trifluoperazine: low risk of weight gain and postural hypotension.
Sulpiride	Nil	1) Sulpiride carries low risk of weight gain. 2) Sulpiride carries low risk of postural hypotension. 4) Sulpiride carries the low risk of sedation.
Clozapine	1) Most common side effect is sedation till the next morning. 2) Second most common side effect is hypersalivation. 3) Clozapine (like olanzapine) carries the highest risk of weight gain.	1) Clozapine is least likely to cause tardive dyskinesia.
2nd Generation		
Risperidone	High risk for EPSE and galactorrhoea compared to other 2nd generation.	1) Risperidone carries low risk of sedation.
Olanzapine	Olanzapine carries the highest risk of weight gain among all antipsychotics.	1) Olanzapine carries low risk of EPSE. 2) Olanzapine carries low risk of hyperprolactinaemia. 3) Olanzapine carries low risk of QTc prolongation.
Quetiapine	1) Quetiapine has high affinity for muscarinic receptors and cause anticholinergic effects.	1) Quetiapine carries the lowest risk for EPSE. 2) Quetiapine carries the lowest risk of sexual dysfunction. 3) Quetiapine carries low risk of hyperprolactinaemia.
Aripiprazole	Nil	1) Aripiprazole carries the lowest risk of QTc prolongation. 2) Aripiprazole carries low risk of sexual dysfunction. 3) Aripiprazole carries low risk for EPSE. 4) Aripiprazole carries low risk of dyslipidaemia, weight gain and glucose intolerance. 5) Aripiprazole carries low risk of hyperprolactinaemia. 6) Aripiprazole carries low risk of postural hypotension.

| | | 7) Aripiprazole carries the low risk of sedation. | |

The first generation antipsychotics

Indications and contraindications	Mechanism of action	Side effects	Drug interactions
Indications: Schizophrenia. Schizoaffective disorder. Substance induced psychosis. Personality disorder with psychotic features. Affective disorders. Tourette's syndrome Huntington's disease Nausea, emesis and hiccups. **Contraindications** Parkinson disease. Lewy body dementia. Elderly who are prone to develop extrapyramidal side effects.	**Symptom control:** mesolimbic dopamine blockade is thought to be the most important for control of positive psychotic symptoms. **Receptor occupancy:** PET studies have shown 65-90% occupancy of brain D2 receptors after normal antipsychotic doses.	**Neurological adverse effects:** **Extrapyramidal symptoms** are due to blockade of D_2 receptors in the basal ganglia. **Tardive dyskinesia** is due to D_2 receptor hypersensitivity. **Drowsiness** is due to antihistamine activity. **Secondary negative symptoms** (indifference to the environment, behavioural inhibition and diminished emotional responsiveness) are due to dopamine antagonism in the mesocortical pathway. **Memory impairments** attributable to antimuscarinic effects and dopamine blockade in the cortex. **Impairments in cognitive and psychomotor functions** occur after acute treatment in both healthy volunteers and patients. Chronic treatment does not cause any significant impairment on psychometric tests or psychomotor performance. **Fine motor incoordination** is due to nigrostriatal blockade. **Hormonal side effects:** **Galactorrhoea** is due to dopamine antagonism in the tuberoinfundibular pathway. Plasma neuroleptic levels correlate with prolactin increases. **Other endocrine effects include:** 1) False pregnancy test 2) Weight gain 3) Secondary amenorrhoea 4) Unilateral gynaecomastia. **Allergic side effects:**	**Drug interactions of first generation antipsychotics e.g. phenothiazines:** They potentiate the depressant action of antihistamine, alcohol, GA and benzodiazepine. They increase analgesic effects of opiates. They cause a marked increase in intracellular lithium. They antagonise the dopaminergic effect of anti-parkinsonian drug. Phenothiazines are protein bound, care must be taken when administered with other highly protein-bound medications (e.g. warfarin, digoxin, theophylline) and potent 2D6 inhibitors (e.g. fluoxetine, paroxetine, cimetidine and erythromycin).

		1) Contact dermatitis 2) Opacities in the lens 3) Cholestatic jaundice 4) Optic neuritis 5) Aplastic anaemia **Haematological side effects:** 1) Transient leucopenia 2) Agranulocytosis	

Chlorpromazine (Phenothiazines with aliphatic side chain) [Dose: 100mg to 800mg/day; 2012 price: $0.12/ mg]

Indications and contraindications	Mechanism of action	Side effects
Chlorpromazine is sedative antipsychotics. indicated for schizophrenia. Chlorpromazine is contraindicated in cholestatic jaundice, Addison's disease, myasthenia gravis, glaucoma and bone marrow depression.	High level of anticholinergic, anti-α-adrenergic, and antihistaminergic actions. At a dose of 100 mg twice daily, chlorpromazine exhibits 80% dopamine D2 receptor occupancy. $T_{1/2}$ = 16-30 hours	Most sedative first generation antipsychotics. Cataract (↑ risk by 3-to-4 fold), miosis, weight gain, increased duration of SWS, galactorrhoea, haemolytic anaemia or agranulocytosis, leucocytosis or leucopenia, cholestatic jaundice (hypersensitivity reaction), quinidine like side effect (prolonged QTc interval, ↓ST and ↓T wave blunting, photosensitive rash and hyperglycaemia. **Priapism:** alpha-receptor antagonism unopposed by cholinergic stimulation may be the underlying mechanism. Summary: Antihistaminergic side effects> anticholinergic side effects = extrapyramidal side effects (EPSE).

Trifluoperazine (Phenothiazines – piperizine side chain) [Dose range: 5mg to 15mg/day; 2012 price: $0.12/mg]

Indications	Mechanism of action	Side effects
Less sedative antipsychotics	Increased D_2 blockade and a lower affinity to muscarinic α-adrenergic and histaminergic receptors. $T_{1/2}$ = 10-20 hours	More likely to cause EPSE. Summary of side effects: EPSE > anticholinergic = antihistaminergic side effects.

Haloperidol (Butyrophenones) [Dose range: 5mg to 10mg/day; 2012 price: $0.10/mg (tablet), Injection 5mg $1.87/mg, Drops $26.66/ ml]

Indications	Mechanism of action	Side effects
Indications: 1) Positive symptoms of schizophrenia. 2) Delirium. 3) Rapid tranquillisation (IM) 4) Available in liquid form (covert medication) 5) Tourette's syndrome	Very potent D_2 blocker Lower level of activity with the nigrostriatal pathway. Little antimuscarinic, antihistaminergic and anti-adrenergic activity.	High doses often lead to EPSE, akathisia and akinesia.

| Contraindications:
1) Parkinson's disease.
2) Lewy body dementia. | $T_{1/2}$ = 10-30hours | |

Sulpiride (Substituted benzamines) [Dose range: 200mg to 800mg/day; 2012 price: $0.24/mg]

Indications and contraindications	Mechanism of action	Side effects
Schizophrenia patients who cannot tolerate EPSE associated with other first generation antipsychotics. Contraindication: renal failure, acute porphyria.	Selective antagonist at D_2 and D_3 receptors. $T_{1/2}$ = 7 hours	Hyperprolactinaemia. Dry mouth, sweating, nausea.

Flupenthixol (Thioxanthenes) [Dose range: IM 20mg once per month; 2012 price: $0.12/mg]

Indications	Mechanism of action	Side effects
Depot antipsychotics For people with schizophrenia who are non–compliant. Not a recommended as a first-line treatment for depression (some patients found it useful).	Antipsychotic effect. Low doses may have antidepressant effects. $T_{1/2}$ = 10-20 hours	Acute dystonia EPSE Long term usage may lead to tardive dyskinesia.

The second generation antipsychotics

Risperidone (Benzisoxazole) [Dose range: 1mg to 6mg/day; 2012 price: $0.40/mg]
Riperdal consta (IM depot) [Dose range: IM 37.5mg to 50mg every 2 weeks; 2012 price: $155.06/ 25 mg]

Indications and contraindications	Mechanism of action	Side effects
Schizophrenia: 1-2mg is the minimum effective dose for the first episode and 4mg for relapse. Bipolar disorder – mania. Behavioural problems in dementia and autism. For covert administration (Risperidone droplets). Contraindications:	High-affinity antagonist of 5-HT_2, D_2-like. α-Adrenergic antagonism It allows more dopaminergic transmission than conventional antipsychotic because the normal inhibitory action of serotonin on dopamine neurons is inhibited due to antagonism of the $5HT_{2A}$ heteroreceptor. The affinity of risperidone for D_2 receptors is approximately 50-fold greater than that of clozapine and 20%-50% of haloperidol.	Hyperprolactinaemia and EPSEs at higher doses. (Reduction of EPSE may be confined to a low dose range, of 2-6 mg, with higher doses giving a profile that approaches that of a first-generation agent). Other side effects include: insomnia dizziness, anxiety, menstrual disturbances and weight gain.

patients reported EPSE with risperidone in the past.	$T_{1/2}$ = 3-20 hours	

Olanzapine (Thienobenzodiazepine) [Dose range: 5mg to 15mg /day; 2012 price: $5.69 / 5mg tablet]

Indications and contraindications	Mechanism of action	Side effects
Indications: Schizophrenia, Schizoaffective disorder. Bipolar mania. **Olanzapine zydis (rapidly dissolvable form) for rapid tranquillisation.** **Contraindications** Stroke patients Patients with diabetes. Obese patients. Narrow angle glaucoma due to anticholinergic effect	High affinity for D_2 and $5HT_{2A}$ but low affinity for D_1 receptors. Similar chemical structure to clozapine. Clozapine selectively binds to many different dopamine receptors, whereas olanzapine is only partially selective for D_2 receptors. $T_{1/2}$ = 21-54 hours	Weight gain. High risk of diabetes. Appetite increase. Sedation. Anticholinergic side effects: dry mouth and constipation. Dizziness. EPSE under olanzapine are not absent all together but if they occur tend to be mild. at relatively high levels of D_2 occupancy. This occurs in association with high anticholinergic effect which may contribute to mitigation of EPSE. The annual rate of tardive dyskinesia is 0.5%.

Quetiapine (Dibenzothiazepine) [Dose range: 200mg to 800mg/day; 2012 price: $2.10 / 200 mg tablet]

Indications	Mechanism of action	Side effects
Schizophrenia patients who cannot tolerate EPSE. Parkinson disease patients with psychotic features after taking levodopa.	Has a high affinity for muscarinic receptors. Quetiapine has high affinity for $5-HT_{1A}$ receptors may increase dopamine levels in the hypoactive mesocortical dopaminergic pathway and improve negative symptoms. It has lower affinity for all receptors than clozapine. $T_{1/2}$ = 6 hours	Sedation (17.5%) Dizziness (10%) Constipation (9%) Postural hypotension. No difference from placebo in terms of EPSE and prolactin level. Less weight gain compared to olanzapine and clozapine (clozapine = olanzapine > risperidone > quetiapine > ziprasidone).

Ziprasidone [Dose range: 20 mg BD to 80mg BD; 2012 price: This drug is not available at NUH]

Indications	Mechanism of action	Side effects
Schizophrenia patients who cannot tolerate weight gain and people with schizophrenia taking warfarin.	Potent $5HT_{2A}$ and D_2 antagonist The effects of ziprasidone on negative symptoms and possibly cognitive symptoms may also be related to its potent antagonism for $5-HT_{2A}$ receptors. It also exhibits $5-HT_{1A}$ agonism and inhibits the reuptake of	The overall effect of ziprasidone on movement disorder is no difference from placebo. Ziprasidone produced only modest weight gain in short term (4- to 6-week) trials, with a median weight gain of 0.5kg. ECGs revealed a modest prolongation with ziprasidone treatment in short-term (4- and 6-week).

| | noradrenaline and serotonin. $T_{1/2}$ = 7 hours | |

Aripiprazole [Dose range: 15mg – 30mg/day; 2012 price: $10.86 / 15mg tablet]

Indications	Mechanism of Action	Side effects
Schizophrenia patients who develop weight gain, metabolic syndrome, galactorrhoea, EPSE, QTc prolongation associated with other antipsychotics.	D_2 and 5-HT_{1A} partial agonist. 5HT_{2A} antagonist. High affinity for D_3 receptors; moderate affinity for D_4, 5HT_{2C}, 5-HT_7, adrenergic, and histaminergic receptors. There are no significant differences in outcomes compared to 1st and 2nd generation antipsychotics. $T_{1/2}$ = 74h to 94h (due to active metabolites)	Excellent safety and tolerability profile. Most common side effects include: Headache Insomnia Agitation Anxiety. Aripiprazole is less likely to cause elevation in prolactin compared to other antipsychotics.

Paliperidone [Dose range: 6mg to 12mg per day; 2012 price: $6.38 / 6 mg tablet]

Indications and contraindications	Mechanism of action	Side effects
Indicated for schizophrenia, schizoaffective disorder especially patients with Tourette's syndrome with liver impairment. Contraindication: renal impairment. This drug is known to be substantially excreted by the kidney.	Potent 5HT_{2A} and D_2 antagonist paliperidone is the major active metabolite of Risperidone $T_{1/2}$ = 23 hours	EPSE Akathisia QTc prolongation Hyperprolactinaemia, metabolic syndrome and increase in risk of seizure

Clozapine (Dibenzodiazepine) [Dose range: 200mg to 450mg daily; 2012 price: $0.26 / 25 mg]

Indications and contraindications	Mechanism of Action	Side effects
Clozapine should be offered to patients whose illness has not responded adequately to treatment despite the sequential use of adequate doses and duration of at least two different antipsychotics (MOH	Higher antagonist affinity for non-dopamine than for dopamine receptor subtypes); D_2 receptors: moderate affinity for D_2 receptors and a high affinity for 5-HT_{2A} receptors.	**Pulmonary embolism:** 1 in 4500 **Myocarditis:** 1 in 1300. **Agranulocytosis:** 1 in 10,000

guidelines 2011). Patients require full blood count on a weekly basis for the first 18 weeks, then fortnightly until the end of the first year. Then patient requires full blood count on a monthly basis after 1 year. **Contraindications:** Potential lethal combinations if patients take clozapine and carbamazepine or sulphonamide. This will lead to blood dyscrasia. Patients take lithium and clozapine increase the risk of seizure, confusion, dyskinesia and neuroleptic malignant syndrome (NMS)	$5\text{-}HT_{1A}$ partial antagonism with D_2-like antagonism this may contribute not only to mitigation of EPSE but also to enhancement of prefrontal dopamine release and putative therapeutic effects. Hypersalivation caused by clozapine is due to antagonism of α_2-adrenergic receptors and agonism at the M_4 receptor. $T_{1/2}$ = 6-26 hours	Neutropenia and agranulocytosis (agranulocytosis is defined as an absolute neutrophil count of < $500/mm^3$). Neutropenia is not dose related and occurs in 1-2 % of patients. If the temperature is over 38.5.°C, consider withholding clozapine until the fever subsides. Weight gain, metabolic syndrome, Hypotension, tachycardia and ST segment changes. Hypersalivation, constipation. and urinary incontinence. The incidence of seizures in people with schizophrenia taking clozapine at more than 600 mg per day is roughly 15%.(first seizure requires a reduction in dose and second seizure requires an addition of anticonvulsant such as sodium valproate)

Selective serotonin reuptake inhibitors (SSRIs)

Indications	Contraindications	Mechanism of action	Side effects
Depressive disorder (first line treatment, less sedative than TCAs). Anxiety disorders Obsessive compulsive disorder Bulimia nervosa	Mania	SSRIs selectively block reuptake of serotonin at presynaptic nerve terminals and increase synaptic serotonin concentrations	Nausea, abdominal pain, diarrhoea, constipation, weight loss. Agitation, tremor or insomnia. Sexual dysfunction

| (fluoxetine) Premature ejaculation. | | | |

Common SSRIs used in Singapore (2012 price in SGD)

Fluoxetine [Prozac] (Dose range: 20 – 60mg/day; $0.46 for 10mg and $0.24 for 20mg)

Special features 1) Non linear elimination kinetics 2) safe in overdose.

Other indications: 1) OCD (>60mg/day); 2) Panic disorder; 3) Bulimia Nervosa; 4) PTSD; 5) Premenstrual dysphoric disorder; 6) Premature ejaculation and 7) Childhood & adolescent depression.

Pharmacokinetics: 1) Fluoxetine inhibits the P450 3A3/4, 2C9, 2C19 & 2D6 and it also inhibits its own metabolism. 2) Due to non-linear pharmacokinetics, higher doses can result in disproportionately high plasma levels and of some side-effects (e.g. sedation) rather late in the course of treatment with this drug. Its metabolite norfluoxetine is much less potent. Long $t_{1/2}$ = 72 hr.

Pharmacodynamics: The serotonin system exerts tonic inhibition on the central dopaminergic system. Thus, fluoxetine might diminish dopaminergic transmission leading to EPSE.

Side effects: 1) Anxiety, agitation 2) Delayed ejaculation/orgasmic impotence 3) Hypersomnolence (high doses) 4) Nausea 5) Dry mouth.

Drug interaction: 1) The washout period for fluoxetine before taking MAOI is 5 weeks. 2) Through inhibition of P450 2D6, fluoxetine may elevate the concentration of other drugs especially those with narrow therapeutic index such as flecainide, quinidine, carbamazepine and TCAs.

Fluvoxamine [Faverin] (Dose range: 50-300mg/day; $0.24 for 50mg)

Special features: 1) Highly selective SSRI 2) FDA approval for OCD 3) Lower volume of distribution, low protein binding and much shorter elimination half-life compared to SSRI.

Other indications: 1) Efficacious for social phobia 2) Panic Disorder 3) PTSD

Pharmacokinetics: 1) Well absorbed 2) $t_{1/2}$ =19 hours 3) metabolized to inactive metabolites. 4) lower volume of distribution and low protein binding 5) maximum plasma concentration is dose dependent. 6) Steady-state levels is 2 to 4-fold higher in children than in adolescents especially females. 7) Well tolerated in old people and in people with mild cardiovascular disease or epilepsy. 8) It offers potent inhibition of P450 1A2.

Pharmacodynamics: The specificity for 5HT re-uptake is greater for other SSRIs. Two neuro-adaptive changes: 1) specific serotonin receptor subtypes that change following presynaptic blockade 2) neurogenesis of hippocampal brain cells occurs and results in changes in behaviour.

Side effect: 1) Nausea – more common than other SSRIs; 2) Sexual side effects with fluvoxamine are similar in frequency to those with other SSRIs; 3) It minimal effects on psychomotor and cognitive function in humans.

Sertraline [Zoloft] (Dose range: 50-200mg/day; $1.92 for 25mg)

Special features: 1) For young women with mood disorders. 2) For mood and anxiety disorders.

Other indications: 1) Premenstrual dysphoric disorder 2) OCD 3) PTSD

Pharmacokinetics: 1) It inhibits P450 2C9, 2C19, 2D6, 3A4. 2) $t_{1/2}$ is 26 – 32 hours 3) More than 95% protein bound 4) Its metabolite, desmethylsertraline, is one-tenth as active as sertraline in blocking reuptake of serotonin.

Pharmacodynamics: The immediate effect of sertraline is to decrease neuronal firing rates. This is followed by normalization and an increase in firing rates, as autoreceptors are desensitized.

Side effects: 1) GI side effects 27%; 2) headache 26%; 3) Insomnia 22%; 4) Dry mouth 15%; 5) Ejaculation failure 14%.

Paroxetine CR [Seroxat CR] (Dose range: 12,5-50mg/day; $2.23 for 12.5mg; $2.17 for 25mg)

Special features: 1) Most sedative and anticholinergic SSRI; 2 Risk of foetal exposure resulting in pulmonary hypertension

Other indications: 1) Mixed anxiety and depression; 2) Panic disorder; 3) social anxiety disorder; 4) generalised anxiety disorder 5) Post-traumatic stress disorder; 6) Premenstrual disorder

Pharmacokinetics: 1) Paroxetine is well absorbed from the GI tract. 2) it is a highly lipidophilic compound. 3) It has a high volume of distribution. 4) It is 95% bound to serum proteins. 4) It undergoes extensive first pass metabolism. 5) Paroxetine CR slows absorption and delay the release for 5 hours. 6) The Short $t_{1/2}$ of original paroxetine leads to discontinuation syndrome 7) It inhibits its own metabolism.

Side effects: 1) <u>Anticholinergic side effects</u> 2) Nausea 3) Sexual side effects emerge in a dose-dependent fashion 4) Closed angle glaucoma (acute). MOH guidelines state that first-trimester paroxetine use should be avoided, as it is associated with increased risk of serious congenital (particularly cardiac) defects.

Drug interaction: 1) Clinically significant interaction: MAOI, TCA, Type 1C antiarrhythmics, 2) Probably significant interaction: β-adrenergic antagonists, antiepileptic agents, cimetidine, typical antipsychotics, warfarin.

Escitalopram [Lexapro] (Dose range: 10-20mg/day; $1.5 for 10mg; $3.8 for 20 mg)

Special features: 1) Most selective SSRI 2) Relatively weak inhibition of liver P450 enzymes. 3) Escitalopram has fewer side effects, more potent, shorter $t_{1/2}$, less likely to inhibit P450 system, more selective than citalopram.

Other indications: 1) OCD 2) Panic Disorder 3) CVA 4) Anxiety with major depression 5) Emotional problems associated with dementia

Pharmacokinetics: 1) Escitalopram is well absorbed after oral administration with high bioavailability. 2) Peak plasma concentration is normally observed 2-4 hours following an oral dose. 3) It is subject to very little first-pass metabolism.

Side effects: Nausea and vomiting (20%), increased sweating (18%), dry mouth & headache (17%), anorgasmia and ejaculatory failure, but no significant effect on cardiac conduction and repolarisation.

Trazodone (Dose range: 150-300mg/day; this drug is not available at NUH)

Special features: trazodone is a mixed serotonin antagonist/agonist.

Pharmacokinetics: 1) Trazodone is well absorbed after oral administration, with peak blood levels occurring about 1 hour after dosing. 2) Elimination is biphasic, consisting of an initial phase ($t_{1/2}$ =4 hrs) followed by a slower phase, with $t_{1/2}$ = 7 hours. 3) Its metabolites, mCPP, is a non-selective serotonin receptor agonist with anxiogenic properties.

Pharmacodynamics: Trazodone antagonises both α_1 and α_2 adrenoceptors but has very weak anticholinergic side-effects.

Side effects: 1) Priapism 2) Orthostatic hypotension 3) Increased libido 4) Sedation 5) Bone marrow suppression.

Noradenaline Specific Serotonin Antidepressant (NaSSa)

Mirtazapine (Dose range: 15-45mg/day; $ 0.60/15mg)

Special features: 1) Mirtazapine blocks negative feedback of noradrenaline on presynaptic α_2 receptors and activates noradrenaline system; 2) Mirtazapine stimulates serotonin neuron and increases noradrenaline activity; 3) Mirtazapine has no effects on seizure threshold or on cardiovascular system. 4) Suitable for patients who cannot tolerate SSRI induced sexual dysfunction..

Other indications: insomnia or poor appetite, dysthymia (40% reduction), PTSD (50% reduction) and chronic pain.

Pharmacokinetics: 1) The peak plasma level is obtained after approximately 2 hours. 2) Linear pharmacokinetics and a steady-state plasma level is obtained after 5 days. 3) The elimination $t_{1/2}$ is 22 hours 4) Metabolised by P450 1A2, 2D6, and 3A4 4) 75% excreted by the kidney and 15% excreted by GI tract.

Pharmacodynamics: 1) Blockade of release-modulating α_2-adrenoceptors leads to enhanced noradrenaline release 2) the released noradrenaline stimulates serotonin neurons via the activation of α_1 adrenoceptors which in turn results in an enhanced noradrenaline effect, together with the selective activation of 5-HT$_{1A}$ receptors, may underlie the antidepressant activity; 3) 5HT$_{1A}$ agonism: antidepressant and anxiolytic effects. 4) 5HT$_{2A}$ antagonism: anxiolytic, sleep restoring and no sexual restoration 5) 5HT$_{2c}$ antagonism: anxiolytic & weight gain 6) 5HT$_3$ antagonism: no nausea, no gastrointestinal side effects. It also blocks histaminergic receptors and results in drowsiness.

Side effects: 1) drowsiness, 2) weight gain, 3) increased appetite, 4) dry mouth, 5) postural hypotension.

Serotonin noradrenaline reuptake inhibitors (SNRIs)

Venlafaxine XR [Efexor XR] (Dose range: 75mg – 375mg/day; $ 0.96/ 75mg tablet)

Special features: Low doses of venlafaxine blocks serotonin reuptake. Moderate doses of venlafaxine block noradrenaline reuptake. High dose of venlafaxine block noradrenaline, dopamine and serotonin reuptake. 2) Metabolised by P450 3A4 to inactive metabolites while P450 2D6 to active metabolites 3) More rapid onset action and enhanced efficacy in severe depression

Other indications: generalised anxiety disorder.

Pharmacokinetics: 1) minimally protein bound (<30%) 2) renal elimination is the primary route of excretion 3) the original version (venlafaxine) has relatively short $t_{1/2}$ = 5-7 hours; 4) prominent discontinuation syndrome (dizziness, dry mouth, insomnia, nausea, sweating, anorexia, diarrhoea, somnolence and sensory disturbance) and hence venlafaxine extended release (XR) is available.

Side effects: 1) nausea (35%); 2) sustained hypertension is dose related and 50% remitted spontaneously 4) dry mouth, constipation, 5) sexual dysfunction

Drug interaction: The toxic interaction with MAOIs, leading to a serotonin syndrome, is the most severe drug interaction involving venlafaxine.

Duloxetine [Cymbalta] (Dose range: 30-120mg/day; $3.8/ 30mg tablet)

Other indications: 1) Depression and chronic pain; 2) Fibromyalgia

Pharmacokinetics: Blood levels of duloxetine are most likely to be increased when it is co-administered with drugs that potently inhibit cytochrome P450 1A2.

Pharmacodynamics: Duloxetine exerts a more marked influence on noradrenaline reuptake than on serotonin reuptake.

Side effects 1) Nausea, dry mouth, dizziness, headache, somnolence, constipation and fatigue are common. 2) A small but significant increase in heart rate was observed 3) Rate of sexual dysfunction is low.

Agomelatin [Valdoxan]

Indications: Major depressive disorder

Special features: Reduced level of sexual side effects as well as discontinuation effects as compared to other antidepressants.

Mechanism of action: Melatonergic agonist and 5-HT2c antagonist.
No effect on monoamine uptake, no affinity for most of the other receptors

Side effects: Contraindicated for patients with renal or hepatic impairment.

Tricyclic antidepressants (TCA)

Indications	Contraindications	Mechanism of action	Side effects	Examples
TCA is an old antidepressant and it is not first-line antidepressant treatment due to potential cardiotoxicity if patient takes an overdose. Depression Anxiety disorder Severe OCD (Clomipramine) Neuropathic pain Migraine prophylaxis Enuresis	Cardiac diseases (e.g. post myocardial infarction, arrhythmias) Epilepsy Severe liver disease Prostate hypertrophy Mania	TCA inhibits the reuptake of both serotonin and noradrenaline and increase the concentration of these neurotransmitters. TCA also blocks histaminergic H_1, α-adrenergic and cholinergic muscarinic receptors on the postsynaptic membrane.	Anticholinergic (e.g. constipation, blurred vision, urinary retention, dry mouth). dizziness, syncope, postural hypotension, sedation). Histaminergic and dopaminergic blockade: nausea, vomiting, weight gain, sedation, Other side effects: sexual	**Amitriptyline** (25mg to 150mg daily) has the most potent anticholinergic effect. **Clomipramine** (100 – 225mg daily): Most potent TCA at D_2 receptors; More selective inhibitor of serotonin reuptake.

| | | | dysfunction, hyponatraemia

Cardiac: arrhythmias, ECG changes (QTc prolongation), tachycardia, heart block TCA overdoses may lead to delayed ventricular conduction time, dilated pupils and acidaemia due to central respiratory depression and a fall in pH reducing protein binding. | |
|---|---|---|---|---|

Monoamine oxidase inhibitors (MAOIs)

Reversible MAOI – Moclobemide (Dose range: 75mg to 225mg daily; $0.45 per 150mg tablet)

Indications	Contraindications	Mechanism of action	Side effects
Atypical depression. Depression with predominantly anxiety symptoms (e.g. social anxiety). Hypochondriasis.	Acute confusional state. Phaeochromocytoma.	Monoamine oxidase A acts on • Noradrenaline • Serotonin • Dopamine • Tyramine •	Visual changes. Headache. Dry mouth. Dizziness. GI symptoms.

The old irreversible MAOIs may lead to hypertensive crisis with food containing tyramine. Irreversible MAOIs are seldom used nowadays. The following food should be avoided :
1) Alcohol: avoid Chianti wine and vermouth but red wine <120 ml has little risk.
2) Banana skin.
3) Bean curds especially fermented bean curds.
4) Cheeses (e.g. Mature Stilton) should be avoided but cream cheese and cottage cheese have low risk.
5) Caviar.
6) Extracts from meats & yeasts should be avoided but fresh meat and yeast.

Other antidepressants

Bupropion [Wellbutrin] (Dose range: 150-300mg/day; $1.6 / 150mg tablet)

Similar efficacy as SSRI but voluntary withdrawal in the US due to induction of seizure at doses if the daily dose is higher than 450mg/day.

Other indications include patients encountering SSRI induced sexual dysfunction, female depressed patients do not want weight gain from medication and smoking cessation.

Pharmacodynamics: blocking dopamine reuptake.

Side effects: agitation, tremor, insomnia, weight loss and seizure. Bupoprion is not associated with sexual side effect

Lithium	Sodium valproate CR [Epilim Chrono]	Carbamazepine CR [Tegretol CR]	Lamotrigine [Lacmatil]
Dose: Oral, start at 400mg. Maximum 1200mg per day. Lithium carbonate CR: $ 0.38 / 400mg tablet.	Dose: Oral, starts with 500mg daily, Maximum 1300mg per day. $0.5 per 300 mg tablet.	Dose: 400mg – 800mg Carbamazepine CR $0.16 / 400mg tablet.	A slow titration is required and this will avoid a serious skin rash: 25 mg/day for 2 weeks doubling the dose every two weeks to a maximum of 400 mg/day. $ 1.09 / 50mg tablet $1.72 / 100mg tablet
Monitoring: Lithium level should be checked every 3 months. (0.4 – 0.8 mmol/L: maintenance range. RFT is checked every 6 months. TFT is checked every year.	LFT is a must before starting a patient on sodium valproate.	* MOH has recently announced and implemented a mandatory genetic test to screen for the gene responsible for developing adverse skin reaction with usage of the above medications	
Properties: • Remains a first line treatment in bipolar disorder. Lithium increases Na/K/ATP-ase activity in patients. It affects serotonin, noradrenaline, dopamine and acetylcholine. Lithium also interferes cAMP (second messenger	•Efficacy is superior to placebo but equal to lithium, haloperidol, and olanzapine. Valproate enhances GABA function and produces neuroinhibitory effects on mania. • Valproate can be combined with antipsychotics in	• • The application of carbamazepine is limited by its properties as an enzyme inducers and side effects such as diplopia, blurred vision, ataxia, somnolence, fatigue, nausea, and blood dyscrasia. • Generally effective in maintenance	• Doubtful efficacy in mania • Effective in bipolar depression in bipolar I patients with clear efficacy at 200 mg/day.

system). • Onset of action: 5-14 days. • Anti-mania effect is proportional to plasma levels. The level is set to be between 0.4 – 0.8 mEq/L for Asian patients. •Lithium is often used with antidepressants and other mood stabilizers in bipolar depression. • Lithium reduces suicidal ideation in bipolar patients Lithium is contraindicated to people with renal failure, thyroid diseases and rapid cycling disorder.	treatment of mania and lower dose of antipsychotic drug is required. • Valproate is effective in maintenance treatment to prevent mood episodes. • Effective plasma levels: 50-99 mg/L but clinical response is more important. Valproate is indicated in patients with renal failure and rapid cycling disorder but contraindicated in people with liver failure.	treatment to prevent mood episodes. • No data on plasma levels and response. Carbamazepine is indicated to patients with bipolar disorder who are concerned about weight gain caused by lithium and valproate because carbamazepine does not cause significant weight gain.	
Side effects: Common side effects: Metallic taste Nausea Polydipsia Polyuria Oedema Weight gain Fine tremors. Long term complications: Hypothyroidism Renal failure	Common side effects: Weight gain Nausea Gastric irritation Diarrhoea Serious side effect: Thrombocytopenia.	Common side effects: Dizziness Somnolence Nausea Dry mouth Oedema Hyponatraemia (due to potentiation of ADH) ↑ALP and ↑GGT. Uncommon side effects: Ataxia Diplopia Nystagmus Serious exfoliative dermatological reactions (3% of patients and requires cessation of carbamazepine. Agranulocytosis Leucopenia Aplastic anaemia.	Common side effects: Dizziness Headache Diplopia Nausea Ataxia. Uncommon side effects If patients develop for lamotrigine-associated rash (10%), hold the next dose and seek immediate medical attention.

www.ingramcontent.com/pod-product-compliance
Lightning Source LLC
Chambersburg PA
CBHW080904170526
45158CB00008B/1980